Geology and landscapes
of Scotland

Good luck!
Oma

Quinag, Assynt, Sutherland, Northwest Highlands: an inselberg (isolated mountain) of Torridonian Sandstone resting unconformably on a pavement of ancient Lewisian gneiss (photograph by Roger Jones).

Cover image: Suilven, Sutherland; see Chapter 3 (photograph by Roger Jones).

Geology and landscapes of Scotland

Con Gillen
University of Edinburgh

TERRA

First published in 2003 by Terra Publishing

Terra Publishing
PO Box 315, Harpenden, Hertfordshire AL5 2ZD,
England
Telephone: +44 (0)1582 762413
Fax: +44 (0)870 055 8105
Website: www.terrapublishing.net
E-mail: publishing@rjpc.demon.co.uk

ISBN: 1-903544-09-2

13 12 11 10 09 08 07 06 05 04
11 10 9 8 7 6 5 4 3 2

British Library Cataloguing-in-Publication Data
A CIP record for this book is available from the
British Library

Library of Congress Cataloging-in-Publication
Data are available

Typeset in Palatino and Helvetica
Printed and bound by Biddles Limited,
Guildford and King's Lynn, England

Contents

Preface

Scotland has been a Mecca for geologists for over 200 years. Indeed, the subject of geology was born here, with the pioneering work of the great James Hutton in the 1780s, and ever since, Scotland has played a key role in the development of geology. And with good reason, for Scotland boasts the most varied geological tapestry of any country of its size in the world.

This book has been designed for non-expert readers, and has been written in such a way that all the key concepts in geology are explained as plainly as possible. Here is an up-to-date account of each of Scotland's unique geological regions, from Shetland and Orkney, through the Western Isles to the Highlands, Central Lowlands and Southern Uplands. We trace movements across the globe from the dawn of time – how Scotland moved from southern ice-bound seas across the Equator into warm and humid coal-forming swamps, to be confronted eventually by icy conditions, this time near the North Pole. The book deals with not just the wide variety of rocks (and the minerals and fossil remains they contain), but also the landforms and types of scenery that owe their origin to the structure and composition of the bedrock. We explore how these landforms developed through time, to be used to great advantage by Scotland's first settlers, and how the mineral riches have been exploited, from iron ores and building stones to coal, oil and gas.

Many of the fundamental ideas in geology were developed on the basis of rocks and structures seen in Scotland. Indeed, Scotland could justifiably be called the cradle of modern geology. For its relatively small size, the country boasts a truly amazing variety of rocks and structures. At first sight, this may appear daunting and even confusing to the novice, but *Geology and landscapes of Scotland* will put you at your ease and allow you to grapple with the most complex of notions, as you travel on a journey, following Scotland's fortunes in time and space.

Con Gillen, Edinburgh, January 2003

Acknowledgments

I am grateful to the British Geological Survey for permission to base the maps and cross sections on BGS work. I thank Stephen Cribb, Colin MacFadyen and Alwyn Scarth for their helpful comments, and those colleagues who allowed me to use their photographs.

I wish to express my gratitude to Roger Jones of Terra Publishing for his unstinting support, professional advice and expertise shown at every stage of production, from the initial idea to the finished work. Secondly, I thank my students who, with their stimulating questions and bubbling enthusiasm inspired me to write down my explanations for the fascinating geological history of Scotland. In order to make the book as readable as possible, direct citations of individual published works – e.g. "(Smith & Jones 1999)" – have been omitted, because many readers may not be comfortable with this academic habit. However, I do wish to acknowledge the contribution of the many geologists whose work has made the writing of this book possible; they are listed in full in the Bibliography. I am extremely grateful to them. I also thank my colleagues for their support. Finally, for their patience and support, I wish to thank my family, to whom this book is dedicated – Patricia, Ann-Marie and Kathleen.

A view south from the summit of Quinag, Assynt, Northwest Highlands: horizontal Torridonian Sandstone on the right; the rounded summit of the middle peak (Spidean Còinich, 764 m) is an outlier of Cambrian Quartzite; Lochan Bealach Cornaidh in the middle distance is a glacial lake; slopes on the left are within the Assynt thrust zone (photograph by Roger Jones).

CHAPTER 1
Geology –
the science of the Earth

Introduction

This book is about the landscape you will see around you as you travel throughout Scotland. What makes up the natural landscape? Its components are part of the natural environment: rocks, soils, relief, climate and vegetation. This natural or physical environment in turn consists of closely interrelated systems working in harmony within cycles of varying duration. The landscape itself is not static, but has been evolving constantly throughout the history of the Earth by the interaction of animals, plants and physical processes, sometimes gradual, often abrupt or catastrophic. And, for at least the past 5000 years, humans have played an important part in changing the landscape, at an ever-accelerating pace. Recent climatic change as a result of the enhanced greenhouse effect is thought to be mainly induced by humans, and the future effects on the landscape could turn out to be profound and possibly irreversible in the scale of our lifetime.

Scotland is famous for its varied and attractive scenery, and for the sharp contrasts between different parts of the country. These scenic differences derive from natural forces acting on bedrock geology that is extremely varied, given the relatively small size of the country (less than 80 000 km^2). We now embark on a journey of exploration through Scotland, looking at the variety of landscapes, and discovering the reasons for the patterns we see at the surface. In human terms, the landscape may seem never changing, save for seacliff collapses, or rockfalls down mountainsides, but over millions of years, it has been changed profoundly by the action of rain, ice, waves and wind. In the more distant past, the rocks of Scotland were formed by processes that originated in the Earth's interior: mountain building, folding of rocks, faulting, earthquakes, volcanic eruptions and deep underground intrusions of molten rock. The Scottish landscape is an intricate set of different landform elements of different ages and origins, resulting from a complex geological fabric and combining to produce an attractive patchwork, characterized by sharp contrasts in the various parts of the country, particularly between the east and the west.

Why do we see such marked contrasts, between high rugged mountains in the northwest, flat, high table lands in the northeast, lowlands in Central Scotland, and rolling hills in the Southern Uplands? The answer, as we shall soon see, lies in the nature of the bedrock that was sculpted by the forces of erosion. In very general terms, Scotland is an upland country, with the oldest rocks in the northwest and a land surface that slopes to the east and southeast, towards the North Sea, and is the direction of flow of most of the main rivers. The fabric of the rocks has strongly influenced the development of the landscape, since some rocks are relatively weak and can be easily sculpted by erosion, whereas others are much tougher and more resistant to erosion, so they can form prominent landscape features. In some parts of the country, such as the Northwest Highlands, the variety of rock types over a small area is considerable, and the result is a complex pattern of landform features that endows the region with a scenic beauty scarcely rivalled in the rest of Britain or Europe.

The rock fabric of Scotland

The Earth is made of three concentric shells: the crust, mantle and core. For our purposes, the crust and the upper part of the mantle are the most important. We live on the crust, the thin, outermost shell, which is 35 km thick on average in the continents and 7 km under the oceans. Continental mountain chains have thicker crust; for example, the crust beneath the Himalayas is 85 km thick. The centre of the Earth lies 6371 km beneath the surface.

The three main rock types that make up the Earth's crust are referred to as:

- igneous, formed by the cooling of molten material from below the crust
- metamorphic, formed when heat and pressure alter rocks that have already been formed
- sedimentary, formed on the surface by the deposition of material carried by wind, water or ice.

In very broad terms, igneous and metamorphic rocks, being crystalline, are harder and more resistant than sedimentary rocks, and so tend to form the higher ground and more rugged landscapes in the highland areas. Most of Central Scotland is made of sedimentary rocks and the land is generally low lying. Higher ground in the central belt is underlain by igneous rocks, such as the lava flows that make up the

2

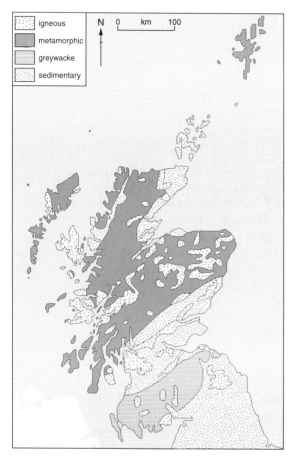

Figure 1.1 Sketch map of Scotland, showing distribution of the main rock types.

Campsie Fells north of Glasgow. The map of Scotland (Fig. 1.1) shows that the greater part of the country is made up of igneous and metamorphic rocks. Rocks are very varied in terms of their colour, the shape and size of crystals and grains, and their texture (how the grains interrelate), and this has given rise to a profusion of names. However, do not be daunted by this, for there are no more than a couple of dozen important rock names, and even fewer are used in this book.

Minerals – the building blocks of rocks

Rocks are made of minerals, which are naturally occurring chemical compounds that grow as crystals in igneous and metamorphic rocks, or exist as fragments in most sedimentary rocks, except limestones and salt deposits which are also crystalline. Individual minerals can be recognized on the basis of their composition, crystal shape and physical properties (i.e. colour, hardness, lustre, cleavage, density, etc.). Talc is the softest mineral and diamond the hardest. There are about 3700 known minerals, and more are being found all the time. Half of them are named after the individuals who discovered them, and a quarter are named after the place where they were found. Because they are chemical compounds, minerals are divided into groups on the basis of their chemical composition. Very few natural elements are able to exist in a pure state, gold, silver and platinum being well known examples of these. As far as the rocks of Scotland are concerned, they are mostly made of silicate minerals. Important non-silicate minerals include the metal ores that were once mined.

Igneous rocks

Igneous rocks are produced when molten magma (hot silicate melt with dissolved gases) cools and solidifies. Interlocking crystals of mainly silicate minerals grow freely together in all directions and in random fashion to create a solid rock. If magma penetrates other rocks underground, the igneous rocks that form are referred to as intrusive. When liquid magma escapes at the surface through fissures and cracks in the crust, or from volcanoes, the resulting lava flows that pour across the landscape are known as extrusive igneous rocks. Depending on the cooling rate, crystals of different sizes can grow, the relationship usually being that slow cooling underground (over possibly up to a million years) results in coarse-grain rocks; rapid cooling, as in surface lavas flows, produces fine-grain rocks. Sometimes cooling from a melt at a temperature of 1000°C or more is so rapid that no crystals have a chance to grow, and instead a dark volcanic glass forms. Large and small crystals can occur side by side in an igneous rock, which normally indicates a two-stage cooling history, larger crystals forming deep within the crust and finer crystals enclosing these as a groundmass or matrix when the rock finally cools at a faster rate on or close to the surface. Differences in mineral composition, related to magma chemistry, result in a great variety of igneous rocks, made all the more complicated by differences in crystal sizes.

Depending mainly on the amount of silica present, igneous rocks are traditionally classified as acidic, intermediate, basic and ultrabasic. In addition to silica (which is silicon plus oxygen), the other major components are aluminium, iron, magnesium, calcium, sodium, potassium and titanium; these elements combine with silica to make the very large family of silicate minerals that form nearly all the crystals in igneous rocks (and in most metamorphic rocks). These are referred to as the rock-forming minerals.

The igneous rocks are then subdivided on the basis of the content of these minerals and the overall chemical composition, and different names are used for coarse and fine varieties (Table 1.1).

Rocks with the same chemical composition as basalt, but with medium-size crystals, are usually called dolerite; the informal quarryman's term "whinstone" is sometimes used in Scotland to denote basalt and dolerite in the field, and refers to the fact that whin (gorse) grows on the very thin, poor soils that form on these rocks.

Acidic rocks are rich in quartz, potassium feldspar (orthoclase) and mica, whereas basic rocks have calcium plagioclase feldspar, pyroxene and olivine. Since quartz and potassium feldspar are pale, acidic rocks tend to be white, pink or grey. On the other hand, pyroxene and olivine are black, so basic rocks are dark and dense, because they are rich in iron and magnesium silicates. Because igneous rocks can vary so much in colour, grain size, texture and mineral content, they have a bewildering variety of hundreds of names. However, it is possible to whittle this list down to a few very common types, and these are the names used in this book. Which particular minerals actually grow in an igneous rock depends on the original composition of the magma, and the order in which they crystallize out of the melt depends on the temperature. Basic and ultrabasic rocks form at about 1200–1300°C and therefore contain minerals with high melting points. On the other hand, acidic rocks start to crystallize at about 700–800°C, and so generally contain minerals with lower melting points. Differences

in rock composition can be picked out by weathering and erosion, to produce contrasting landforms. One of the best examples where this can be seen is on Skye, where the black gabbro Cuillin Hills, with their rough surface and jagged peaks, look completely different from the nearby Red Hills, made of granite, which are lower, more rounded and subdued.

Depending on the conditions in which igneous rocks have formed, a variety of landforms can result. When erupted at the surface, lava forms flat flows, each one usually 1–3 m thick, that create their own landform features. Excellent examples of horizontal piles of lava flows can be seen in the western part of the Midland Valley, such as the Campsie Fells, and in the Hebridean islands of Mull and Skye. The volcanoes that produced these lavas have long ago been removed by erosion, but the hard igneous rocks that formed in the necks of these volcanoes now stand up as prominent "plugs", which are particularly abundant across the Midland Valley – Dumgoyne, Dumfoyne and Dumbarton Rock in the west, Arthur's Seat (Edinburgh), Dundee Law, the Lomond Hills and Largo Law in the east, to name just a few. Material broken off the walls of a volcanic vent by the force of an eruption will often fall back down into the vent and pile up as jumbled angular rubble known as volcanic agglomerate. Examples of this can be seen very clearly at the Arthur's Seat volcano in Edinburgh and at St Abb's Head in the Scottish Borders.

In the Inner Hebrides, many of the eruptions took place in narrow fissures, fed from below as lava rose to the surface, and these fissures now appear as straight vertical walls of solidified rock, known as dykes (Fig. 1.2). They are clearly visible around the coasts of the islands, especially on the south of Arran. When tongues of lava are squeezed underground between sedimentary layers, the result is a "sandwich" or sheet of igneous rock known as a sill. Salisbury Crags in Edinburgh's Holyrood Park are Scotland's most famous example of a sill; another is the Midland Valley Sill, on which Stirling Castle sits. During relatively rapid cooling at or near the surface, lavas and sills shrink and form cracks known as cooling joints, which often form patterns of hexagonal columns resembling organ pipes. The joints of Fingal's Cave on Staffa are world famous, other examples including Samson's Ribs in Holyrood Park and the

Table 1.1 Igneous rocks: common names and compositions.

Rock type	Acidic	Intermediate	Basic	Ultrabasic
Silica content	Over 66%	66–52%	52–44%	Under 44%
Coarse grain	Granite	Syenite	Gabbro	Peridotite
Fine grain	Rhyolite	Andesite	Basalt	None

Figure 1.2 A narrow Tertiary igneous dyke (black) cutting older metamorphic quartzite; the dyke is cut in turn by a small fault. Rum, Inner Hebrides.

Giant's Causeway in Antrim, in the north of Ireland.

Intrusive igneous rocks formed underground and are now seen at the surface only because the overlying rocks have been stripped away by erosion. They are sometimes referred to as plutonic rocks, having formed in large intrusions called plutons. Most large bodies of granite, gabbro and peridotite would have been intruded at depths of 5–10 km and have cooled extremely slowly, possibly taking almost a million years to solidify completely. The reason for the slow cooling is that the surrounding rocks are good insulators and are themselves quite hot because of their depth of burial. As they become exposed by erosion, the pile of rocks above is gradually reduced, and intrusive rocks, particularly granite, will develop cracks or joints, both horizontal and vertical. This can often result in roughly cubic or rectangular blocks at the surface. In the case of granite, chemical weathering in formerly humid climates has created tors (from the Gaelic tòrr: a heap of rocks), a rare landscape feature in Scotland, but relatively well developed in the Cairngorms and on Bennachie in Aberdeenshire and, of course, a well known feature of Dartmoor.

During cooling, crystals form within molten magma and are able to grow freely in all directions. For this reason, igneous rocks generally have a random texture, wherein the crystals interlock to form strong three-dimensional frameworks. This property makes them particularly useful as roadstone. In the case of ultrabasic and basic rocks (e.g. peridotite and gabbro), the earliest-formed crystals of minerals such as iron ore, chrome ore and olivine are very dense and may sink through the magma to form crystal layers at the bottom of an intrusion. Such layered bodies are known from Skye, Rum and Aberdeenshire, at Huntly, Belhelvie, Insch and Haddo House. One intrusion at Unst in the north of Shetland has layers of chromite that were formerly mined.

Igneous rocks of several different ages are widespread in Scotland and give rise to many prominent landscape features in the Highlands, Lowlands, Southern Uplands and all the island groups, from the Cheviot Hills along the English border to the seacliffs of Shetland and the Uig Hills of Lewis. Scotland's highest mountain, Ben Nevis, is granite topped by lava, and the highest plateau in Britain, the Cairngorms, is part of an extensive granite body. Within the Caledonian Mountains, igneous rocks constitute up to 25–30 per cent of the chain.

In addition to the silicate content, magmas have other components such as water, carbon dioxide, sulphur dioxide and various materials that are released during volcanic eruptions as clouds of gas, often spectacularly and violently. These gas products are called volatiles and are easily driven off by boiling when the pressure is reduced suddenly during an eruption. Gas bubbles within lavas rise to the surface and form porous honeycomb skins at the top of a flow. With time, the cavities are filled with minerals deposited from hot solutions circulating in the lava pile, and the resultant infillings are referred to as vesicles. Many of the lava flows in Mull, Skye, the Campsies and other areas are vesicular, and the minerals they contain are collectors' items. Larger cavities are often filled with colour-banded silica, to form agates. Granitic magmas tend to contain more volatiles than basic magmas do, and, when granites are nearly totally crystallized, water and the other volatiles will accumulate in the remaining portion of magma, creating conditions in which the last crystals can form freely in volatile-rich portions such that very large crystals can grow. The resulting rock is known as pegmatite, which will usually form veins and patches near the top of a granite intrusion, or as dykes cutting across into the surrounding country rocks. Pegmatites typically contain large crystals of quartz, feldspar and mica, and sometimes rather less-common minerals such as black tourmaline (schorl – a beryllium silicate) and possibly ore minerals of metal such as tin, tungsten and molybdenum, although these types are very rare in Scotland.

Heat from intrusions is lost only very gradually, and large bodies of molten magma, with initial temperatures above 1200°C, will affect the surrounding rocks by heating and baking them, causing new minerals to form as the existing components of the country rocks recrystallize. The larger the intrusion, the wider the area around it that will be heated (called a thermal aureole). When lava flows are erupted and small bodies such as sills and dykes are intruded near the surface, their edges chill rapidly on contact with the colder country rocks. The result is a thin edge of finer-grain rock – a chilled margin – with crystals slightly larger in the interior. Whenever lavas are directly erupted onto the sea floor during deep submarine volcanic activity, the effect of sudden chilling by cold sea water produces a skin, which is then punctured by the pressure of lava within, and a sequence of pillow shapes is gradually built up. Pillow lavas are conspicuous at Stonehaven and Downan Point, Ballantrae, near Girvan; other examples include the famous Tayvallich Volcanics in the Dalradian rocks of Appin, Argyllshire.

Sedimentary rocks
Rocks at the surface of the Earth are exposed to rain, frost, wind, and the action of plants and animals. Over time, all rocks are broken down physically into smaller particles that can be carried away, or they may rot because of chemical and biological activity. The minerals in igneous rocks, which formed at high temperature, are unstable at the surface and will break down readily. When the fragmented material is removed by streams or the wind, it is transported elsewhere and eventually deposited as loose sediment (e.g. sand and gravel). Some components may be dissolved in running water and remain in solution until chemical reactions cause material to precipitate out of solution to form, for example, iron or salt deposits.

Plants growing in soils on the surface can themselves form sediment after death if they are buried quickly enough (coal is an example), whereas corals and seashells living in shallow water can form rocks (limestones) in their own right. Rocks that form at the surface by these processes are known as sedimentary rocks, and they are normally laid down in layers or beds (also called strata) separated by bedding planes (Fig. 1.3). Well bedded rocks showing horizontal bedding planes are spectacular in the Torridonian sandstone of the Northwest Highlands around Torridon, Applecross and Ullapool (see Ch. 3); mountains such as Liathach and Ben More Coigach are the best examples. The Caithness Flagstones forming the cliffs and seastacks at Duncansby Head near John o'Groats are equally magnificent. As more sediment piles up, the lowest beds sink and are progressively buried, and eventually the separate grains are compacted more closely and bonded by a cement, so that unconsolidated sediment is transformed into a sedimentary rock, which has strength and cohesion (a process known as lithification), and the changes that take place during the growth of a cement are together called diagenesis. The cement forms from material that has been dissolved in groundwater and which subsequently fills pore spaces and grows along boundaries between individual grains. In general terms, sedimentary rocks are not as strong as igneous rocks, where the crystals have intergrown randomly in three dimensions. The grain size of sedimentary rocks is directly related to the energy of the water or wind that transported loose material across the surface. High-energy environments, such as fast-flowing mountain streams, can carry larger fragments than rivers meandering gently across flat valley floors. As particles move down stream, they are worn down by colliding or being dragged and crushed, and they will usually become smaller and more rounded by these actions. Softer and weaker minerals are ground down more readily, so the composition of the sediment load in a stream changes as it winds its way down hill towards the sea. Coarse sedimentary rock containing rounded pebbles and cobbles of rock, with finer sand between, is known as conglomerate and was generally deposited by fast-flowing streams close to the source of its components. The Old Red Sandstone, for example, contains large pebbles and some boulders

Figure 1.3 Bedding planes in Old Red Sandstone cliff; a wave-cut platform is in the foreground. New Aberdour, Morayshire.

along the Highland Boundary Fault, because streams were transporting debris from the high Caledonian Mountains of the north into the Midland Valley. This is very clear at Aberfoyle, which lies almost on the Highland Boundary Fault.

One of the most durable minerals is quartz, which is very hard and is resistant to chemical and biological attack. For this reason, the quartz content of sediments tends to increase as a river moves towards the sea. The eventual product of quartz-rich sand is sandstone, a type of rock in which the quartz particles (or grains) may be cemented by various materials, including most commonly silica, lime (calcium carbonate) or iron salts (hæmatite, which stains the rock red). Very fine particles such as clay minerals can be transported much farther than coarser heavy grains, and the fine light flakes will settle only very slowly to form mud on the sea floor, which is then compacted to give mudstone, or shale if more mica is present. Whenever sediment of mixed grain size is deposited suddenly, larger grains, which are heavier, are deposited first, followed by finer material; this creates a structure known as graded bedding, with an increase in grain size from top to bottom. Organic matter is often found in sediment, and such rocks may be black from the carbon content. One of the main constituents of mudrocks is clay (actually a large group of flat platy silicate minerals), the end product of the alteration of many silicate minerals, particularly feldspars. Clay minerals tend to stick together rather well, so that mudrocks are often quite impermeable to water.

Those sedimentary rocks that have been formed as a result of physical breakdown, transport and deposition generally have a fragmental texture and are classified as clastic (Greek: to break) rocks. Normally they have been laid down as beds, bounded by bedding planes, which can often be marked by slightly finer material. Although most beds are formed horizontally, sediment can be laid down at an angle by fast-flowing currents, which, if they change direction, will give rise to cross bedding in which each successive bed lies at an angle different to the one beneath. In the Old Red Sandstone, cross bedding is a very common feature of the sandstones that were laid down in braided rivers criss-crossing over their floodplains in a semi-arid environment. This can be compared with the New Red Sandstone rocks, which were formed on a desert floor, and the cross beds are of a m size, because they represent fossil sand d are clearly visible at Hopeman near El

ROGER JONES

Figure 1.4 Ripple marks on bedding planes of siltstone in Torridonian rocks, Stoer Peninsula, northwest Sutherland; straight cracks running across the picture are joints.

(calcium carbonate) or dolomite (calcium–magnesium carbonate). Fossils are common in limestone, and indeed some limestone beds such as those at Barn's Ness in East Lothian are made entirely of fossil coral reefs. Chalk is a type of pure limestone formed from plankton skeletons; it is very rare in Scotland. Non-carbonate sediments include coal from fossil plant remains; flint or chert, which is banded sedimentary silica from the remains of minute floating plankton; ironstone, which formed in stagnant lagoons; rock salt (sodium chloride, common salt) and gypsum (calcium sulphate), deposited by the evaporation of sea water in shallow coastal lagoons during extremely hot climatic conditions. Limestone, salt and gypsum have crystals of minerals that grew in sea water, and so their texture is crystalline. Table 1.2 summarizes the main characteristics of the commonest sedimentary rocks you are likely to see in Scotland; some are more important than others, notably sandstone, conglomerate and shale. Greywacke is restricted to the Southern Uplands. Sedimentary rocks are generally softer and more easily eroded than igneous rocks, and usually produce more subdued landforms. However, certain differences can occur in landforms produced in areas of sedimentary rocks, depending on the composition and strength of the particles and of the cement. Shales and mudstones are easily worn away and form low ground, as is the case in much of Central Scotland. Coal also is very soft; it occurs on low ground in the Midland Valley, but very little is exposed at the surface. Sandstones are variable and can form upland areas, especially if the cement is strong, as is the case with silica cement. Carbonate cement is readily attacked by rainwater and may dissolve out, leaving quartz grains loose and liable to fall out, so that such sandstones may develop pockmarked surfaces; this is especially seen at the coast. In the far Northwest Highlands, the pure white Cambrian quartzite on Beinn Eighe, Foinaven and Arkle is in fact clean sandstone deposited in shallow water and cemented by silica. Compaction by overlying rocks, now removed, forced the grains to become almost welded together, producing a very tough durable rock. Frost action on the mountain tops has given rise to the spectacular white screes. Some of the higher ground in the Midland Valley, close to the Highland Boundary Fault, has Old Red Sandstone conglomerate, as seen at Aberfoyle

Moray Firth coast, at Brodick on Arran, and in the old sandstone quarries at Mauchline and Sanquhar in Southwest Scotland. Individual sand grains in dune-bedded sandstones are normally perfectly smooth spheres, rounded by the wind.

Particular kinds of structures can form on the surface of sediment after it has been deposited, such as ripples made by water currents (Fig. 1.4), mud cracks whenever the surface dries up, trails made by animals, burrowing tubes made by worms and molluscs. In some instances, these structures may be preserved during diagenesis and they are invaluable in unravelling the secrets of past environments.

In addition to clastic rocks – conglomerate, sandstone, siltstone, shale and mudstone formed from the erosion of rocks from the continents, there are sedimentary rocks that originated by other processes. One group is made up of limestones, rocks made of calcite

Table 1.2 Common sedimentary rocks.

Rock name	Composition	Grain size	Characteristics
Conglomerate	Pebbles of various rocks	Very coarse; pebbles rounded	Formed by fast-flowing rivers
Breccia	Angular rock fragments	Very coarse	Fossil scree deposit
Sandstone	Quartz grains	Medium Grit has angular grains	Cemented by silica, calcite or iron ores; may sometimes contain fossils
Arkose	Quartz and feldspar fragments, sometimes angular	Medium to coarse	A type of Torridonian sandstone formed from gneiss and granite fragments; red or brown
Greywacke	Quartz, clay and igneous rock fragments	Medium	Well cemented, tough, breaks in angular pieces
Siltstone	Quartz grains	Fine	Finely bedded, like flagstone; tough
Shale	Quartz, mica, clay	Very fine	Splits easily, very soft
Mudstone	Clay minerals, carbon	Very fine, black	Sticky, impervious
Limestone	Calcite or dolomite	Medium to fine	White or grey, often rich in fossils, crystalline
Chalk	Calcite	Medium to fine; pure white	Soft; rounded particles of fossil algae and plankton
Coal	Carbon (plant remains)	Fine, black, shiny or dull	May be banded; plant fossils sometimes seen

and Stonehaven. The Maiden Pap in Caithness and the prominent Knockfarril Hill near Dingwall, with its iron-age fort on top, are also Old Red Sandstone conglomerate.

Limestone occurs in two main areas of Scotland. From Durness on the northwest coast of Sutherland southwards through Assynt to Skye is the narrow band of Durness Limestone, formed 500 million years ago in shallow tropical seas. Characteristic landforms in limestone terrain are very clear around Smoo Cave near Durness. Limestone is soluble in rainwater, and features such as disappearing streams, dry valleys, swallow holes, underground cave systems and limestone pavements are seen in Assynt, albeit on a small scale, especially around Elphin and Inchnadamph. The other area of limestone development is in Central Scotland, where the Carboniferous limestones have been used extensively in the manufacture of lime for agricultural use. At Dunbar in East Lothian, the limestone has been worked for decades in the production of high-quality cement.

However, most sedimentary rocks in Scotland are found in the Midland Valley, with smaller areas in the Borders, Caithness and Orkney, and around the Moray Firth. Offshore, nearly all the rocks in the North Sea consist of very thick piles (sequences) of sedimentary rocks, accumulated over a long time by erosion of continental land areas, mostly of igneous and metamorphic rocks formed during the building of the Caledonian Mountains 500–400 million years ago.

Metamorphic rocks

Metamorphic rocks form by the action of heat, pressure and fluids on pre-existing rocks, within mountain belts and at depths in the Earth's crust of up to 20–25 km. In Scotland, metamorphic rocks are very abundant and form parts of the Outer Hebrides, the Northern and Northwest Highlands, Shetland and the Grampian Highlands. Most of the rocks in the Southern Uplands are weakly metamorphosed sedimentary rocks. During the compressive events involved in mountain building, pre-existing rocks are folded, flattened, squeezed, sheared and heated up, to the extent that the original constituents are recrystallized. Temperatures during metamorphism can be 600°C or more. Beyond that, especially if water is present (or minerals containing water in their crystal structure, such as clays and micas), rocks will reach the point where they can begin to melt and therefore form igneous rocks. Since compression and shearing are prevalent, most metamorphic rocks will take on a structure in which the minerals grow parallel to each other, at right angles to the direction of compression. This is especially the case with minerals such as mica, which have a natural tendency to grow as thin flat sheets, with a very strong cleavage or plane of weakness. Mica-rich metamorphic rocks then take on a schistosity, which is the ability to split easily into slabs. The best example of this phenomenon is seen in slate, which is metamorphosed shale or mudstone. Clay minerals recrystallize into micas at right angles to the direction of maximum compression, and the result is an extremely well developed slaty cleavage. Metamorphic changes take place slowly in the solid state by recrystallization and reorientation of existing particles; no wholesale melting is involved. Where metamorphism has resulted mainly from the action of heat, for example around large igneous intrusions, the metamorphic rocks in the heat-affected zone (aureole) will usually have a more random texture, in contrast to the foliated metamorphic rocks. Hard splintery rocks in metamorphic aureoles are known as hornfels, frequently having dark spots representing clusters of new minerals that crystallized in response to higher temperature. Apart from water and carbon dioxide, very little enters or leaves rocks during metamorphism; the combined action of heat, pressure and fluid activity is enough to cause new minerals to grow. However, the type of activity in zones around large granite bodies that have intruded limestones is a special case. Calcite (calcium carbonate) and dolomite (calcium–magnesium carbonate) are highly reactive, and water-rich fluids expelled from granite induce chemical reactions in limestones to produce rocks that show the effects of hydrothermal alteration (from the effects of hot water-rich solutions). The changes that occur are described as metasomatic (Greek: altered body), meaning that there has been a change to the overall body of the rock. Metasomatic rocks can be characterized by their containing many different minerals, arranged in streaks, patches and bands. Good examples in Scotland are the Ledmore marble in Assynt, or the Broadford marble in Skye, both produced by granite intrusions that metamorphosed Durness Limestone.

The varieties of metamorphic rocks depend, first,

on the composition of the original rock, and secondly on the pressure, temperature, fluid composition and fluid-pressure conditions during the metamorphism. Sedimentary rocks such as sandstone usually consist mainly of grains of quartz (silica) cemented by silica, and when this recrystallizes a quartzite is produced, made solely of quartz. In the case of mudrocks, the large variety of particles, the very fine grain size and the presence of fluids create a recipe that produces many different minerals by chemical reactions and grain growth, as pressure and temperature progressively increase during the process of metamorphism and deformation. As a result, higher metamorphic grades are produced. The term grade is used to refer to the relative temperature and pressure conditions (temperature being the more important) to which a rock has been subjected during metamorphism. Slate is described as being of low grade, mica schist medium grade, and gneiss (pronounced "nice") is a high-grade rock. The highest grades would commonly reflect conditions at the deepest levels of the crust. Grain size also increases progressively with grade: fine in slate, medium in schist and coarse in gneiss, and is usually also banded into separate mineral layers. Certain minerals will grow at different pressure and temperature conditions, and these will characterize the different metamorphic rocks.

At the beginning of the twentieth century, George Barrow first described the progressive sequence of minerals in metamorphic rocks while he was mapping the Dalradian rocks of the Southwest Highlands, just north of the Highland Boundary Fault.The zones he identified gave Barrow an indication of the pressure and temperature conditions within the Caledonian mountain belt. Barrow's zones are used world

The early geologists

George Barrow, who worked for the Geological Survey, was strongly influenced by the ideas of James Hutton, who had proposed over a century earlier that metamorphic rocks resulted from the effects of heat deep within the Earth. The field area where Barrow worked is just south of the extensive Aberdeenshire granites, and Barrow assumed that the metamorphism arose from the effects of these granites. Later this was shown to be an oversimplification.

As an interesting historical aside, Hutton's colleague Sir James Hall (1761–1832) provided him with proof of the origins of metamorphic rocks by conducting some of the first geological experiments in 1798. He filled an iron cylinder with powdered chalk and sealed it to

prevent the escape of carbon dioxide gas and to simulate high-pressure conditions. The cylinder was then placed in a blast furnace and heated; when it was opened, crystalline marble was found, instead of the lime that would normally be produced when chalk is heated under surface conditions. In another experiment, he created an artificial igneous rock by mixing quartz and feldspar in the proportions found in natural granite, and melted these together. He made the important observation that the melting temperature of the mixture was lower than that of either the quartz or feldspar. Such experiments disproved the notions of the Neptunists, who believed that all rocks were the result of deposition during the Biblical Flood.

Table 1.3 The commonest types of metamorphic rock in Scotland.

Original rock	Metamorphic rock produced
Sandstone, siltstone	Quartzite
Shale, mudstone	Slate, phyllite, schist, gneiss
Limestone, dolomite	Marble
Basalt lava, volcanic ash	Hornblende schist (greenstone)

wide in describing metamorphic rocks in mountain belts in which the main original rocks were clay-rich mudstones and shales; the term "Barrovian metamorphism" is sometimes used. Where the original composition is different, then various other metamorphic rocks will form, including quartzite. Table 1.3 lists the main types of metamorphic rocks.

Although most of the minerals found in metamorphic rocks – quartz, feldspar, mica – also occur in igneous and sedimentary rocks, a few are much more common, such as chlorite and hornblende, and some are unique to metamorphic rocks, garnet being the most common of these. Other minerals found only in metamorphic rocks are the aluminium silicates andalusite, kyanite, sillimanite, cordierite and staurolite. It was the identification of these minerals in the Highland schists that enabled George Barrow to establish his zones of regional metamorphism. Intrusive igneous rocks, in particular basic and ultrabasic varieties such as gabbro and peridotite, will often not show many signs of change during metamorphism. The reason is that the minerals (pyroxene and olivine in the main) in such rocks crystallized at very high temperature, in excess of what is usually achieved during metamorphism. Metamorphosed basic lavas in the Dalradian belt of southwest Argyllshire are known as greenstone from the presence of chlorite and epidote, two metamorphic minerals (Fig. 1.5). Granite is sometimes an exception, since many of the minerals began forming at temperatures of about 700°C, which can be attained in high-grade metamorphic conditions. The result is that granite may partially melt or recrystallize to form a banded granitic gneiss, or augen gneiss if the feldspars recrystallize into large flattened crystals ("augen" is German for "eyes"); the Inchbae augen gneiss is a fine example of this rock type.

In Scotland, the various metamorphic belts show pronounced differences as a result of the original rock compositions and the pressure and temperature conditions during the mountain building that produced

Figure 1.5 Top: Kildalton Cross, Islay (eighth century), made of local greenstone, a weakly metamorphosed Dalradian basic lava. Foot: Detail, showing white feldspar crystals standing proud of matrix, because of differential weathering and hence loss of detail.

the metamorphic rocks. The oldest rocks in the country are the Lewisian gneisses of the Outer Hebrides and Northwest Highlands, which formed 2–3 billion years ago, at the base of the crust, some 30 km or so below the surface. These are very high-grade rocks that were affected by many metamorphic events over an exceedingly long period of time, producing the Lewisian Complex. Temperatures during metamorphism were 550–650°C, and some of the rocks became partially molten to form granites and pegmatites, seen most spectacularly around Loch Laxford in Sutherland. Dolerite dykes were sheared along their margins, and pyroxene recrystallized as amphibole, creating hornblende schist, in which the hornblende needles are aligned in parallel to form a mineral lineation. Metamorphic rocks in the Moine schist belt of the Northern Highlands are typically quartz–mica schist, reflecting the fact that the original rocks were mainly sandstones with occasional mud and shale layers, which accumulated in shallow sedimentary basins about a billion years ago. Dalradian rocks, in contrast, display much greater variety as a result of the original pre-metamorphic rocks having formed in contrasting environments. Mudrocks produced slate, schist and gneiss in places; limestones gave rise to marble; basaltic lavas, tuffs and ash formed greenstones and hornblende schist; sandstone was converted to quartzite, or to grit if the original sedimentary rock was a gravel. Despite folding, metamorphism and complete recrystallization, it is still possible to identify signs of the original nature of some of these metamorphic rocks. For example, quartzites in the Moine and Dalradian rather frequently show cross bedding and graded bedding, features that are useful in the field, to identify the top and bottom of a folded sequence of rocks, and therefore whether the rocks are right way up or upside down. This technique, using what became known as "way-up criteria", was first developed and used in the Dalradian of Scotland by Robert Shackleton (1910–2001) in the 1960s, and allowed huge strides to be made in our understanding of the evolution of the complex Caledonian mountain belt; the technique is employed world wide. By using way-up criteria in the Southern Uplands, the geological structure of the region was completely reinterpreted in the 1980s, and early notions of the origin of the belt as being some

sort of concertina of simple folds was abandoned in favour of a model that involved intense folding and thrusting of slivers of rock piled up one above the other adjacent to a subduction zone, where oceanic crust is returned to the mantle (see p. 14). The Southern Uplands were described above as being made of weakly metamorphosed rocks, the low metamorphic grade being evident in the growth of chlorite and mica, indicating relatively low-temperature conditions. Shales and mudstones became slates, whereas the angular sandstones containing abundant clay particles and fragments of igneous rocks formed greywacke, a tough, strong, durable rock that breaks with an angular fracture, owing to the cement having recrystallized to form a much more closely knit bond than in the original rock. (Incidentally, greywacke is often referred to locally as "whinstone", but is quite different in origin from the igneous dolerite that is also more commonly called whinstone).

Metamorphic rocks that have been formed over large areas of a continent, in mountain belts, are referred to as regional metamorphic rocks, as opposed to more local effects of thermal metamorphism in aureoles around igneous intrusions. In Scotland, the Buchan area of Aberdeenshire is noted for the unusual conditions that existed during the metamorphism of the Dalradian rocks. During the early stages of folding, the rocks were intruded by large bodies of basic and ultrabasic magma, so that a considerable amount of additional heat was introduced from the upper mantle into rocks at the base of the crust that were being deformed and metamorphosed. The unusually hot conditions produced a broad zone of thermally metamorphosed rocks in the Buchan area, and the mineral associations that formed in these conditions are unique, to the extent that this type of metamorphism is called "Buchan type", to distinguish it from Barrow's metamorphic zones in the Highland border area to the south ("Barrovian type").

The rock cycle

It is possible to show how rocks of all three major groups are interrelated by connecting them in the rock cycle, a concept that was first put forward in 1785 by James Hutton (1726–97), born in Edinburgh, who is credited as being the founder of modern geology. Hutton used field observations of rocks, soils and natural processes to conclude that rocks form and are recycled by erosion and deposition over a very long period of time, by processes that operated in the past in much the same way that they operate today. At the time, this was a major advance in scientific thinking, and was in direct opposition to the contemporary philosophers' view that the Earth had formed suddenly in a series of brief catastrophes.

The rock cycle can be envisaged as beginning with the formation of new crust of igneous intrusions from the top of the Earth's mantle into the lower continental crust, or directly onto the ocean floors to create new basaltic crust. Intrusions may be accompanied at the surface by volcanic activity, with the production of lava and ash. Rocks on the surface are subjected to weathering and erosion by water, wind, ice, gravity, chemical and biological attack, and particles are removed, transported and deposited to form sediments. Deep burial leads to compaction and cementing of loose surface sediments, and sedimentary rocks eventually form, often added to by life itself, as limestone from corals and shells, or coal from trees, soil and peat. Movements of continents drifting around the Earth will sooner or later lead to collisions, and rocks caught up in such zones will be folded, thickened and metamorphosed deep within the crust. Once temperature and pressure rise sufficiently, the rocks will be metamorphosed and some will approach melting conditions. Those that do melt in part then form igneous rocks, to be injected into the lower crust. Uplift and denudation by weathering and erosion of the resulting mountain chain creates the conditions for sedimentary rocks to form once more, and the cycle is complete. Mountain building typically takes 200–300 million years to complete, and since the oldest rocks so far dated are a little under 4000 million years old, it is clear that there has been more than enough time for the rock cycle to have operated many times over. Other cycles operate within the rock cycle, at different scales and over different time periods. Important among these is the hydrological or water cycle, connecting the oceans, seas, lakes, rivers, atmosphere, groundwater and ice; it is the operation of this cycle that makes life possible for humans and most other animals and plants. Secondly, there is the carbon cycle, which shows the complex interrelations between animals and plants in the seas and on land, between groundwater, soils and life on land, and carbonate sediments (limestones) in the seas. Parts of these two major cycles operate over days, years, or millions of years in the case of the formation of limestone and its eventual appearance at the surface, to be attacked by rainwater and dissolved, then returned to the ocean.

Structure of the Earth

Planet Earth has a clearly defined layered structure. We live on the thinnest, outermost layer, the crust, below which lies the mantle, which extends to 2900 km from the surface. The mantle is made of peridotite, a rock consisting of the dense iron–magnesium silicates olivine and pyroxene. The innermost part of the Earth is the core, solid iron–nickel alloy at the centre, and surrounded by a liquid outer layer made of iron and some sulphur. In total, the core makes up a third of the Earth's mass, and the fact that it is made of iron–nickel alloy is responsible for giving the Earth a magnetic field, which acts as a protective shield, deflecting harmful cosmic radiation from outer space, and preventing it from reaching the surface. It is possible to identify two types of crust, based on composition: continental and oceanic. Continental crust is 35 km thick on average, but up to 75 km thick beneath young mountain chains, such as the Himalayas. It is made of a wide variety of rocks – igneous, sedimentary and metamorphic – which vary greatly in age, the oldest being close to 4 billion years old. These very old rocks are found only in the central part of stable areas of the continents, often near the centre, and typically are surrounded by progressively younger mountain belts. Oceanic crust on the other hand has a quite different structure. It is composed of basalt and is typically 5–10 km thick. New oceanic crust is continually being formed at ocean ridges as basalt emerges from the mantle beneath and is added to the crust along clearly defined narrow submarine volcanic zones. Iceland sits on top of one of these, the Mid-Atlantic Ridge. Because of the way in which oceanic crust is constructed, the youngest parts are at the mid-ocean ridges and the oldest parts are farthest away, at the edges: new material pushes older crust away from the ridges. A second major contrast with the continents is that the oldest oceanic crust is nowhere older than about 180 million years, and the age pattern is regular and symmetrical on either side of the ocean ridges. Ocean ridges slope away from a high active central zone towards deeper and flatter areas known as abyssal plains. Around the Pacific Ocean, the boundary between continents and the ocean is often marked by very deep ocean trenches, down to 10 km or so. Being made of basalt, oceanic crust is able to record changes in the Earth's magnetic field: basalt is relatively rich in iron and, as lava cools, the iron in the form of magnetic iron ore (magnetite) crystallizes out to form small needles, which line up along the magnetic field. This direction is then effectively frozen in the rock when the basalt solidifies. However, the magnetic field does not have a constant orientation. The present orientation, where the magnetic North Pole is close to the geographical North Pole, is said to be "normal", but in the past the magnetic North and South Poles have been switched around by 180°, giving rise to reverse magnetism. This switching of the poles happens suddenly and unpredictably, every few hundred thousand years, for reasons that are still uncertain, but probably related to current motions in the liquid outer iron core. As a result, the ocean floors have a highly characteristic striped arrangement of normally and reversely magnetized basalt, with stripes of the same age and magnetism being parallel to the ridges (Fig. 1.6). The magnetic reversals for the past 200 million years have been recorded in the geomagnetic polarity timescale.

Plate tectonics
It was the discovery in the middle of the twentieth century of the alternating magnetic stripes on the ocean floors that gave rise to the notion of plate tectonics, a theory that has been considerably developed and which attempts to explain many interconnected patterns and features of the composition, age and structure and of the crust. In broad terms, the plate tectonic theory envisages the crust and upper mantle as being divided into large rigid plates, meeting at plate boundaries where interactions between adjacent plates give rise to features such as volcanoes, earthquakes, ocean trenches and continental mountain belts (Figs 1.6, 1.7). Tectonic activity is thus mainly concentrated in narrow zones, which form an interconnected system around the globe.

The plates are in constant motion, at various speeds in different directions, and this motion is driven by the Earth's internal heat. Plates are able to move because large slabs of solid material – the lithosphere (the crust and top part of the mantle) – are riding above a weak zone in the upper mantle, known as the asthenosphere, which is in a partially molten state and can move by plastic deformation. Hot mantle material rising upwards melts beneath the crust to form basaltic magma, which is extruded along ocean ridges, forcing the crust to expand sideways. Ridges are known as constructive margins, where new oceanic crust is formed. Since the Earth is a sphere of constant size, creation of new crust must be compensated for by the destruction of old crust. Where this happens, at destructive plate margins, old oceanic crust sinks down into the mantle along subduction zones, and is effectively recycled in a continuous conveyor-belt process. At subduction zones, where plates are colliding, the forces are enormous and give rise to powerful earthquakes. In three dimensions, slabs of oceanic crust dip down at 45° on average beneath continents or other ocean slabs and, as they are partially melted, lighter material emerges at the surface to form volcanic belts.

Collision zones may occur between two oceanic plates, creating island arcs (e.g. Japan), or between ocean and continent, forming mountain belts on the continent adjacent to an ocean trench, and characterized by earthquakes and volcanoes (e.g. the Andes); or between two continental blocks, where any intervening oceanic crust is subducted beneath one or both continents. The final result in the last case is a continental mountain chain, like the Himalayas. On a global scale, subduction effectively pulls slabs of oceanic crust into the mantle, and the growth of ridges pushes plates away, down the slope of the expanded hotter material beneath the ridge (Fig. 1.7). Individual plates move at different speeds, depending on the length of ridge or trench at the boundary, and the amount of continent relative to ocean. Today, the Pacific plate is moving fastest, at speeds of up to 100 mm per year. The African plate on the other hand is effectively stationary, since it has a large area of very thick

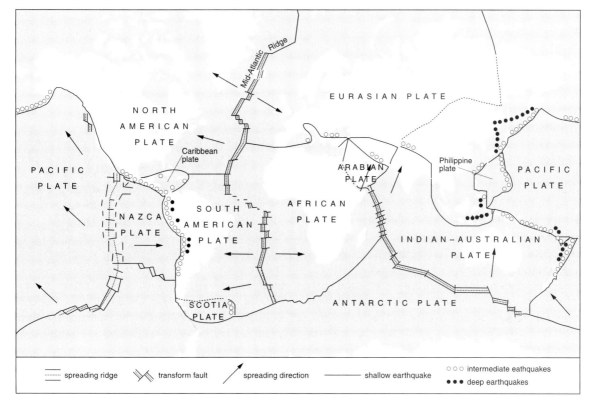

Figure 1.6 Sketch map showing the Earth's tectonic plates.

continental lithosphere and no currently active subduction zones at any of its margins. Plate-tectonic theory has been a powerful tool in explaining the evolution of Scotland throughout geological time, and this theme will be developed in later chapters.

Prior to the establishment of plate tectonics, a controversy had raged for decades about continental drift, whereby previously joined continents (e.g. Africa and South America) were thought to have rifted apart and separated. No mechanism was forthcoming that

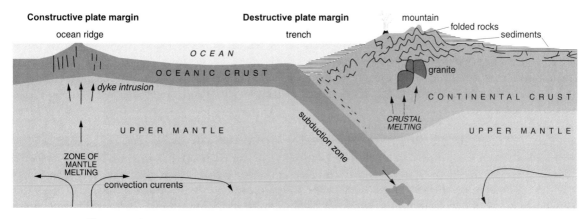

Figure 1.7 Cross section through plate boundaries (mid-ocean ridge and trench, with subduction zone).

would convince physicists, who claimed that continents could not move across solid oceanic crust. Many geologists also refused to accept continental drift, at least until the 1950s. However, Arthur Holmes (1890–1965), who was professor of geology at the University of Edinburgh from 1943 until 1956, suggested a mechanism for continental drift whereby continental crust was able to move as rafts on top of partially molten mantle that was moving in a series of convection currents, with hot material rising, spreading out horizontally beneath the continents, then finally returning down into the upper mantle. It was only in the year he died that convincing evidence began to be accumulated from the ocean ridges and the striped magnetic pattern of oceanic crust. However, the basic notion that the continents had shifted position over geological time was rather older, and was first hinted at by Sir Charles Lyell (1797–1875), a geologist born near Kirriemuir in Angus. He published his highly influential work, *Principles of geology*, in the early 1830s. Lyell maintained that changes in the geological past had been caused by forces still operating today, and that instantaneous catastrophes were not necessary to explain the natural features on the surface.

Since the oldest oceans are less than 200 million years old and the age of the Earth is 4560 million years, there has been more than enough time for many cycles of plate movements to have occurred. It is thought that plate tectonics may have operated more rapidly in the early history of the Earth, when the thin oceanic crust was forming and being rapidly destroyed and recycled by convection currents and meteorite impact damage. Once continental crust began to form, a different regime of plate tectonic activity became established, and inevitably continents began to collide and become welded together to form larger masses. Continental crust is lighter, thicker and more buoyant that oceanic crust and therefore cannot be subducted into the mantle or destroyed, so that the amount of continental crust has increased with time. On several occasions, several continents became attached (or accreted) to form extremely large supercontinents. The last such event happened 220 million years ago when all the continents had collided to form a landmass, named Pangaea (Greek: the whole Earth). Alfred Wegener (1880–1930) coined the term in 1915 in his book, which attempted to explain the origin of the continents, and it was this theory of continental drift that was supported later by Arthur Holmes. Wegener proposed that Pangaea split apart along a series of rifts and that the continents then moved apart. In plate tectonic theory, these rift zones are the mid-ocean ridges. The main rifting of Pangaea and the separation of the continents took place from about 180 million years ago, which explains why the oldest oceanic crust is that age.

Folds, faults and shear zones

We have seen above (p. 8) that sedimentary rocks have structures in them that reflect the processes at the time of formation, such as bedding, graded bedding, cross bedding, ripple marks, mud cracks and dune bedding. Likewise, igneous rocks may have features that are original and relate to the way in which the liquid magma cooled, such as crystal layering in large ultrabasic intrusions, flow banding and bubble cavities in surface lavas, pillow structures in lavas erupted under water, bedded ash, or irregular agglomerate blocks inside ancient volcanic vents. Cooling joints in igneous rocks relate to the development of shrinkage cracks that formed while a lava flow, a dyke or a sill was cooling and crystallizing relatively quickly. However, the rocks in Scotland have at many times in their long history been subjected to forces that have altered the shape of the original rocks and caused them to deform. This has mainly been the result of plate tectonic movements, when rocks were either compressed by collision or pulled into tension by thinning and rifting of the crust. The study of the effects of these forces is known as structural geology.

When rocks are stressed by being subjected to force, they are said to deform, and the amount of deformation – which is a change in the shape or position of a rock – is known as the strain: stress produces strain. Externally applied forces can affect rocks in three ways: compression (flattening, thickening or squashing), tension (thinning or stretching) and shear (sideways slip or gliding). Rocks will behave or react in different ways to stress, depending usually on depth of burial (and therefore pressure and temperature conditions at different levels in the crust), the presence of fluids and the rate at which the forces are applied.

So the deformation may be elastic if the rock goes back to its original state when the force is removed; or plastic, if the rock becomes ductile and flows into a new permanent shape; or brittle, if the strength of the rock is exceeded – it breaks or fractures, which is also permanent deformation.

Folds

Folds develop when thin and variable beds are compressed at deep levels in the crust. This is plastic behaviour and the folds represent permanent deformation – it is impossible to "unfold" a rock. However, rocks may be folded and re-folded several times during deformation. In the centre of mountain belts such as the Caledonian Highlands of Scotland, metamorphism and folding have gone on simultaneously, so that schists and gneisses formed, in which recrystallization has produced minerals different from the original starting materials. Folds are described according to their shape and orientation. The sides of a fold are called the limbs (Figs 1.8, 1.9); when the limbs dip in towards the bend or hinge, the fold is a V-shape syncline, and younger rocks are found in the middle. If the limbs dip away from the hinge, the fold is described as an anticline; here, older rocks are at the centre, in the core of the fold. In mountain chains, deformation can be so intense that folds may be bent right over on themselves, or may be re-folded several times. Large-scale crumpled folds, like those that form when a tablecloth is pushed across a table, are known as nappes (French: sheet or tablecloth). The base of a nappe fold will frequently become detached from the underlying rocks, resulting in huge slabs being pushed (or displaced) almost horizontally for great distances. Such features are seen to great effect in the Assynt district of northwest Sutherland, at the edge of the Caledonian Mountains.

Faults

Faults occur in brittle rocks when a break occurs and blocks shift past each other along a fault plane (also called a fracture surface). The total amount of movement is known as the fault displacement. The vertical movement is called the throw on a fault, so that faults have relative downthrow and upthrow sides. When rocks are pulled apart by tension, normal faults result, whereas reverse faults result from compression, and

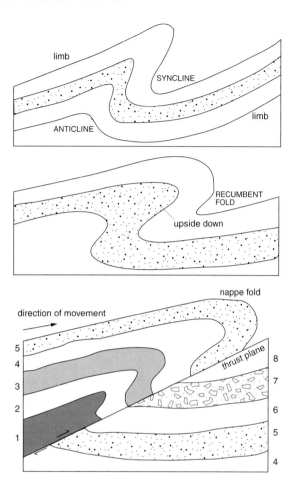

Figure 1.8 Sketches of fold types.

one block is upfaulted against its neighbour (Fig. 1.10). When the angle of a reverse fault is low, the fault is referred to as a thrust. In the case of vertical fault planes, one block will be pushed sideways past its neighbour on the opposite side of the fault line, to create a tear fault (also called a strike-slip fault; pronounced as in "Terra"). Movement may be either leftwards (sinistral) or rightwards (dextral). A block of crust downfaulted between two parallel normal faults is known as a graben (German: furrow or trench) or a rift valley. Large-scale examples of all these fault types are well known in Scotland:

- the Highland Boundary Fault and the Southern Uplands Fault are normal faults
- the Midland Valley is a graben or rift valley between these two faults

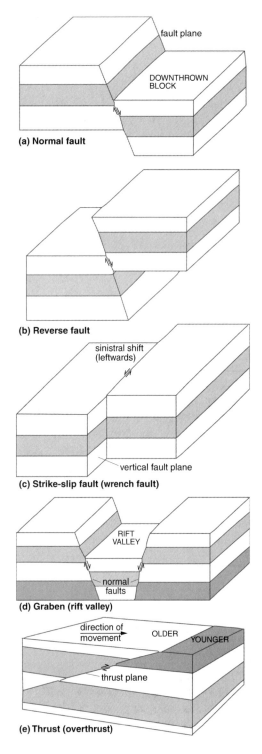

(a) Normal fault

(b) Reverse fault

(c) Strike-slip fault (wrench fault)

(d) Graben (rift valley)

(e) Thrust (overthrust)

Figure 1.9 Sketches of fault types.

- the Great Glen Fault is a sinistral tear fault, as is the Minch Fault
- the Moine Thrust and Outer Isles Thrust are examples of thrust faults
- the Pentland Fault in the east of the Midland Valley is a reverse fault.

Most of these major structures have long histories and have moved in different directions over their lifetimes. For example, the Highland Boundary Fault and Southern Uplands Fault were both initially sinistral strike-slip faults when they were initiated during the Caledonian mountain-building episode, but they subsequently acted as normal faults. The Great Glen Fault moved mainly as a sinistral strike-slip fault (i.e. the north side slid down the Great Glen to the southwest), but later moved as a dextral fault, and still later as a normal fault (Fig. 1.11). Large-scale structures such as these, which may penetrate the entire thickness of the crust, and which all formed 420–400 million years ago towards the end of the Caledonian orogeny (see Ch. 4), are long-lived features that controlled much of the subsequent development of the crust in Scotland.

Physical movement of one block past another will frequently produce fault breccia (broken rock) in a zone adjacent to a fault plane, and slickensides (Old

Figure 1.10 Folds in quartz-rich Moine schist, Monar, Central Highlands (outcrop at Monar Dam, Glen Strathfarrar, Inverness-shire).

Figure 1.11 Urquhart Castle (built of Old Red Sandstone) on Loch Ness, looking along the length of the Great Glen Fault, with its very straight shoreline. Nearby, rocks on Loch Ness shore are granite, pegmatite and gneiss, all probably of Moinian age.

English: to smooth or to polish), which are scratch-marks on rocks in the fault plane. In some cases, scratching and grinding are so intense that the fault planes become polished. Many larger faults have mineral veins, caused by solutions circulating in the fractured rocks within a fault zone, and quartz, calcite or hæmatite may be deposited in narrow fissures and cracks.

Shear zones

Faults form in the upper 1–10 km of the crust in the brittle zone, but below about 10 km the temperature is about 300°C and confining pressure is very high. Here we enter the plastic zone, where rocks behave quite differently and respond by flow and recrystallization, rather than fracturing. In the plastic zone, the combined action of depth of burial, high compressive stress, high temperature and fluid activity leads to the formation of shear zones, which are relatively narrow zones of high deformation that contain pronounced features such as schistosity or foliation parallel to the walls of the shear zone, and caused by platy minerals such as mica growing in alignment. Closely spaced small-scale folds may also appear to have their limbs

parallel; flattening and stretching have caused the folds to be streaked out into parallel alignment. Where rocks have been intensely crushed, ground down and recrystallized, the resulting banded streaky rock is referred to as mylonite (Greek: milled), a term first introduced in the Assynt district to describe the rocks at the base of the Caledonian nappes in the Moine Thrust zone. Linear arrangements of either needle-shape minerals such as hornblende, or clusters of deformed quartz and feldspar, or streaked-out fold noses, may produce lineations that penetrate the entire width of a shear zone. In some cases, circulating fluids have become concentrated along shear zones, and these may be the sites for ore deposits, including gold. Shear zones are formed only in the deeper crust, and they may be regarded as the equivalent of faults, but in which the movement of material was ductile, rather than brittle. At the surface, faults are relatively narrow structures, whereas shear zones may be considerably wider. Strain tends to be concentrated in narrow belts and thus rocks on either side of and between shear zones may be quite undeformed.

Deposition of first beds

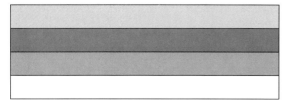

Folding and uplift

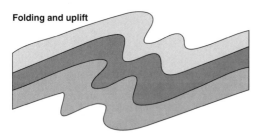

Weathering and erosion

Deposition

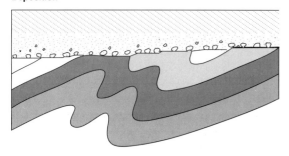

Figure 1.12 The formation of an unconformity.

Time and life

The concept of the vastness of geological time was first introduced by James Hutton, whose ideas still form the basis for modern geology. Hutton based his views on field observations and noted that natural laws must have operated over a long period of time to produce the rocks he could see. For Hutton, rocks had been produced by erosion from the continents,

deposition of the fragments in the sea, and their burial and lithification, followed eventually by uplift to form new continents, on which the cyclical processes immediately began to operate once more, a cyclicity that forced him to conclude that the Earth had "no vestige of a beginning, no prospect of an end". Hutton's ideas sharply contradicted the prevailing notion that all landforms had been produced instantaneously as a result of a catastrophic flood, from the waters of which all rocks formed as precipitates, first granite, then schist above, followed by sandstone, limestone and coal and finally the loose sand and gravel on top. Hutton went out around Scotland and found localities where the relationships between the various rocks utterly disproved this theory. He established the concept of unconformities in geology: early-formed beds are uplifted, eroded and eventually buried beneath younger rocks (Fig. 1.12). We now know that the Earth formed 4560 million years ago – so there is in fact a vestige of a beginning – and the end will come when the Earth's heat has finally been dissipated; then the lithosphere will thicken to such an extent that plate tectonics will fail to operate, with no volcanic belts and no subduction zones. The Earth will one day cease to be a dynamic planet, and life too will cease whenever the Sun runs out of energy.

Geological time is divided up into units of varying lengths. The divisions were established in the nineteenth century, mostly in Britain, and were based on extinctions of important fossils, usually of animals that lived in shallow seas, and the sudden appearance of new forms. An immediate difficulty arose: rocks without fossils were difficult to date, and their age could be judged only from the age of rocks above or below, presumed to be younger or older respectively. This is the relative method of dating, and it was not until the 1950s that absolute dating of rocks was made possible, following the discovery of radioactive decay of elements in rock-forming minerals. Arthur Holmes was the greatest pioneer of this approach.

Animals with hard shells that could be fossilized appeared quite suddenly about 550 million years ago, but the vast majority of older rocks, representing nearly 90 per cent of geological time, contain very few fossils, and most of those were soft-body creatures such as worms and jellyfish, or imprints of seaweed and algae. The first fossiliferous rocks to be described and defined were those belonging to the Cambrian period (545–495 million years ago), named by Adam Sedgwick in 1835 after the Latin name for Wales. The sudden appearance of shelly fossils at 545 million years ago has been described as the Cambrian "explosion" in evolution. Rocks lying beneath the Cambrian (i.e. those older than 545 million years), are called Precambrian. In Scotland, almost all the rocks of the Highlands are Precambrian in age. It is possible to subdivide all the periods into much smaller units, based on their fossil content, and these subdivisions can be refined further, particularly as we approach more recent periods. However, for the sake of simplicity, only these broad subdivisions are used in this book (Table 1.4).

Rocks and geological time had initially been divided into Primary, Secondary and Tertiary by the Italian Giovanni Arduino in 1759. To this was added the fourth or Quaternary period by Adolphe von Morlot in 1854. Only Tertiary and Quaternary are still in use today in Britain. Broad subdivisions of Precambrian time are Hadean (4560–4000 million years ago; from Hades, the Greek underworld), Archaean (Greek: ancient) (4000–2500 million years ago) and Proterozoic (2500–545 million years ago; Greek: first life). These subdivisions are not used in this book, and instead reference is made to the metamorphic complexes found in Scotland, which are the Lewisian, Moinian and Dalradian. The Torridonian is a Precambrian sedimentary succession. The remaining segment of time from the Cambrian to the present day is referred to as the Phanerozoic eon, meaning the time when life was visible and preserved in rocks as fossils.

Based on the fossil content of sedimentary rocks, it is possible to give relative ages of rocks and to match them up (or correlate) in different parts of the country. This method was introduced in Britain by William Smith (1769–1839), a civil engineer who was involved in building canals and who made the first geological map of England and southern Scotland in 1815. The study of rocks and the fossil sequences they contain is known as stratigraphy (from "strata" or "layers"). The method of correlating rocks has its limitations, though, because animals and plants are preserved after death only rarely. Most will rot or be scavenged, or be crushed by falling debris; there are not many environments conducive to fossilization. The fossil record is therefore incomplete and is biased to what happens to be preserved in certain conditions and then later found by geologists. Organic remains are most likely to be preserved in fine-grain sediments, provided that they are buried quickly enough to avoid decay or scavenging. Soft tissue usually decays rapidly and is almost never preserved. One important exception to this general rule is the Rhynie Chert from Aberdeenshire, which has exquisitely preserved fossil plants, among the earliest land plants found anywhere. They were suddenly engulfed by hot water emerging from volcanic springs, and the internal structures of the mosses and reeds were replaced by silica to form chert. Hard parts – bones, shells, and teeth – are slowly replaced during lithification, as rock forms from unconsolidated sediment. Land animals and plants are rarely preserved, nor are those remains that fall onto coarse pebbles and are smashed up by wave action. Some animals leave behind other evidence of their existence: trace fossils, which include footprints, burrows and feeding trails. The animals responsible for these marks are often never found.

As scientists were building up the geological column and the defining time periods within it, many such as William Smith observed that each separate period had its own unique set or assemblage of fossils that changed or evolved with time. Life probably made its first appearance on Earth by 4 billion years ago, once the surface had cooled enough to allow

Table 1.4 Major divisions of geological time (millions of years).

Time period	Age
Quaternary	1.8–0
Tertiary	65–1.8
Cretaceous	142–65
Jurassic	206–142
Triassic	248–206
Permian	290–248
Carboniferous	354–290
Devonian	417–354
Silurian	443–417
Ordovician	495–443
Cambrian	545–495
Precambrian	Older than 545

Table 1.5 The evolution of life on Earth (millions of years).

Age	Significant events in evolution
4560	Origin of the Earth; meteorite bombardment for 700 million years
4300	Oldest dated individual mineral grain (zircon from granite)
4030	Oldest dated rocks (granite)
3800	Primitive bacteria – life begins
3500	Cyanobacteria – first photosynthesizing organisms; oldest known fossils
	Stromatolites (algal mats) in shallow warm seas
2200	Single-cell plankton in seas; atmospheric oxygen increases
1300	Multicellular algae in seas
680	Soft-body animals – worms, jellyfish in seas
545	Animals with shells appear, in all varieties (invertebrates)
500	Jawless fish – first animals with backbones (vertebrates)
420	First land plants – club mosses, lichens, ferns, reeds in bogs; insects on land – first land animals; bony fish appear
370	Amphibians appear – land and sea
350	Reptiles appear – land, sea and air; dominant forms, including dinosaurs
300	Conifers; giant club mosses, etc., form first forests on land
150	Birds appear
100	Flowering plants appear and spread rapidly, with grasses to form prairies
65	Dinosaurs die out; mammals dominant on land
5	First human ancestors appear

Figure 1.13 Graptolites on slate.

water to remain as a liquid. Most life requires energy from sunlight, but some forms are able to exist deep on the ocean floor, surrounding hot black smokers – volcanic pipes that pump sulphide-rich hot water onto the floor. Animals living there use chemical energy to survive. Earliest life-forms were primitive bacteria that may have lived in such conditions. The first forms to use the Sun's energy to produce food (by photosynthesis) were the cyanobacteria or blue-green algae. During photosynthesis, oxygen is produced, and the action of cyanobacteria began to change the Earth's atmosphere and thereby enable other forms to evolve. In broad terms, the main stages in the evolution of life on Earth can be summarized as in Table 1.5.

Invertebrate fossils

Because of their relative abundance, fossil invertebrates – animals without backbones – have been the most useful group in correlating sedimentary rocks. Many groups exist, some of which are now totally extinct, but others have descendants alive today. Figure 1.13 shows graptolite fossils, and Figure 1.14 ammonites, both now extinct, but very important for correlating Silurian and Jurassic rocks respectively.

The main groups of invertebrates are sponges, with carbonate or silica skeletons – usually only their needle-shape spines are preserved; corals, which can build reefs and form limestone; bryozoans (sea mats); the extinct trilobites, shallow-water arthropods related to crabs, with a shell divided into head shield, segmented body, and tail; brachiopods (lampshells), with two shells hinged with teeth and sockets; molluscs, soft-body animals surrounded by a carbonate shell. The three main groups of molluscs are all very widespread in time and place: bivalves (cockles, mussels, scallops, oysters, razor shells, with two opening shells), gastropods (snails, limpets, whelks, with a single shell, often coiled into a spiral), cephalopods (octopus, squid, nautiloids, belemnites and the extinct ammonites with flat coiled shells and internal chambers); echinoderms (sea urchins, star fish, crinoids (sea lilies), with skeletons made of calcite plates, and

Figure 1.14 Jurassic ammonite fossils.

usually with fivefold symmetry); graptolites, an extinct group of floating colonial animals, with tiny polyps that lived in cups, attached together along branching arms; they resemble pencil marks on slate or black mudstone.

Microfossils

Of great importance to the oil industry are the micro-fossils found in drill cores from sedimentary rocks in the North Sea. These are the remains of tiny planktonic animals – radiolaria, foraminifera and conodonts – and seeds, pollen and spores from plants carried by the wind. Radiolaria have silica-rich skeletons in the form of intricate lattices; their remains can give rise to radiolarian chert, a microcrystalline form of silica. Foraminifera have chambered shells consisting usually of calcite, and they can be the main constituent of

some types of limestone, including chalk, which also contains calcareous algae. Remains of foraminifera and radiolaria contribute to deep-sea oozes found on the floor of the ocean. Conodonts are tiny tooth-shape fossil elements made of phosphate; they were produced by floating and swimming marine animals, and may be part of the feeding apparatus of types of small worms or eels that are now extinct. The first finds of conodonts were in the Carboniferous rocks of Scotland. They are so abundant, widely distributed and varied that it has been possible to devise 140 separate zones, making them one of the most useful fossils in stratigraphy. Ostracods are tiny shrimp-like crustaceans, often less than 1 mm long, with a hinged bivalve shell made of carbonate. Some forms are still found today, living in marine and freshwater environments. Other microfossils include diatoms, from single-cell floating algae; coccoliths from marine planktonic algae, one of the main constituents of the Chalk of southern England; and pollen grains and spores from land plants (trees, mosses, ferns) deposited in lakes and peat bogs. As their name suggests, microfossils are minute; they are rarely seen with the naked eye and can be identified only by means of a microscope. But their importance in correlation and dating of sedimentary rocks cannot be overstated, as well as their role in helping to unravel past environments. Their widespread geographical distribution is a key factor here.

Vertebrates

Vertebrates – animals with backbones – first appeared in the Cambrian period, but these earliest forms had no hard parts. The first forms with skeletons appeared in the late Ordovician and early Silurian as jawless fish that lived in shallow water, feeding on decaying matter that fell to the sea floor. In the succeeding Devonian period, armour-plated fish with jaws became very abundant and varied. One of the first geologists to study fossil fish in great detail was Hugh Miller (1802–1856), from Cromarty on the Moray Firth. Evolution since the Devonian period has successively given rise to sharks (with cartilage skeletons – only the teeth are preserved), amphibians (frogs and newts) at the end of the Devonian; reptiles (including dinosaurs, snakes, lizards, crocodiles) towards the end of the Carboniferous, culminating in the age of the dinosaurs in the Jurassic and Cretaceous, a period of 150 million years. The first bird appeared 150 million years ago, late in the Jurassic period. Although the oldest mammals originated at the same time as the dinosaurs, they remained small, the size of mice, and it was not until the extinction of the large reptiles at the end of the Cretaceous period 65 million years ago that the mammals rapidly evolved into many diverse forms occupying new environments, including open prairies and grasslands. Primitive human fossils are known from Ethiopia and have been dated at 4.4 million years old. Our own true ancestors appeared 2.4 million years ago, and *Homo sapiens*, our particular species, finally evolved 100 000 years ago.

The pattern of vertebrate evolution is such that in terms of size, variety and number, each of the main groups dominated for a particular span of geological time, giving rise to such popular terms as the Age of Fishes, Age of Reptiles, and so on. Other classes of vertebrates had certainly appeared earlier, but remained relatively insignificant until circumstances changed, and earlier dominant forms declined by extinction, leaving environmental niches to be filled by new groups. The geological timetable (also called the stratigraphical column) was devised on the basis of the changing assemblages of fossils, each period marking quite abrupt extinction of many families. Extinction of animals and plants has been a continuous process in the history of life on Earth, but, against this background of gradual change, the geological column is punctuated by a few mass extinction events. The most important of these occurred at the end of the Ordovician, Devonian, Permian, Triassic and Cretaceous periods. At least 95 per cent of all species then in existence became extinct at the end of the Permian, and 77 per cent of those in existence at the end of the Cretaceous, which is when the dinosaurs finally died out. This end-Cretaceous mass extinction at 65 million years ago is thought to be associated with the impact of a meteorite, 10 km in diameter, which hit Mexico and produced a dust cloud that encircled the Earth, blocking out sunlight for many years. However, the Deccan Plateau basalt lavas were erupting in India at this time, and may have contributed to global cooling. Mass extinctions were probably induced by changes in global climate brought on by changes in the size, shape and relative position of continents and shallow seas because of plate movements. Most animals live in shallow waters around continental shelf areas and, if the sea retreats from these areas, the loss of habitat is pronounced, with serious consequences for life-forms. During the past 250 million years, there have been 12 evenly spaced extinction events. Some scientists interpret this periodicity as indicating extraterrestrial events, such as giant meteorite impacts, as the Earth passes through meteorite showers.

Fossil plants

From an economic point of view, plants have been more important than fossil animals, since the coal deposits of the Midland Valley, and the oil and gas in the North Sea Basin, originated from plants. Microscopic algae are among the oldest known fossils on Earth. Land plants first appeared in the late Silurian period (420 million years ago), and the earliest forms had no true roots or leaves. Club mosses appeared in the Devonian and reached their maximum development 300 million years ago in the Carboniferous, when they rapidly spread to form the first rainforests, together with horsetails, tree ferns, seed ferns and primitive conifers. Their remains were rapidly buried in the stagnant swamps where they grew, and were compacted to form peat and then coal. The climate at the time was warm and exceptionally humid, and these coal swamps developed when Central Scotland lay at or near the Equator. At Fossil Grove in Glasgow there is a fragment of one of these forest environments, complete with petrified tree trunks in their growing position. A less extensive episode of coal formation occurred later, in the Jurassic, to produce seams at Brora and Helmsdale. Where coals (of whatever age) are deeply buried, they may act as the source for natural gas, which can migrate upwards into younger rocks. This is the case in the southern part of the North Sea, where the Carboniferous coals have produced gas. Most of the oil in Jurassic sedimentary rocks of the North Sea were derived from marine planktonic algae that lived in great abundance 140 million years ago in tropical seas. Their remains fell to the bottom and were attacked in the still, dark, stagnant conditions by bacteria. This organic-rich mud was buried, heated and altered to form fats, waxes and oils. Flowering plants did not appear until the middle of the Cretaceous period, 100 million years ago, then spread rapidly in the Tertiary, from 65 million years

Origin of the Earth

It is now commonly accepted that our Solar System – the Sun, planets and asteroids – began over 5 billion years ago as a cloud of gas and dust contracted under its own gravity. Swirling fragments collided and amalgamated together to form progressively larger bodies in a process known as accretion. Gravitational attraction meant that larger bodies grew by pulling in and capturing smaller particles and the energy of this process was converted to heat. In the space of perhaps a few million years, the Sun and planets had formed from the Solar Nebula, i.e. the gas and dust cloud, with the rocky Earth-like planets (Mercury, Venus, Earth, Mars) closest to the Sun, followed outwards by the larger, lighter, gas-rich planets (Jupiter, Saturn, Uranus, Neptune; Pluto may be a small remnant planetesimal left over from the formation of the Solar System). Heat from accretion, together with radioactive heat from the decay of elements, ensured that the young Earth was much hotter than it is today, and may have melted completely before cooling and forming the layered or differentiated planet – core, mantle, crust – that we now have. Bombardment from outer space is likely to have been intense, and a glancing blow from a giant body the size of Mars or the Moon may have caused a fragment of the Earth to be ejected to form the Moon at 4500 million years ago, i.e. shortly after the Earth had formed. Continuous meteorite impacts may have prevented the growth of the continental crust into plates of any significant size. High heat flow (up to three times present levels), a very hot mantle which may have been convecting more rapidly than at present, extensive volcanic activity and meteorite bombardment of a relatively thin crust combined to ensure that small continental blocks were being continually recycled. It may have taken about 100 million years before continental crust could remain stable, at about 4300 million years ago. These timespans are immense, of course, but relative to the age of the Earth, all the main events were completed in quite a short timespan – about the first 4 per cent of geological time.

Once the crust had stabilized, continents were able to grow. And since there is evidence that the earliest rocks contain material that was derived from pre-existing formations, we can safely assume that there was water on the surface, so the temperature had dropped relatively rapidly to allow this. With an atmosphere, geological processes on the early Earth began to operate in ways familiar to us today – rivers, waves, wind and rain did their bit in weathering and transporting materials across the surface. One immediate result was the destruction of evidence for meteorite impacts. The Moon's surface is clearly pitted with countless craters, preserved for billions of years because the Moon has no atmosphere, no volcanoes and no moving plates. In contrast, the Earth is a dynamic planet, with plates in constant motion and new rocks forming (and older rocks being destroyed) continuously.

ago; this coincides with the spread of the newly evolved mammals. Well preserved leaves from Tertiary plants are found at Ardtun on the Ross of Mull, and trunks in their growing position are known from the nearby Wilderness area on the Burg Peninsula (National Trust). These represent part of a Tertiary forest that was engulfed by lava during an eruption of the Mull volcano.

Dating of rocks

Although stratigraphy has allowed rocks to be dated relatively (i.e. such and such a rock is older or younger than some other one), particularly for fossil-bearing sedimentary rocks, geologists have long sought to give absolute figures in millions of years to the age when a rock formed. Lord Kelvin (William Thomson, 1824–1907), Professor of Physics at Glasgow University for 53 years, attempted to calculate the age of the Earth by assuming that the crust was once molten. In 1863, he arrived at a figure of 100 million years, which his contemporary, Charles Darwin, concluded was far too short a timespan to have allowed all the known animals and plants to have evolved. Kelvin had no idea then that much of the Earth's heat is produced by the radioactive decay of isotopes, and so his basic assumption was incorrect, and the Earth is much older. Using his knowledge of isotopic decay, Arthur Holmes was the first to pioneer the radiometric-age dating technique for minerals and rocks, from 1913 onwards. The various techniques are briefly outlined in Chapter 3 (see p. 52). An important development of the technique is the comparison between radiometric ages (sometimes also known as absolute ages) with indirect methods used to date sedimentary rocks, for example the strontium-87–strontium-86 method on carbonate rocks can give information about mountain building and climatic events recorded by marine animals, which incorporate some strontium into their shells. In Tertiary and Quaternary sediments, oxygen-isotope studies give us useful information about water temperature, and therefore past climates. Light oxygen (oxygen-16, ^{16}O) is stored in ice during glacial periods, hence the ratio of oxygen-18 (normal oxygen, ^{18}O, heavier than the 16 isotope) to oxygen-16 in animal shells can give a measure of temperature. Actual ages can be correlated with known events; for example, volcanic eruptions on Iceland have left ash deposits in the Greenland ice, so it is possible to draw up an accurate timetable of events.

Skye, Inner Hebrides, showing contrasting landscapes in Tertiary igneous rocks: Black Cuillin (right), Red Cuillin (left) and basalt lava (foreground) (photograph by Pat & Angus Macdonald).

CHAPTER 2
Geological regions of Scotland

Introduction

Although a small country, Scotland has an unrivalled diversity of rocks and landscapes. Most periods in the long history of the Earth are represented here, from the oldest hard crystalline rocks of the Outer Hebrides and Northwest Highlands to the younger and softer sedimentary rocks of central and southern Scotland. It is thanks to the underlying geology and the varying impact of different external processes that we have such varied landscapes, and the rocks have had a profound influence on the development of soils, land use, industry, transport and economic development.

Scotland owes the great diversity of its geology to the fact that it has always been at a sort of "geological crossroads", at the edge of a large continent. We now know that continents have been continually moving across the surface of the Earth, colliding together to form great mountain chains, or being split apart as new oceans are born. These processes have been taking place throughout billions of years, all over the Earth, but in different regions at different times. Once the activity ceased at any one place, the forces of erosion immediately began to erode the newly formed mountains to produce thick layers of sediment on low ground (or in "basins"), which sank slowly to accommodate the increased mass of material transported. Scotland's most recent mountain-building period took place 400–500 million years ago, to produce the Caledonian Highlands that stretch from Argyll to Shetland. What we have here, though, is only a small part of a much more extensive chain of mountains that runs from the Appalachians in North America, through Ireland, Wales and Scotland, then on northwards to eastern Greenland and western Norway (see Fig. 4.1). This mountain chain, although mostly worn down by erosion, forms the backbone of Scotland and is responsible for the northeast–southwest "grain" of the country. Almost the last event in the formation of these mountains was the faulting created when smaller portions of the land were forced against one another. Like the mountain chain, the faults also run roughly northeast–southwest, and now they divide Scotland into geological regions. From north to south the faults are the Outer Isles Thrust, the Moine Thrust, the Great Glen Fault (and the Walls Boundary Fault, its extension in Shetland), the Highland Boundary

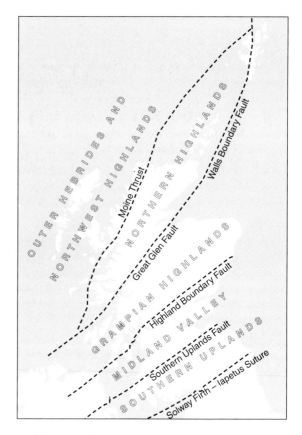

Figure 2.1   Sketch map of geological regions of Scotland.

Fault, the Southern Uplands Fault, and the Solway Fault, which more or less coincides with the Scotland/England boundary (Fig. 2.1). We shall shortly see that Scotland and England used to be parts of two quite separate continents. The union came about 420 million years ago during a collision that produced the Caledonian Mountains. Once welded together, they have remained in that position; so the political boundary is really a geological one too. Figure 2.1 shows how these large faults have controlled the coastline, particularly the eastern seaboard of the Outer Hebrides, the east coast of Caithness and Sutherland, the long inlet of Loch Linnhe off Mull, and the Solway Firth. Most of the industrial development took place in Central Scotland, between the Highland Boundary Fault and the Southern Uplands Fault. The Highlands were once almost impenetrable north of the Highland Boundary Fault, and the Great Glen Fault has always acted as the main transport route in northern Scotland.

Figure 2.2 Scourie Bay, northwest Sutherland. Lewisian gneiss in the foreground, Arkle (left, Cambrian Quartzite) and Ben Stack (right, Lewisian) in the background. Rocks in the foreground are Scourian gneisses.

Geological regions

Later chapters in this book are divided up along geological lines, the boundaries of which are these faults. The oldest rocks are found in the Outer Hebrides and the Northwest Highlands, to the west of the Moine Thrust zone (Fig. 2.2). Between the Moine Thrust zone and the Great Glen Fault lie the Northern Highlands, with rocks folded and re-folded several times, on which rest the Old Red Sandstone sediments of Devonian age in Orkney, Caithness and around the Moray Firth. From Argyll to Shetland is an immense tract of metamorphic rocks, punctured by the granites of Etive, Monadhliath, Cairngorm and Aberdeenshire. This belt of ground, the Grampian Highlands, has as its margins the Great Glen Fault and the Highland Boundary Fault (Fig. 2.3). Central Scotland, or the Midland Valley to give it its geological name, is a block of crust that occupies the low ground between a

parallel pair of faults, the Highland Boundary Fault and the Southern Uplands Fault. Within the Midland Valley are mostly young sediments, including Old Red Sandstone and the Carboniferous rocks, which include the coal, limestone and sandstone on which much of Scotland's early industrial development depended. Volcanic lavas and old eroded volcanic plugs abound in Central Scotland, especially in Fife, the Lothians and around the River Clyde (Fig. 2.4). Southwest Scotland and the Borders, between the Southern Uplands Fault and the Solway Fault, consist mainly of folded sediments laid down in an ancient ocean, and punctured by a few granites in Dumfries and Galloway (Fig. 2.5). The border itself is marked in part by the Cheviot Hills, made of old lavas and the granite massif of the Cheviot. It is almost as though these once-molten rocks stitch Scotland and England together. Subsequent events in the 250-million-year interval between the Carboniferous coal swamps and

Figure 2.3 Dalradian rocks, Argyllshire highlands, from Greenock looking north beyond the Highland Boundary Fault across Firth of Clyde.

Figure 2.4 Stirling, looking from the Wallace Monument across the River Forth meanders and flood plain, low ground of Carboniferous sedimentary rocks covered by glacial and postglacial sediments.

the Tertiary volcanic eruptions have left only relatively minor traces in the coastal fringes of the northeast and the Inner Hebrides, and the low-lying valleys of the southwest.

Our journey through the rocks will finally, and perhaps surprisingly, take us back up to the northwest, to the Inner Hebrides, for Skye, Rum, the Small Isles, Ardnamurchan (St Kilda and Ailsa Craig too) have the youngest solid rocks in Scotland. About 60–55 million years ago, this region was an active volcanic zone. The Cuillins of Skye and Rum, and Ben More on Mull, are the now-eroded stumps of powerful volcanoes that poured out lava across the western seaboard for two or three million years. Goatfell on Arran (Fig. 2.6) and the islet of Ailsa Craig are granite bodies, whereas the rocks of the St Kilda Group are in many respects closer in type to the Cuillin of Skye. Vertical walls of rock, called dykes, radiate out from these volcanic centres, and some of them, especially those from the Mull volcanic centre, stretch great distances to the southeast into northeast England, cutting across older rocks. They must represent cracks in the Earth's crust that were filled with molten rock as it made its way to the surface to feed the lava flows. While volcanoes were erupting on the west coast during the Tertiary, the land surface was tilted to the east and the main river pattern was established. These rivers removed prodigious amounts of weathered rock debris, which found its way into the North Sea. Global cooling during the late Tertiary, from 10 million years ago, associated with tectonic movements that greatly affected ocean current patterns and resulted in the clustering of large continental masses around the North Pole, coupled with changes in the Sun's heat, led inexorably to the establishment of glacial conditions. Ice sheets waxed and waned many times over in the past two million years, leaving their mark on the landscape, as

Figure 2.5 The Scottish Borders: plateau and rolling hills of greywacke, river terraces in foreground, in Tweed catchment area.

Figure 2.6 Goat Fell, Arran: Tertiary granite intrusion, cutting Dalradian schist (lower slopes); New Red Sandstone in foreground, Brodick Bay in middle distance; from Claonaig, Kintyre; the deep Kilbrannan Sound is in the foreground.

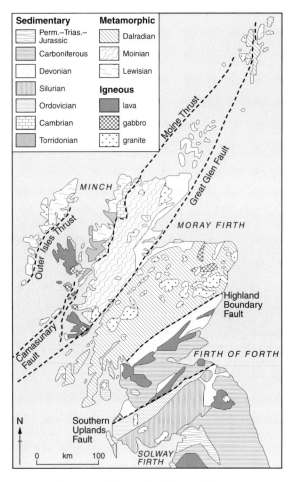

Figure 2.7 Outline geological map of Scotland.

the fundamental pieces having been assembled about 400 million years ago. The great event that produced the basic structure was the Caledonian mountain-building episode (called an orogeny). During this orogeny, Scotland and England, which had previously been parts of different continents on opposite sides of a wide ocean that geologists have called the Iapetus, were forced together in a collision of the Earth's plates. The collision caused the Iapetus Ocean to disappear, and the sediments that had been laid down in the ocean were crumpled into folds, and then faulted. As the collision proceeded, the crust continued to thicken while Scotland and England were being welded together. Energy produced by the collision and thickening of the crust led to a steep temperature gradient and therefore metamorphism of the rocks caught up in the folding. At many locations, the heat was intense enough to cause some of the rocks near the base of the thickened crust to melt and form granites. Because granite is lighter than the surrounding rocks, it moved upwards to higher levels in the crust, thereby contributing to the elevation of the new mountain chain. At the very end of the orogeny, the major faults became active, and here unfolds a story of incredible complexity. Enormously powerful forces were responsible for sliding segments of the crust into place along these faults. Westward-directed forces carried the folded rocks of the Northern Highlands out in the direction of the Hebrides along the gently dipping Moine Thrust, while the Grampian Highlands slid into place along the Great Glen Fault. Central Scotland formed as a block of crust sank between the Highland Boundary Fault and the Southern Uplands Fault to create the Midland Valley.

Rocky deserts

These events did not take place overnight. In fact, the Caledonian orogeny lasted about 100 million years, immense by human standards. However, when you realize that the Earth's plates are still moving today, some of them at a rate of 10 cm a year, then you can work out that in 100 million years one part of the crust has plenty of time to move around the globe. When continents move, they pass through different climatic zones, and Scotland was no exception. Once the Caledonian Mountains had formed, Scotland lay somewhere south of the Equator and well away from

both erosional and depositional features. The Hebridean volcanic province has a north–south orientation and thus cuts sharply across the northeast–southwest grain of the country, which was established at the end of the Caledonian orogeny.

Figure 2.7 shows in general terms that the old hard metamorphic and igneous rocks are mainly in the north and west of the country, and the younger, softer sedimentary rocks are found mainly in the south.

Geological evolution

Caledonian Mountains

In terms of its underlying structure then, Scotland is something of a geological jigsaw puzzle, with most of

Figure 2.8 Graemsay, Orkney: Old Red Sandstone conglomerate (fossil scree) resting on Precambrian granite basement.

the moist effects of any ocean, just the sort of place one would expect to find desert conditions. And as the young, high mountains were rapidly worn down, the debris was carried by floods and short-lived rivers into upland valleys and onto the low ground of Central Scotland – the Midland Valley – and what was to become a large, shallow, land-locked lake in the northeast, around the area that is now Orkney, Caithness and the Moray Firth Basin. These rocks are the famous Old Red Sandstone, of Devonian age (Fig. 2.8).

As Scotland continued its drift northwards, it moved into tropical regions, where there were warm shallow seas, and eventually it crossed the Equator about 315 million years ago during the Carboniferous period. Volcanoes were erupting in Central Scotland at this time. Rivers continued to erode the Caledonian Mountains, bringing mud, silt and sand into the tropical sea, which was gradually clogged up with sediment. River deltas and coastal swamps formed, to be invaded by the Earth's first dense rainforests, whose rotting remains piled up and were buried to make coal. Continuing its drift, Scotland entered desert conditions once more, 280–200 million years ago, this time north of the Equator. The sediments formed then (in the Permo-Triassic) are the New Red Sandstone. Reptiles roamed the land, probably around waterholes, their fossil footprints well preserved, for example, in sandstones at Hopeman near Elgin.

During subsequent periods of geological time, sea level was higher and most of Scotland was flooded, apart from the mountains. Sand and mud, later chalk, were laid down in warm tropical seas, swarming with abundant life. Rocks formed then still exist in the

North Sea Basin and on land around the Moray Firth and Inner Hebrides; otherwise, much of this 140-million-year timespan has been lost to erosion.

By 65 million years ago, Scotland was coming close to its present latitude and was mostly above water, but with a warmer and even wetter climate than today. The North Atlantic Ocean was then opening up, and a north–south chain of volcanoes running from Skye to Ailsa Craig formed. Such was the force of these volcanoes that the crust in the west was pushed up and Scotland was tilted to the east. It was at this time that most of the major rivers formed, all of which flow generally eastwards into the North Sea or south- and northeastwards into the Moray Firth. Erosion in the subtropical climate was deep and rapid, and vast quantities of rock were removed from the land and carried east to fill the North Sea Basin. It was at this time that most of the major rivers formed, almost all of which flow into the North Sea, or northeastwards and southeastwards into the Moray Firth. It was during this long 50-million-year period of weathering and erosion in Tertiary times that the essential features of the Scottish landscape developed.

Scotland in the Ice Age

About two million years ago in the Pleistocene period, the climate suddenly deteriorated and Scotland was plunged into the most recent ice age. Continental glaciers formed rapidly, and grew to a thickness of over 2000 m, completely covering the land surfaces and mountain tops with a slowly moving ice sheet. As the ice made its way from high ground to low, the powerful grinding and scouring action of the glaciers removed more rock. Glacial erosive processes effectively smoothed the features developed in the Tertiary period. Then, by 10 000 years ago, the ice melted more swiftly than it had appeared. The resulting meltwaters dumped enormous quantities of sand and gravel, which had been trapped and carried in the ice, over the lower ground and in the waters around Scotland.

Today we await the return of the ice. Meanwhile, rivers are still doing their work of transporting sediment to the sea, and the sea itself is constantly attacking the coast and wearing away the cliffs. Chemical weathering and physical erosion continue, albeit at a slower pace than during the warmer and wetter Tertiary period. The Earth forces that created Scotland are continuing their work today, shaping the land surface and modifying the landscape. Additional changes are now also being made by our own actions, as we extract ores, rocks and fossil fuels, dump the waste, farm the land, build roads, bridges, dams and towns, attempt to shape rivers and divert their paths, and defend the coastline. But Nature is more powerful than human beings, and nothing that we do can stop the Earth's processes from continuing to shape the planet. This book, therefore, represents a journey through space and time, explaining how Scotland was created and how the interplay between the bedrock and the forces of erosion have created the landscapes we see today.

So much, then, for the broad outline, which is relatively straightforward, and is summarized in Table 2.1. In order to follow the continuous narrative, the table begins with the oldest rocks at the top.

Lewisian
Europe's oldest rocks are found in the Outer Hebrides and in a narrow strip on the northwest coast, from Cape Wrath to Skye, Coll, Tiree and Iona. These are the Lewisian gneisses, high-grade metamorphic rocks, originating over 3 billion years ago, and intensely folded and deformed in two main episodes at 2900 and 1800 million years ago. The rocks were metamorphosed at temperatures of up to 900°C and pressures equivalent to a depth of 30 km in the continental crust. At 2200 million years ago, stretching of the ancient crust led to continental rifting and the intrusion of many basic dykes and lava outpourings. Thick sediments were also deposited then (the Loch Maree Group), including limestone (metamorphosed to marble), sandstone (metamorphosed to quartzite), shale (metamorphosed to mica schist) and banded ironstone. Iron ore was extracted and smelted locally. Gold is also present, but was never mined. The Lewisian Gneiss Complex represents a fragment of deep-level continental crust, belonging to Laurentia, that was involved in collisions, as great crustal blocks slid past each other, forming zones of intense thinning and high deformation – shear zones, along which water, carbon dioxide, and heat-producing radioactive elements (uranium, thorium and potassium) were removed, to make the Lewisian crust effectively dry. Significant quantities of granite were formed at 1800 million years ago by melting of the lower crust.

All of the dominant features in the Lewisian block – folds, shear belts, basic dykes and granite sheets – run northwest–southeast; this has been accentuated by erosion, so that the coastline has the deep indentations of fjords or sea lochs aligned consistently with that trend. Scotland then was part of Laurentia, which included Canada and Greenland. There are many areas of Lewisian gneiss in the Northern Highlands to the east of the Moine Thrust, within the Caledonian mountain belt, and the Lewisian is probably present as the basement in all of northern Scotland. On Islay there are metamorphosed igneous rocks 1800 million years old, derived from the mantle, that superficially resemble some Lewisian gneisses, but are quite distinct. These, together with some overlying rocks of uncertain age on Colonsay, probably belong to a block of crust that includes northwest Ireland, wedged between offshoots of the Great Glen Fault. The metasediments in the east of Iona resemble the Colonsay and Islay rocks, whereas the western half is made of Lewisian gneiss and marble.

Torridonian
By 1200 million years ago, uplift and erosion of the crust resulted in the Lewisian basement rocks being exposed at the surface. Large easterly flowing rivers became established in the region of Greenland and Canada, which then lay much closer to Scotland. These rivers deposited thick continental redbeds in fault-bounded basins in the northwest of Scotland. The sedimentary rocks that formed then are known as the Torridonian, which are the oldest sedimentary rock sequences in Britain (i.e. which have not been metamorphosed). Boulders of Lewisian gneiss are found among these deposits, which fill in hollows and cover low rounded hills of the basement (e.g. on Slioch). Recent erosion has therefore brought to life narrow zones of a billion-year-old landscape, although only a few isolated mountains of Torridonian sandstone now remain of what was probably an extensive sheet of rocks covering the Lewisian. The red colour is explained by the presence of feldspar derived from the Lewisian. Temporary lakes were formed, in which red muddy limestones are found draping Lewisian boulders; these represent mounds of algae that grew in warm shallow water. Beds of volcanic ash are also present, but no volcanic centres have yet been found.

Table 2.1   A brief history of the geology of Scotland (ages in millions of years).

Geological period and age	Latitude and climate	Main tectonic events in Scotland	Geological setting and environments
Lewisian 3300–1750	Position unknown; part of Laurentia	Metamorphism and igneous intrusion deep in crust (20–30 km)	Volcanic and sedimentary rocks at surface, on granitic crust; banded ironstone, limestone, basic lavas formed in sea
Moinian 1100–800	At edge of Laurentia; temperate	Folding and metamorphism at 780 and 450	Shallow-water sediments (sand and shale) deposited on Lewisian basement; birth of Iapetus Ocean
Torridonian 1050–850	Moved from north of Equator to south; arid	Rifting in the Minches to create fault-bounded basins	Rivers brought coarse sediment from eroded gneiss terrain (Canada and Greenland) to form continental red-beds
Dalradian 750–550	Close to South Pole; temperate, then cool; glaciations	Thinning and rifting of continental crust to create Iapetus Ocean crust	Sediments in sea, shallow then deeper, in separate fault-bounded basins: sandstone, shale, mudstone, limestone; basalt lavas on ocean floor; glacial deposits
Cambrian 545–495	30°S; tropical, warm, humid	Iapetus Ocean opens to its maximum extent (4000–5000 km possibly)	Sandstones then limestones deposited in warm shallow shelf sea; first fossils with hard shells – explosion in evolution
Ordovician 495–443	Moved from 30°S to 15°S; warm	Iapetus Ocean starts to close as southern continent (Avalonia) approaches; oceanic-island arcs	Limestones form on shelf of warm, shallow sea in NW; volcanic islands in ocean; deepwater muds in south
Silurian 443–417	Around 10–15° S; warm to hot, tropical	Iapetus Ocean closed at 420, Laurentia and Avalonia collide; growth of Caledonian Mountains; faulting	Land area in Highlands; deepwater greywackes in south scraped off trench and folded; sediments and igneous rocks on site of future Midland Valley
Devonian 417–354	Moved from 10°S to 5°S; hot, arid desert	Uplift and rapid erosion of mountains; lavas and granites at 400; opening of a sea in the far south (Devon and Ireland)	Scotland in interior of large continent; desert redbeds (Old Red Sandstone) in south; lake deposits (flagstones) in NE, with abundant fish; primitive plants around hot volcanic springs in north
Carboniferous 354–290	5°S to 15°N, crossing the Equator; arid, then humid, then semi-arid	Volcanoes in Midland Valley, and shallow seas, then deltas and swamps; Variscan mountains form in south	Coral limestones form in clear warm shallow seas; sandstone deposited by rivers, then deltas silted up, invaded by dense forests, coal swamps at sea level
Permian 290–248	15–25° N; hot, arid desert	Volcanoes in Midland Valley; erosion of uplands; Pangaea forms	Dune-bedded red desert sandstones form in valleys; evaporites form in shallow inland Zechstein Sea to the east
Triassic 248–206	Northward drift, 25°N to 35°N; hot arid desert	Stable conditions; Pangaea supercontinent remains in place	Desert sandstones, occasional shallow lakes; evaporite deposits; Elgin reptiles abound, living on dunes around oases
Jurassic 206–142	Moved from 35° to 40°N; subtropical then warm and humid	Break-up of Pangaea; rapid rise in sea level; lowlands flooded by shallow sea	Shallow-water limestones, then sandstones and organic-rich shales (oil source rocks); reptiles abundant; rift valleys in North Sea region filled with sediment
Cretaceous 142–65	Around 40–45° N; warm	Scotland mostly upland, then parts of west coast drowned	Clear warm shallow seas in later part; chalk deposited, very thick in North Sea region; dinosaurs become extinct
Tertiary 65–2	Around 50°N; subtropical, warm, humid, becoming cold and dry later	Scotland an upland region; volcanic centres intruded in west, land tilted to east and southeast; opening of North Atlantic Ocean; Scotland separates from North America	Lavas poured over land surface, 60–55 million years ago; occasional shallow lakes between eruptions; growth of dense forests, low-grade coal forms; vast amount of weathering and erosion, material carried to North Sea; establishment of the main river pattern; sudden climatic cooling at end of period
Quaternary 2 million years ago to end of the Ice Age	57°N; glacial conditions with interglacials	Continents clustered around North Pole; ocean-current pattern altered	Ice sheets grow and melt many times over; erosion, then deposition from ice and from meltwater streams
Holocene 10 000 years ago to present day	57°N; interglacial conditions	Rapid climatic change	Last ice melted; sea level rose, then land rebounded; forests grew then died, peat forms; human impact increases

Moinian

The Northern Highlands, lying between the Moine Thrust zone and the Great Glen Fault, are made of what is often described as a monotonous series of pale grey schists, the Moine rocks (or Moinian), representing metamorphosed shallow-water sandstones and shales, laid down in a slowly subsiding basin 1000–870 million years ago. These thick sedimentary rocks were folded and metamorphosed in an episode of mountain building termed the Grenville orogeny, one of a series of continental-collision events that gave rise to the large supercontinent of Rodinia. The shallow-water basins represent the beginnings of a new ocean, the Iapetus, as Rodinia began to break up 750–600 million years ago. Although there is some overlap in the age of the Torridonian and the Moinian, it is no longer believed that the two sequences actually formed beside each other, or that the Moinian represents metamorphosed Torridonian.

Dalradian

Dalradian rocks form a wide belt from Shetland through the Grampian Highlands to Argyll, and include rocks at the base that resemble the older Moine rocks in many respects, suggesting continuity of deposition. However, the Dalradian rocks are, at 25 km thick, much thicker and more varied than the Moinian, at least in the upper units, reflecting deposition in separate fault-bounded basins that sank to accommodate increasing amounts of sediment. In addition to schists and quartzites, there are also limestones, slates, phyllites, grits and volcanic rocks, as well as glacial deposits that indicate a sequence of glaciations 650 million years ago when Scotland was far to the south. This period of time (800–570 million years ago) has been referred to as Icehouse Earth, because glacial deposits (tillites, including the Port Askaig tillite of Islay, which is correlated with the Varanger Tillite of Norway) are rather widespread, and it is possible that continental ice sheets stretched much closer to the Equator than at any other time. The reasons for this apparently severe period of glaciations are not entirely clear, but are probably related to the amount of carbon dioxide in the atmosphere, which in turn depends on the amount of biological activity, the positions of continents, the height of mountain chains, and the intensity of weathering and erosion. It seems probable that the sedimentary rocks were laid down in separate adjacent basins over a period of 200 million years or more, because a complete 25 km thickness is not seen at any one place, and in any case this would represent a most unusual thickness for a sedimentary succession. The Tayvallich Volcanics, at about 600 million years old, represent the rifting of continental crust and the final opening of the Iapetus Ocean.

Cambrian to Silurian

In Northwest Scotland, Cambrian and Ordovician sediments (sandstone and limestone, with the first ever shelly fossils) were deposited in shallow shelf seas at the edge of the Iapetus Ocean, while shales and muds were laid down on top of Dalradian rocks (or possibly in one or more separate basins) along the Highland Border, at the edge of the future Midland Valley. Farther to the south, the Iapetus Ocean was much deeper, and thick accumulations of sand, grit and mud were carried down into the depths to form greywackes, sandstones that resulted from rapid turbidity currents. Slow sedimentation of fine mud on top of the greywackes allowed the delicate graptolite fossils to be preserved. These little animals assumed exceptional importance in the dating and correlating of the Ordovician and Silurian rocks in the Southern Uplands (see p. 95). Then, as the Iapetus Ocean began to close, chains of volcanic islands were thrown up, and their products formed not only pillow lavas on the ocean floor but also debris that was transported as sediment into the deeper ocean basin. While the ocean was closing at an ever-increasing rate, the sediments forming above an ocean trench in the area destined to become the Southern Uplands were scraped off and thrust northwards, and the island arcs were progressively swept up and crumpled between two advancing continents: the Iapetus was doomed. Very early in the original Geological Survey work, it was noted that the fossils in Scotland resembled those in Canada in rocks of the same age, but were entirely different from those in Wales, suggesting a wide gulf between the two regions. The gulf was the Iapetus Ocean. Inliers (areas of older rocks surrounded by younger) of Ordovician and Silurian rocks in the Midland Valley represent remnants of volcanic and sedimentary arcs that existed in the Iapetus and were swept up during continental collision, to become attached to Scotland.

The Caledonian Mountains

What happened next, 400 million years ago, was to shape the remainder of Scotland's geological evolution. These events resulted in the collision of Scotland and England, and the complete closure of the Iapetus; all that remains is the suture line, marking the place where they are joined. All the rocks of the Highlands and Southern Uplands were strongly deformed by folding and thrusting. Metamorphism was more intense in the north, and grades increase into the core of the mountain chain northwards away from the Highland Boundary Fault, while in the Southern Uplands the grade remained very low. The Dalradian rocks were converted to slate and schist at the same time that they were folded into immense tablecloth folds (nappes) pushed outwards and over to the northwest and southeast. Near the Highland Boundary Fault, the nappe was bent over into a steep structure, and over much of the Grampian Highlands the Dalradian rocks are upside down in a flat belt. In a final phase of upheaval, the nappes were pushed across to the northwest, many kilometres, such that older rocks now lie on top of younger formations in the Northwest Highlands – a situation that can be produced only by the forces involved in continental collision. The Moine Thrust and other such faults in Assynt are famous throughout the world, and have been of fundamental importance in helping to unravel the evolution of mountain belts. As a result of the Caledonian orogeny, Scotland was assembled as a geological jigsaw, mainly as several segments were moved sideways into place along faults trending northeast–southwest, including the Great Glen Fault, Highland Boundary Fault and the Southern Uplands Fault. The last two faults later moved again, as a pair of normal faults, so that the crust between them sank to create a rift valley, later to become the Midland Valley. Thus, Scotland was born: mountains to the north, hills to the south, a low-lying basin in the middle.

Devonian

At the ending of the Caledonian orogeny 400 million years ago, the mountain chain was uplifted by large volumes of granite being intruded from the base of the crust to much higher levels. A great swathe of granite outcrops runs from Peterhead and Aberdeenshire in the northeast, through the Cairngorm Mountains to Rannoch Moor and eventually the Ross of Mull. The granites in the mountain chain were derived by partial melting near the base of the thickened continental crust. Dalradian rocks in the Grampian Highlands consist mainly of schists, originally derived from shale, mud and silt. Such rocks are rich in hydrous minerals (mica and clay in particular) and in heat-producing elements; they are susceptible to melting within fold mountain chains, such as the Caledonian.

Granite magma is the first product of such melting and, being a low-density liquid, it makes its way up to higher levels in the crust. Volcanoes at the surface are an expression of this igneous activity, and examples of this can be seen in the lava flows of the Ochil, Sidlaw and Pentland hills in the Midland Valley, as well as around Glen Coe and the Lorne Plateau near Oban. Some of these huge volcanoes were involved in cauldron collapse around circular faults. The Glen Coe cauldron subsidence appears to have been controlled by deep faults in the crust, with collapse taking place several times over a long period, resulting in a volcanic caldera at the surface. Igneous activity was intensely explosive, a feature common to volcanoes in continental settings, where the magma is much richer in silica than the basalts of the ocean floor. In the northeast, hot volcanic springs were active and those at Rhynie have preserved the world's oldest land-based ecosystem of primitive plants, arthropods and crustaceans, petrified in their living position.

Following the collision between Laurentia and Avalonia, Scotland was landlocked and surrounded by mountains, and some 5–15° south of the Equator, in an arid climatic zone. Trees and grasses had not yet evolved, and the bare rocky landscape was rapidly eroded as screes from mountainsides were transported in flash floods to the valley floors and spread out in the Midland Valley along the edges of the mountains (i.e. near the Highland Boundary Fault and Southern Uplands Fault). These are the Old Red Sandstone deposits, which are about 10 km thick in total. At the same time, a large inland sea developed in the northeast, stretching from Shetland, through Orkney, Caithness and the Moray Firth, a sea (or possibly a northerly embayment of the ocean that existed to the south) that expanded and shrank as the climate changed. The fine silty sediments deposited in this basin, Lake Orcadie, were compacted to form the

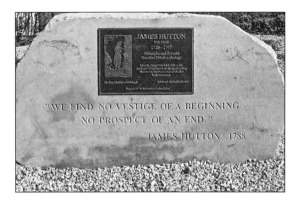

Figure 2.9 The plaque to James Hutton (1726–97); St John's Hill, Edinburgh, at the site of Hutton's house.

Figure 2.10 Siccar Point, Berwickshire: Hutton's unconformity; Old Red Sandstone at top, dipping gently, above vertical Silurian greywacke and shale.

Caithness Flagstones. Occasional incursions of the sea brought a great diversity of fish communities into the lake. During times of drought or the growth of algal blooms, the fish were killed off in mass-mortality events to create fish beds; these fossils were intensely studied by Hugh Miller (see p. 116) among many others, and have provided important insights into the evolution of the fishes. Far to the south, where Devon is now, a new ocean appeared.

Evidence for the intensity of erosion in the Devonian is to be seen in the conglomerates containing boulders and pebbles of granite, resting on the very granites from which they were derived. These granite bodies had been intruded into the crust at a depth of some 5 km only a few million years previously. The ancient screes, which these deposits represent, are especially clear around the Moray Firth, suggesting a fossil landscape of granite hills encircling Lake Orcadie. These granites were the source of the uranium found in flagstones. James Hutton was the first geologist to identify and interpret the situation of eroded material having been deposited on the remnants of older rocks beneath; he discovered his famous unconformities at Siccar Point, Jedburgh and Arran in the 1780s (Figs 2.9, 2.10; see also p. 20).

Carboniferous

During the period 350–300 million years ago, Scotland gradually moved from below the Equator to just north of it, in a monsoon climate, where equatorial rainforests thrived as never before. To begin with, volcanic activity was widespread in the Midland Valley,

as the crust was stretched and the rift valley subsided. Basaltic lavas were poured out from a thousand fissures and vents across the northern part of the Midland Valley. Landforms then created account for the most prominent features of Central Scotland – the Campsie Fells and Dumbarton Rock in the west and the eroded vents of Arthur's Seat, North Berwick Law and the Bass Rock in the east. Shallow tropical seas encroached on the Midland Valley, depositing limestones with a rich and diverse fauna of corals, brachiopods, bivalves and sharks. Lagoons and inland lakes continued to exist as the volcanoes gradually subsided, and in these environments varied communities of amphibians, early ancestors of the reptiles, and insects (including 2 m-long centipedes) thrived. The oldest known reptile, the fossil Lizzie, was found in West Lothian in rocks of this environment (see p. 122). Renewed uplift in the now-denuded hills produced sand that was transported by large rivers and laid down on deltas that advanced across the coral seas, killing the reefs and burying them under thick beds of sandstone. As the deltas silted up, dense tropical forests colonized the swampy ground around sea level. The trees of the rainforests were killed by forest fires and floods. Plant debris piled up rapidly and the peat beds were buried under new layers of sand, the organic matter being eventually transformed to coal. Lagoons that received floating masses of algae and rotting vegetation among the fine muddy and silty sediments were the source of the Carboniferous oil shales. Coal and oil shales ensured Scotland's place in the industrial revolution, enhanced by ironstones of the same age that formed in the stagnant and brackish-water swamps and lagoons. Relatively gentle folding at the end of the Carboniferous (the Hercynian and Variscan movements of northern and central Europe) caused the Carboniferous rocks to be folded into three more or less north–south synclines, which coincide with the Ayrshire, Central and Fife–Midlothian coal-fields. Scotland continued to drift northwards into the arid climatic belt of 15–25° north of the Equator.

Permian and Triassic
Following the compression that produced the north–south folds of the coal basins, the crust responded by relaxing, and as a result, an important set of east–west dolerite dykes were intruded into the Midland Valley. These are now exploited as valuable sources of crushed rock. Desert conditions became established once more, both in the wake of the Variscan earth movements and because of Scotland's position in arid latitudes. Windblown red sandstones piled up thickly in valleys around the west (Ayrshire and Arran) and along the southern shores of the Moray Firth, as well as in offshore basins in the North Sea and the Firth of Clyde. Reptile footprints on fossil sand dunes and various reptile skeletal remains are known from the Elgin–Hopeman area. To the east lay a salty shallow inland sea, the Zechstein Sea, source of thick evaporites that were deposited from the briny waters in the baking heat. The Permian and Triassic sandstones are referred to as the New Red Sandstone, and they form a curious parallel to the Old Red Sandstone of Devonian age, which formed when Scotland was at a similar latitude but to the south of the Equator, 100 million years earlier, in the lee of the Caledonian Mountains. Rift valleys began to open up in the North Sea area, early precursors of events that ultimately created the North Atlantic Ocean. At the end of the Permian period, 250 million years ago, there was a sudden extinction event that caused the disappearance of 95 per cent of all species and 61 per cent of all life on Earth, the greatest such event in history. This was probably caused by the amalgamation of continents into the newly formed supercontinent of Pangaea, when shallow shelf seas (where the majority of animals and plants lived) were greatly reduced in number and area. Eruption of flood basalts in Siberia caused global cooling and acid rain at precisely that time, and this could have been a contributory factor. A giant meteorite strike is another possibility. New life-forms rapidly invaded the vacated ecological niches in the succeeding periods of the Mesozoic era, including the first mammals, birds, dinosaurs and flowering plants. Primitive life had now gone forever, as the Palaeozoic gave way to the Mesozoic.

Jurassic and Cretaceous
Stretching and thinning of the crust, and the eventual rifting and break-up of Pangaea, continued from 250 million years ago onwards, into the Jurassic period, which in Scotland is marked by limestones, sand-stones, shales and mudstones around the Inner Hebrides and Moray Firth shorelines (Fig. 2.11), but more especially in the North Sea Basins, which subsided as more and more sediment was piling in from what remained of the Caledonian Mountains. The warm shallow tropical seas were home to abundant life; organic productivity was exceptionally high, and the result was a rich soup of decaying matter on the floor of the dark lifeless sea. Continued sinking of the sea floor ensured that the dead remains of planktonic animals, plants, algae and bacteria were buried, tightly packed and effectively transformed in a natural pressure cooker into oil and gas. The warm moist climatic conditions lasted for 140 million years, during which time the crust in Scotland remained relatively stable. Dinosaurs of mid-Jurassic age thrived in and around the estuarine parts of the Hebrides Basin, feeding on each other or on the vegetation; fossil remains were found on Skye in late 2002. Cretaceous chalk is rare on the mainland, but occurs within the North Sea Basins. At the end of the Cretaceous, 65 million years ago, there was another mass-extinction event, perhaps associated with the Chicxulub meteorite impact in the Yucatán Peninsula of Mexico, and the eruption of the Deccan plateau basalt lavas in India. Many species were lost, and the dinosaurs disappeared then, to be replaced by the mammals during the Tertiary period.

Figure 2.11 Laig Bay, Isle of Eigg: iron-rich concretionary limestone (top, dark beds) above sandstone containing cannon ball concretions; the large boulder on the beach is a concretion weathered out of the cliff face; Jurassic age.

Figure 2.12 Helmsdale, east Sutherland: folded Jurassic shales adjacent to the Helmsdale Fault; the three headlands with semi-circular rock outcrops at sea level represent these folds.

Tertiary

The quiet conditions were halted 60 million years ago by the sudden eruption of ash on the west coast, followed by outpourings of basalt lavas over the land surface of what is now the Inner Hebrides. These basalts emerged from fissures in the ground that brought magma from the upper mantle and caused the formation of lava plateaus at the surface, up to

4 km thick in places. This event was the final opening of the North Atlantic Ocean, as Scotland separated from North America. Iceland was eventually born in the middle of the Atlantic, 24 million years ago. As soon as the lavas stopped 55 million years ago, a series of large intrusions forced their way into the crust, in a north–south belt from Skye to Rum, Ardnamurchan, Mull and Arran, each one at the intersection of an earlier Caledonian fault and a north–south line of weakness in the crust caused by the emplacement of an elongate mantle plume. This massive addition to the West of Scotland crust caused the land surface to be pushed up and tilted to the east. The climate was warm, varying from humid to semi-arid, as revealed by the red fossil laterite soil horizons between the lava flows, testimony to deep tropical weathering. Scotland's river system was formed then, and immense quantities of weathered debris were transported east and deposited in the North Sea Basins, over more than 50 million years. That event was almost the last episode in the formation of Scotland's landscape.

Quaternary

By 2 million years ago, Scotland had reached its present position, 56° north. Climatic cooling, triggered in part by plate-tectonic movements, heralded the start of the great Ice Age, which was actually a sequence of 20 rapid climatic fluctuations from glacial to warm and back. Continental ice sheets spread across Scotland and helped to smooth off any rough edges left by the much more extensive Tertiary weathering. Upon melting, the glaciers left thick blankets of boulder clay, sand and gravel in their wake. After the ice had melted, the land surface rebounded in the past 10 000 years in response to the unloading of the heavy ice cover. Forests quickly blanketed the land, but this cover has since been removed during the past 5000 years, mainly as a result of our own efforts. Rivers continue to flow, the coasts are still being battered by the sea, and landslips in the mountains indicate that weathering and erosion are continuing unabated to shape the landscape. We may be speeding up the process as we burn fossil fuels; the greenhouse gases are contributing to global warming, which is implicated in climatic change that manifests itself in more storms and floods. The coal was formed 300 million years ago, and we will have used it all up in 300 years since the industry began. Oil and gas formed 140 million years ago, and we will have burned the lot in a mere 140 years of production. The rock cycle will need to operate for another 140 million years before any more oil will form or 300 million years before any more coal can be produced. Just a thought.

What next?

Predicting the future is a hazardous occupation, especially when dealing with long-term processes, not all of which are fully understood, and some of which are completely random. Meteorite impacts are unpredictable, the last major one being 65 million years ago at the end of the Cretaceous, which hit the Yucatán Peninsula in Mexico but had a global effect on the climate. Upwellings of mantle plumes that create giant volcanic provinces happen in a seemingly haphazard fashion. The last one to affect Scotland was responsible for the Tertiary volcanic events of 60–55 million years ago. Mantle plumes are thought to originate

because slabs of dense subducted oceanic crust pile up at 700 km depth, then suddenly collapse to the core/mantle boundary at 2900 km depth and cause a plume to rise to the upper mantle. Major meteorite impacts and giant volcanic plumes occur roughly once every 100 million years, but not (usually) in the same place twice. We are therefore left with the known speed and direction of plate movements to provide a framework within which it is possible to make some reasonable guesses about the future location of Scotland, though even here it is not easy to predict where new subduction zones might appear.

Plate-tectonic cycles take about 200 million years to complete, and there is every indication that the Atlantic and Pacific oceans will continue to spread and widen for 20–50 million years, until new subduction zones form, possibly down the east coast of North and South America. According to the model formulated by Professor Chris Scotese,[*] by 250 million years from now all the continents will have again re-assembled into a new supercontinent – Pangaea Ultima – with Scotland lying at the very edge, rather close to the North Pole.

The weather might worsen and stay that way for a long time. Of more immediate concern is the fact that we are presently in an interglacial stage, and the pattern is such that we will probably enter another glacial stage within a few thousand years. Global warming caused by the enhanced greenhouse effect may delay or obscure the effects temporarily, but present predictions of a 2°C rise in global temperature mean that sea level will rise by more than 500 mm before the end of the twenty-first century, because of melting ice sheets in Greenland and Antarctica, and climatic change will have hugely variable effects in different regions of the globe. In the longer term, the Earth will recover from the addition of so much carbon dioxide into the atmosphere, but for humans this will not happen quickly enough to stave off possible natural disasters such as flooding, coastal erosion and landslides.

---

[*] Of the University of Texas, and author of the PALEOMAP Project (www.scotese.com).

Suilven, Sutherland, Northwest Highlands: an isolated, steep-sided mountain of Torridonian Sandstone, showing horizontal bedding planes and vertical joints (eroded into gullies) rising above a low plateau of ancient Lewisian gneiss, showing knock and lochan topography (photograph by Roger Jones).

CHAPTER 3

Scotland's oldest rocks – the far northwest

Introduction

Our journey starts way back in time, with Europe's oldest rocks, 3 billion years old, which are found in a narrow strip of the Northwest Highlands and in the Outer Hebrides (Fig. 3.1). Taken together, the Outer Hebrides and Northwest Highlands comprise the Hebridean Terrane, one of five such blocks of crust that make up Scotland. These ancient rocks are known as the Lewisian, after the Isle of Lewis. Except for some Triassic sandstone and conglomerate around Stornoway, all the islands of the Outer Hebrides are made of Lewisian rocks, from the Butt of Lewis to Barra Head, and they are wonderfully exposed, particularly in the southern isles and in South Harris (Fig. 3.2). On the mainland, Lewisian rocks are found along the coast of Sutherland and Wester Ross, from Cape Wrath, through Assynt and Gairloch, towards Rona, northern Raasay and southeastern Skye, passing Dornie and Glenelg, and finally via Rum on to Coll, Tiree and Iona (Fig. 3.3). As in the Western Isles, on the mainland you will see the bare rocky knolls and lochans or peat bogs between (Fig. 3.4), or occasionally soaring cliffs, as at Cape Wrath (Fig. 3.5). Lewisian rocks occur almost exclusively to the west of the Moine Thrust zone, which really represents the western margin of the Caledonian mountain belt. For this reason, the old rocks of the Northwest Highlands are sometimes referred to as belonging to the foreland or basement, that is, a stable segment of the crust that was largely unaffected by the later events of the Caledonian orogeny. At the end of this orogeny, the now-folded metamorphic rocks of the Northern Highlands were pushed many kilometres to the west, over the

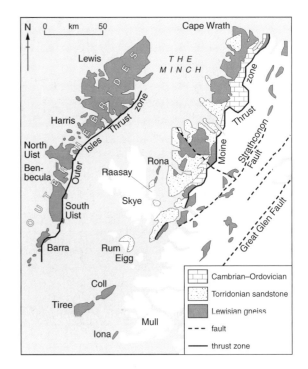

Figure 3.1 The Northwest Highlands and Outer Hebrides.

top of the old basement rocks. This movement took place along a series of low-angle faults, known as thrusts, including the Moine Thrust zone (Fig. 3.6). Imagine pushing a stack of cards across a table top: the cards separate and are moved away from you, each one a little more than the ones below. If you add up all the small amounts of movement, you end up with quite a significant total, or displacement – at least 80 km, and possibly much more. And so it is in the Northwest Highlands, the Moine Thrust zone is just

Figure 3.2 South Harris Igneous Complex: high rounded hills of resistant rock.

Figure 3.3 Iona: Lewisian gneiss and machair sands in foreground; the islands in the background (Dutchman's Cap and Staffa) are of Tertiary basalt lavas.

Figure 3.4 Ben Stack, near Loch Laxford, Sutherland: knock and lochan topography in Lewisian gneiss. Ben Stack is made of many sheets of Lewisian granite and pegmatite dykes in the Laxford shear zone.

Figure 3.5 Cape Wrath, Lewisian gneiss cliffs, showing folds.

Figure 3.6 The Moine Thrust at Knockan Cliff, Assynt, Sutherland: Moine schist (older) at top, Durness Limestone (younger) below; guide book resting on thrust plane (overhang).

the top one of a series of thrusts, each piled on top of the others, and each one carrying a segment of folded rocks above for at least 20 km each. This is especially clear around Loch Assynt and Loch Glencoul (Figs 3.7, 3.8). Lying on top of the Lewisian are rocks that give Northwest Scotland its eye-catching mountain scenery – the Torridonian sandstone and the sparkling white Cambrian quartzite and Durness Limestone (Figs 3.9–3.12).

The Northwest Highlands

Northwest Scotland must surely have the grandest scenery in Britain, and among the most inspiring in Europe. For well over a century, the region has been a Mecca for geologists, students and professionals alike, who have come to study the superbly exposed rock sections on the mountains, cliffs and rocky shores. Here we find rocks that are over 3 billion years old, the oldest in Europe. Many hundreds of doctoral theses have been written about the rocks here, and more remains to be discovered as analytical techniques and equipment continue to improve. The first map of the region by the pioneering Geological Survey geologists

Figure 3.7 Glencoul Thrust, Loch Glencoul, Assynt, Sutherland: Lewisian gneiss at top and bottom of picture, Cambrian–Ordovician rocks (bedded quartzite and limestone) forming distinct beds in centre of hill and dipping down to the right.

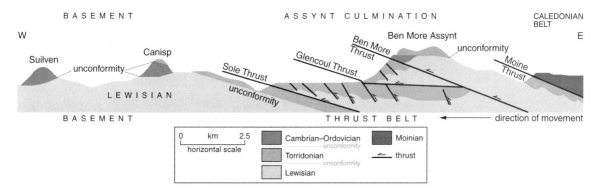

Figure 3.8 Sketch cross sections through the Assynt Thrust belt.

Ben Peach and John Horne was published in 1907 with an accompanying 680-page memoir. This work was astonishing in its detail and clarity, and it has stood the test of time. Much of the progress in understanding the complexities of areas of highly deformed basement gneisses around the world began in Scotland with the work of Peach & Horne. Remember that this work was completed when road transport and accurate contour maps hardly existed, to say nothing of laboratories and equipment.

Lewisian gneiss

Along the west coast, from Cape Wrath to Gairloch, the basement rocks belong to the Lewisian Complex, or Lewisian Gneiss Complex, named by Dr John MacCulloch in his famous 1819 work *A description of the western islands of Scotland*. This part of Scotland is regarded by many as one of the classic areas of geology in the world, and the Lewisian rocks have been studied more intensively than rocks of similar age anywhere else. These are Precambrian (3300–1750 million years old) crystalline rocks, banded in white, pink and black, everywhere folded and intensely metamorphosed deep within the crust. Fossils are completely absent, and age dating has been carried out exclusively by radioactive-isotope methods (see p. 52). They often contain black or very dark green, coarsely crystalline "pods" or rugby football-shape bodies within the banding (Fig. 3.13). Pods such as these consist of very large crystals of the minerals

Figure 3.9 Canisp (left) and Suilven (right): isolated Torridonian sandstone hills above a basement of glacially scoured Lewisian gneiss.

Figure 3.10 Stac Pollaidh (Stack Polly), Sutherland: Torridonian sandstone. **(a)** Deep crevasses at summit, weathered in vertical joints, and heather-clad scree on lower slopes. The castellated summit is the result of frost shattering. **(b)** Part of the summit, showing horizontal bedding planes in Torridonian sandstone, and vertical crevasses along joints.

augite (a pyroxene) and hornblende (an amphibole), and they probably represent basic or ultrabasic igneous material that was stretched, thinned, pinched out and possibly rotated during the formation of the gneisses. Frequently too the gneisses are cut across by thin veins, sheets, dykes and also irregular patches of bright pink granite (Fig. 3.14). This granite, which in places is present in considerable volumes, was produced by the melting of the quartz and feldspar-rich gneiss. Granite melts in the continental crust at 650–700°C, and is then injected at higher levels because it is lighter than the surrounding rocks. However, the

48

Figure 3.11 Foinaven (915 m), near Laxford Bridge, Sutherland: Cambrian quartzite (white scree) above Lewisian gneiss; Lewisian gneiss in the foreground, showing knock and lochan topography.

presence of so much patchy granite within the banding suggests that the entire rock mass must have been in that sort of temperature range.

Patches of intensely altered Lewisian rocks also occur to the east of the Moine Thrust zone in the Northern Highlands, within the Caledonian fold belt. These are very similar to the rocks found on the Sutherland coast, except that their margins are usually highly sheared because they have been caught up in large-scale folding and thrusting. We shall return to these blocks, which are termed inliers because they are old material within younger belts, in Chapter 4.

The deceptively innocent-looking low rocky knolls all the way down the coast disguise an immensely lengthy history of formation and a highly complex geological structure. This little piece of basement in Northwest Scotland is a relatively small fragment of a much more extensive region of Precambrian continental crust in the North Atlantic, usually referred to as Laurentia, which includes the Canadian Shield and Greenland (see Fig. 4.2). Northwest Scotland became detached from Laurentia 100 million years ago when the North Atlantic Ocean opened, and the area is now separated from Scandinavia by the Caledonian fold belt and the rifted basins of the northern North Sea. Therefore, the Lewisian Complex forms part of a

Figure 3.12 Durness Limestone (light colour), Elphin, Assynt, Sutherland; Ben More Assynt (Moine schist) in background.

Metamorphism and metamorphic rocks

Metamorphism relates to changes that take place in rocks as a response to changes in pressure, temperature and the action of fluids. Any rock can be metamorphosed and the resulting metamorphic rocks are crystalline, with interlocking grains. Sometimes the crystals are in regular arrangements – lines (needle-shape crystals such as hornblende are orientated to give a lineation), or planes (platy minerals such as micas lie in parallel sheets to form a foliation). During metamorphism, original grains become recrystallized in the solid state: the rocks do not melt. However, if the temperature is high enough and water-rich fluids are present, then some types of deeply buried rocks can approach melting point and may partially melt. In most cases, the resulting partial melt has a granitic composition, since the components of granite (feldspars and quartz) have the lowest melting point. This is why granite sheets and granitic pegmatite dykes and veins are common in many parts of the Lewisian, particularly the Laxford gneisses and migmatites. These would have formed at a temperature of about 650–700°C and a pressure equivalent to 20–25 km depth of burial in the crust.

Regional metamorphism can be described as being of low, medium or high grade, depending on the pressure and temperature conditions. The following table summarizes this.

Original rock	Metamorphic equivalent
Sandstone	Quartzite
Limestone	Marble
Basalt or dolerite	Amphibolite (or greenstone or hornblende schist)
Mudstone or shale	Slate → phyllite → schist → gneiss → migmatite (in presence of water-rich fluids) or → granulite (in "dry" conditions) – depends on grade
Granite	Metagranite or migmatite
Basalt	Eclogite (at very high pressures, e.g. in subduction zones)

broad conditions of temperature and pressure. Rocks of the same original composition will have the same minerals in them if they belong to the same metamorphic facies. In metamorphic belts, the commonest facies are:

- greenschist facies, 275–400°C, depth 5–15 km
- amphibolite facies, 400–700°C, depth 10–25 km
- granulite facies, 700–900°C, depth 20–40 km

(Actual temperature and pressure values are approximate, and there is some overlap.)

Another concept, used when discussing the Dalradian rocks of the Southwest Highlands (Ch. 4), is that of "metamorphic zones". This was first used to describe the schists just north of the Highland Boundary Fault in Glen Esk. Zones are characterized by the presence of a particular mineral that reflects metamorphic grade. These key minerals are called "index minerals", and are usually easily identified in the field. Metamorphic zones were originally based on minerals found in metasediments that started out as clay-rich rocks, i.e. shale and mudstone. The following table summarizes grades and zones for such rocks.

Metamorphic grade	Rock name	Grain size	Rock texture
Low	Slate	Very fine	Slaty cleavage; splits easily
Medium	Phyllite	Fine	Thin foliation, wavy surface, high sheen
Medium	Schist	Medium	Schistosity; mica common on surface
High	Gneiss	Coarse	Mineral banding, rough foliation
High	Migmatite	Very coarse	Irregular, patchy with veins and streaks of quartz and feldspar
Very high	Granulite	Coarse	Random crystals, granular, may be compositional layering

Notice from the table above that grain size increases with increasing grade: more energy was available at higher grades to allow larger crystals to form.

Since any rock can be metamorphosed, it follows that the resulting metamorphic rock will have a composition that reflects the original rock composition, because very little enters or leaves a rock during metamorphism except for water and carbon dioxide. The following table gives the metamorphic equivalents of some common rocks.

The term "metamorphic facies" is used in the description of the Lewisian gneiss. This is a technical term used to describe sets of minerals occurring in metamorphic rocks that formed under the same

Grade	Rock	Zone	Minerals present
Low	Slate	Chlorite	Chlorite,* quartz, muscovite, plagioclase
Medium	Phyllite	Chlorite	Chlorite,* quartz, muscovite, plagioclase
Medium	Mica schist	Biotite	Biotite,* quartz, plagioclase
Medium	Garnet schist	Garnet	Garnet,* mica, quartz, plagioclase
Medium	Schist or gneiss	Staurolite	Staurolite,* mica, garnet, quartz, plagioclase
High	Kyanite gneiss	Kyanite	Kyanite,* mica, garnet, quartz, plagioclase
High	Sillimanite gneiss	Sillimanite	Sillimanite,* garnet, mica, quartz, plagioclase feldspar

* Index mineral

Figure 3.13 Black ultrabasic pods (of igneous origin) in Lewisian gneiss, Lochinver, Sutherland.

Figure 3.15 Scourie dyke (dark, on right) cutting gneiss in the Laxford shear zone, Rubh an Tiompain, Scourie, Sutherland. The rocks on the left are strongly sheared gneisses.

Precambrian shield area (also termed "craton"; Greek: strong; a Precambrian continental shield area that has remained undisturbed since its formation), similar to those that are found in other continents, but has become separated from the main mass in Canada.

The Lewisian outcrop has been traditionally subdivided into three main regions: the northern region from Cape Wrath to Loch Laxford, consisting of pink quartz–feldspar–mica gneiss with pegmatites and granite sheets and veins (the Laxford gneisses); the central region from Scourie to Gruinard Bay; and the southern region, which extends to Gairloch, Loch Torridon, Rona and Raasay. Both the central and southern regions are made up mainly of grey banded gneiss and granulite (the Scourian gneisses), with layered bodies of basic and ultrabasic igneous rocks – gabbro, peridotite and anorthosite. Around Scourie in particular, there are many vertical and relatively narrow

northwest–southeast-trending basic dykes, of varying ages (2240–2200 million years old), but generally grouped together as the Scourie Dykes (Fig. 3.15). The dykes are recognizably igneous intrusions, which we can see from the way in which they cut across the banding of the gneisses and the fact that the minerals interlock in a random fashion – a typical igneous texture. In places the margins of the dykes were affected by shearing and metamorphism, so that hornblende–garnet schist is found at the edge, whereas in the interior the minerals there is pyroxene and feldspar in random orientation. Some steep, narrow northwest–southeast zones of very high deformation, referred to as shear zones, cut the gneisses, giving the rocks a more platy schistose character. These may represent deep-seated faults, movement along which brought different parts of the Lewisian together. The most important of these are the Laxford shear zone at Loch Laxford (Figs 3.15–3.17), the Gairloch shear zone, the Canisp shear zone near Lochinver, responsible for the inlet at Achmelvich Bay, and the Diabaig shear zone at Loch Torridon. The highly deformed rocks in the near-vertical shear zones are weaker than the gneisses with their flatter banding, so that glacial and marine erosion have exploited this difference, the inlet of Loch Laxford being a particularly good Scottish example of a fjord. The Laxford shear zone marks the boundary between the northern and central regions of the Lewisian outcrop, and runs from the aptly named Red Point (Rubha Ruadh), with its distinctive bright pink and red granite, to Ben Stack (Fig. 3.17), the

Figure 3.14 Folded gneiss, intruded by granite veins and pegmatite sheets in Lewisian gneiss, Loch Laxford, Sutherland.

Age dating of rocks

There are two methods of determining the age of rocks: relative and absolute. Using fossils and knowing something about the evolution of animals and plants, geologists have always been able to work out the relative ages of sedimentary rocks containing fossils. Generally, the more primitive organisms are found in older rocks, whereas those that resemble modern forms more closely appear in younger rocks. Another general principle at work here is that younger rocks are found on top of older rocks: superposition. Until the discovery of radioactivity in 1896 and its first application in geology in the 1950s, the relative method was the only one available.

Attempts to calculate the age of the Earth include Lord Kelvin's calculation based on the rate of loss of heat and the time taken for the Earth to cool from a completely molten state. Others tried to calculate the time to deposit all the known thickness of sedimentary rocks, or the time taken to make the sea as salty as it is today, starting with fresh water. But all these methods used assumptions that turned out to be invalid. Archbishop James Ussher of Dublin has the honour of having established in 1654 that the Earth was created at 9 a.m. on 26 October 4004 BC, based on adding up the ages of tribes in the Old Testament.

Sensitive instruments called mass spectrometers are able to detect minute quantities of different isotopes present in rock and mineral samples. Isotopes used in geology have to be relatively abundant, have recognizable decay products and have half lives roughly the same as the age of the rock or mineral to be dated. A selection of the most common currently used schemes is shown in the table, but several others exist.

Modern instruments and techniques now allow us to determine the exact ages of rocks very precisely, in millions of years; this is absolute dating, and it usually refers to the timing of a particular event in Earth history, such as a volcanic eruption, a metamorphic episode, or the intrusion of an igneous body. The technique that uses radioactive materials occurring naturally in rocks is known as radiometric age dating or isotopic dating. The principle behind this method is that certain elements are radioactive. Matter is made up of distinct elements, and each element is made of atoms, the building blocks of matter. Atoms have a nucleus of particles, called protons and neutrons, each with a mass of one unit. Surrounding the nucleus are clouds of virtually weightless electrons. Many elements have different isotopes, that is, they have the same number of protons but different numbers of neutrons, and hence they have different masses, but they have the same basic properties. Isotopes that are unstable decay over time, and change into isotopes of other elements. The original element is known as the parent and the decay product is called the daughter. Individual isotopes of a particular element decay at a constant rate, which can be measured in the laboratory. The time taken for an isotope to decay to half its initial mass is known as the half life, and for geology the useful isotopes for age-dating purposes have half lives measured in millions of years. During one half life, half the original isotope will have decayed to daughter isotopes, and so only half the original mass of the parent isotope now remains.

Small samples of minerals and rocks are treated chemically and placed into a mass spectrometer, which can measure the amount of parent and daughter isotopes to an incredible degree of accuracy by separating the different isotopes using strong magnetic fields. The age of the rock or mineral is found by comparing the relative amounts of parent material and decay product; the numbers are put into a fairly simple mathematical formula, together with the half life, and a figure is calculated. Obviously, much can happen to a rock in its lifetime, such as being heated, reheated, folded and metamorphosed, and it is possible to unravel much of this history by finding the ages of different heating episodes by using different minerals and different isotope schemes. Nowadays, ages of rocks as old as 3800 million years can be found to an accuracy of one million years or better. Regarding radiocarbon dating (i.e. using carbon-14, the radioactive isotope), we have to be aware that, because of the relatively short half life of carbon-14, this method is used for dating only very recent events (i.e. younger than 30 000 years), and can be used only for materials containing organic carbon, such as peat, wood, shells and bones. The method is used widely in archaeology, but it does not work well with old rocks or with crystalline rocks. The rubidium–strontium method is widely used for dating igneous and metamorphic rocks, and in addition can give useful information about the source of a rock, that is, whether an igneous intrusion originated from melting of the lower crust or was directly injected into the crust from the upper mantle.

Parent isotope	Daughter isotope	Half life (mill. yr)	Minerals used	Rocks dated (commonly)
^{238}Uranium	^{206}Lead	4470	Zircon, apatite, sphene	Granite, basalt, gneiss
^{235}Uranium	^{207}Lead	704		
^{232}Thorium	^{208}Lead	14 000		
^{40}Potassium	^{40}Argon	1400	Muscovite, biotite, hornblende	Igneous rocks, some sedimentary rocks
^{87}Rubidium	^{87}Strontium	48 900	Muscovite, biotite, feldspar	Gneiss, granite
^{147}Samarium	^{143}Neodymium	106 000	Monazite	Igneous rocks
^{14}Carbon	^{14}Nitrogen	5730	Graphite or carbon	Peat, wood, bone

triangular peak that is the highest hill of Lewisian gneiss on the mainland, at 719 m. All around Laxford Bridge you will see many thick sheets of granite and pegmatite cutting the pale striped gneisses. Ben Stack owes its shape to the fact that a huge concentration of these steep southwesterly dipping sheets meet together at the summit. Another clearly visible major feature in the northern region is the Strath Dionard anticline, a huge dome structure with an ice-polished surface that can be seen clearly from the roadside on

Figure 3.16 Lewisian gneiss cut by amphibolite (black, basic dyke or sheet), granite and pegmatite (light), Loch na Fiacail, north of Loch Laxford.

the journey north between Rhiconich and Durness, at Gualin House. Metamorphic rocks such as these in the northern region of the Lewisian, with their abundant pegmatites and granite sheets cutting across and invading striped and banded gneisses, are generally known as migmatites (Greek: mixed rocks). Migmatites form by partial melting of the gneisses, and indicate that the Laxford gneisses were close to melting point in places when they formed (650–700°C). Migmatites frequently contain large pods of gneiss

Figure 3.17 Ben Stack, Loch Laxford, Sutherland: several thick Lewisian pegmatites and granite sheets meet together to form the conical peak of the hill, within the Laxford shear zone. The bare knolls are Lewisian gneiss.

strung out along the banding. These bulging patches are referred to as boudins, after the French word for black pudding, and result from thinning, flattening and stretching of the gneisses, such that the different compositions behave differently in response to the deformation. Pegmatites are very common in the Laxford gneisses, and one prominent body at Beinn Ceannabeinne near Durness was investigated as a possible commercial source of feldspar for use in the manufacture of ceramics. Next time you drive from Durness to Kylesku, note that the gneisses in the north are pinkish, and that the banding is frequently quite steep, but between Scourie and Kylesku they are very dark, nearly black, and look rather greasy. They are also much denser than the Laxford gneisses, because of the greater abundance of heavy minerals containing iron, magnesium and calcium, whereas the pink granitic gneisses are much richer in the lighter potassium, aluminium and silicon. Another feature is that the banding in these dark rocks in the southern belt is very often nearly horizontal. Table 3.1 outlines the main events in the history of the Lewisian Complex.

Occasional patches of metamorphosed sedimentary rocks (often referred to as metasediments) – quartzite, marble and schist – represent surface deposits (sandstone, limestone, shale, ironstone, mudstone) laid down in an ancient shallow sea and prove that water existed on the Earth at least as long ago as 3 billion

Table 3.1   Formation of the Lewisian Complex (ages in millions of years).

Age	Main rock-forming events in the Lewisian Complex
1200	Uplift and exposure of Lewisian at surface, prior to deposition of Torridonian rocks, starting at 1200 million years ago
1750	Intrusion of granites and pegmatites, formation of migmatites, and folding; formation of major NW–SE shear zones; amphibolite facies metamorphism
1850	Metamorphism at amphibolite facies in Northern region
2000	Lavas and sediments of Loch Maree Group
2200–2400	Intrusion of Scourie Dykes – various phases, cross cutting
2500	Metamorphism and folding
2700	Granulite facies metamorphism and folding (high pressure)
2800	Formation of gneisses of northern region (Laxford gneisses)
3000	Formation of gneisses of central and southern regions (Scourian gneisses) – intrusion of tonalite, folding, thrusting, high-grade metamorphism
3000–3300	Formation of sediments and layered basic–ultrabasic igneous rocks, early phase of metamorphism

years (and actually much longer than that, as we shall soon see). Amphibolite sheets (black and white striped hornblende–feldspar schist) associated with these metamorphosed sediments probably represent basaltic lavas and tuffs from volcanic zones, subsequently metamorphosed along with the sediments. Frequently associated with the metasediments are layered ultrabasic to basic igneous intrusions. These have olivine at the base, followed upwards by olivine and pyroxene, with plagioclase entering higher up; finally, feldspar-rich layers (anorthosite) occur at the top. This layering is believed to be explained by gravity stratification of large igneous bodies – olivine is dense and crystallized early, so fell to the bottom, whereas feldspar crystallized last and is relatively light, so it formed the topmost layers. During folding and thrusting, these intrusions frequently became dismembered, and separate units of ultrabasic rock, gabbro (basic) and anorthosite can now be found as quite distinct bodies within the Lewisian, with their tops and bottoms being parallel to the banding in the surrounding gneisses. Although metasediments are quite rare, at Loch Maree and Gairloch there is a more important outcrop of metamorphosed sediments, termed the Loch Maree Group, covering 130 km², and with highly deformed boundaries with the gneisses. The rocks consist of thick units of amphibolite, probably volcanic in origin, with quartzite, mica schist, marble, banded ironstone and graphitic schist. The Loch Maree sediments are 2000 million years old; the abundant amphibolites could represent outpourings of ocean-floor basalts during a period of continental break-up, with the marbles, schists and quartzites representing continental shelf deposits and the ironstones being related to hydrothermal activity (mineral-rich hot water) on the sea floor. Some of the metasediments at Gairloch contain beds of massive copper-sulphide and iron-sulphide deposits, in which there are small inclusions (particles contained within another mineral) of gold and silver within the mineral chalcopyrite (copper–iron sulphide). The host rocks for this low-grade copper–zinc deposit are banded ironstones and what were originally deepwater, impure, muddy sandstones (greywacke), surrounded by amphibolites (which probably represent metamorphosed basalt lavas). Exploration companies have investigated this ore-mineral deposit, discovered in 1978, but it was found to be uneconomic and no extraction was ever carried out.

From studying the minerals in the gneisses, we know that the Lewisian formed deep within the Earth's crust, possibly as far down as 30 km or more. Yet today, geophysical surveys tell us that the crust in that part of the world is still about 35 km thick. This implies that during the Precambrian the crust may have been as much as 60–70 km thick, comparable with that in younger fold-mountain chains, in particular the Himalayas. Erosion therefore removed 30 km of rock during the time that the Lewisian was being lifted to the surface. In other words, the Lewisian Gneiss Complex formed within an ancient mountain belt by folding, compression, metamorphism, partial melting and also granite intrusion. Movements along shear belts brought together the Scourian and Laxford gneisses, which may have formed in distinctly separate parts of this mountain belt. However, comparing the thickness of the crust with that under the Himalayas does not imply that mountains as high as Mt Everest need necessarily have existed in Scotland 3 billion years ago. The land surface on the northwest coast, where Lewisian rocks are exposed, is about 100 m above sea level, say. All we know is that some 30 km of crust was removed before the deposition of the Torridonian sandstone, but we can say almost nothing about the heights of any mountains that may have existed at the surface of the Earth while the Lewisian rocks were being formed at depth. Nor is it entirely certain how the metasediments that formed at the surface of the Earth were taken down to these depths – if indeed they were. Scotland, then at the edge of Laurentia, lay somewhere in the Southern Hemisphere.

Torridonian rocks

Between about 2 billion and 1600 million years ago, various continental blocks are thought to have converged to form a large supercontinent in what is now the North Atlantic region, but before the present ocean opened. This supercontinent included Laurentia and Baltica (Scandinavia), with Scotland being wedged between them at the eastern margin of Laurentia (see Fig. 3.15). About 1400 million years ago, this huge continental mass began to break up during a period of rifting, and Laurentia and Baltica split apart. The Lewisian gneiss was exposed at the surface of this land mass. Associated with this rifting was the deposition of the Stoer Group (named after the Stoer Peninsula in Sutherland) of rocks in Northwest Scotland, some 2000 m of river-borne sediments and wind-blown sands. These sediments were laid down in a desert environment, directly on top of the Lewisian, which must therefore have been at the surface then, and many of the pebbles and boulders at the base of the Stoer Group are of Lewisian gneiss. Above the coarse material, river-lain sandstones occur, as well as shales deposited in shallow lakes and a volcanic unit that represents windblown ash or a mudflow caused by an eruption. It contains broken crystals and ash pellets shaped like peas. This is a useful marker horizon

(a bed with distinctive features that can be used as a point of reference over a wide area), since it is easily recognizable in the field. No source for the volcanic rocks has yet been found. At Enard Bay there is a red mudstone mound draped over Lewisian gneiss that may have been formed by the growth of algae on the sea bed. Volcanic activity is interpreted as evidence for the crust being thinned and split apart at the time, with sedimentation going on as faulting occurred, thus allowing the 2000 m thickness to build up. Some of the shales contain tiny fossil algae, the oldest fossils in Scotland, but dating of these in not precise, and the Stoer Group is believed to be about 1200 million years old. Other indicators of a sedimentary environment are ripple marks and mudcracks on shale surfaces.

Lying on top of the Stoer Group rocks, and about 200 million years younger, we find the Torridon Group, which makes the spectacular scenery of the northwest seaboard (Figs 3.18–3.20). Although the name derives from the area around Loch Torridon, these reddish-brown well bedded sandstones are

Figure 3.18 Slioch, Loch Maree, Wester Ross. Torridonian sandstone (top) filling in hollows and draping over former hills in Lewisian gneiss (at the loch shore): an example of an ancient topography being exhumed by erosion. The very straight edge of the loch is the result of erosion along the Loch Maree Fault.

Figure 3.19 Liathach, Wester Ross: Torridonian sandstone.

this is responsible for the striking colour. Such sandstones rich in feldspar are called arkoses. Normally, feldspar weathers rapidly and breaks down to fine clay minerals. The presence of so much fresh-looking feldspar in the Torridonian is taken to indicate that they must have been eroded from an area close by. Altogether there is a thickness of about 6000 m of these rocks, which on many maps are simply referred to as the Torridonian sandstone. In fact there is another 3500 m-thick group lying underneath, known as the Sleat Group, from the Sleat Peninsula in southeastern Skye. The rocks were mostly deposited by large rivers flowing from the west and northwest across the exposed Lewisian gneiss, which became buried in the process, and evidence shows that Scotland was located at 30–40°N during the formation of the Torridonian, 1000–800 million years ago. Microscopic plant spores indicate an age of 800–900 million years. Everywhere there is a pronounced unconformity (i.e. young sedimentary rocks resting on old metamorphic rocks) with the underlying Lewisian gneiss, which formed the basement to the Torridonian. This is especially clear at Loch Assynt and, more impressively, along the shores of Loch Maree, where Slioch (980 m high) towers over the knobbly Lewisian at shore level (Fig. 3.18). Other Torridonian mountains include Suilven, Liathach, Ben Alligin, Quinag, Stac Pollaidh and An Teallach (Figs 3.20, 3.21). In other words, the Torridonian covers an old land surface, which in places had a relative relief of 600 m, and erosion has exposed this surface. Beneath the Torridonian Northwest Scotland

found all the way from near Cape Wrath to Applecross, Skye and Rum, and they form the sea bed from there for another 125 km to the Great Glen Fault. The coarse red sandstones are rich in feldspar fragments, derived from the underlying Lewisian gneiss, and

Figure 3.20 Ben Alligin, by Loch Torridon, Wester Ross: Torridonian sandstone.

Figure 3.21 Quinag, between Loch Glencoul and Loch Assynt, Sutherland: Torridonian sandstone (high ground) on top of Lewisian gneiss (foreground), with Cambrian quartzite above Torridonian (far left).

must have looked the same today as it did about a billion years ago: a fossil landscape unearthed. The present low plateau is the result of Tertiary erosion. Considering the huge thickness of the Torridonian and north–south extent of the outcrop, the total amount of sediment brought in by the rivers was phenomenal. Clearly, there must have been large powerful rivers that would have been eroding a vast area to the west of the present Northwest Highlands, for Scotland as it is today could not have provided enough source material. It is most probable that these rivers transported sediment for 500 km from Greenland and Canada, which then lay much closer to Scotland.

Between the deposition of the Stoer Group and the Sleat and Torridon groups, in the period just over a billion years ago, another supercontinent formed by the collision of Laurentia, Baltica and Siberia in the north, and Gondwanaland (the southern continents). This supercontinent has been named Rodinia (from the Russian word for "motherland"), and the orogenic belt that appeared in Laurentia as a result of these collision events is referred to as the Grenville belt (after a province of that name in Canada). We shall return to the Grenville orogeny later (p. 76).

Landscape features of the northwest coast are related in large measure to the structure of the Lewisian basement and the overlying Torridonian sediments, modified by extensive Tertiary weathering and erosion, and by subsequent glacial erosion.

Major folds, faults, shear belts, dyke swarms and granite sheets in the Lewisian Complex tend to be aligned northwest–southeast, with the result that differences in weathering resistance have been picked out and accentuated by rivers, then the ice and now by the sea. What we see today are long, narrow inlets, sea lochs and fjords, such as Loch Inchard, Loch Laxford, Loch Glencoul and Loch Inver. Many inlets also cut deeply into Torridonian rocks, such as Loch Broom, Little Loch Broom, Loch Ewe and Loch Torridon. Headlands of flat-lying Torridonian form prominent landforms, such as that adjacent to the Old Man of Stoer. Handa Island (opposite Scourie) and the Summer Isles (opposite Achiltibuie) are of Torridonian sandstone, as confirmed by their almost flat tops. Loch Maree is an inland loch that sits along the Loch Maree Fault, which has been excavated into the wonderful southeast–northwest-trending U-shape glacial valley of Glen Docherty (Fig. 3.22).

Cambrian and Ordovician rocks
Some of the most impressive landforms in the Northwest Highlands are associated with rocks younger than, and lying above, the Lewisian and Torridonian. These include the white scree aprons around the summits of Foinaven, Arkle and Ben Eighe (Figs 3.11, 3.23, 3.24) and the spectacular inlet of Smoo Cave at Durness (Fig. 3.25). The screes are made of sharp, angular fragments of white quartzite, a sedimentary rock that

Figure 3.22   Glen Docherty, Loch Maree Fault, Wester Ross: a northwest–southeast fault excavated by ice; Loch Maree is in the distance.

was deposited as pure sand in a shallow sea some 530 million years ago early in the Cambrian period. Over time, the silica cement of this sandstone was recrystal-lized to form a sedimentary quartzite. In many places, including on Quinag (see Fig. 3.21), the quartzite rests on top of Torridonian sandstone at an angle of 20°, but, clearly, being a beach deposit, it must have been

laid down almost horizontally. This means that the Torridonian rocks were tilted and eroded before the sea came in across the landscape and drowned the Torridonian. Later movements have uplifted all these rocks and caused the Torridonian to be tilted back to the horizontal. Good outcrops of these Cambrian rocks can be seen at Loch Eriboll and around Loch

Figure 3.23   Ben Arkle from Loch Stack, Sutherland: Cambrian quartzite (white scree at top) above ice-scoured Lewisian gneiss basement.

Figure 3.24 Ben Eighe, Wester Ross: Cambrian Quartzite (white scree) above Torridonian sandstone.

Assynt, and also at the Knockan visitor centre. At Skiag Bridge there is a splendid outcrop of the famous Pipe Rock (Fig. 3.26), containing fossil worm burrows cutting vertically across the bedding planes of the quartzite. Although mostly lying unconformably on top of the Torridonian sandstone, in places, such as on the west side of Loch Glencoul, the Cambrian quartzite rests directly on Lewisian gneiss, having overstepped across the Torridonian. The overstepping happened because the Cambrian Sea extended in area and the sands that eventually were compacted to form the quartzite were deposited first on top of Torridonian rocks, then on top of the underlying Lewisian in areas where the Torridonian had been eroded away. Hence, the Cambrian rests unconformably on both Torridonian and Lewisian rocks. Above the quartzite are sandy and lime-rich shales and grit beds, some of which contain fossils, including worm traces and

Figure 3.25 Smoo Cave, Durness, Sutherland: sea cave in Durness Limestone, with swallow hole and disappearing stream above.

trilobites, which are now completely extinct; these lived at the edge of a shallow sea and fed by scavenging on the muddy floor. From Durness to Elphin and again at Broadford in Skye is the outcrop of the Durness Limestone, a thin deposit of marine dolomite (calcium–magnesium carbonate) formed by the action of algae in a shallow continental shelf environment over 500 million years ago. Splendid outcrops can be seen in the Stronchrubie Cliffs at Inchnadamph. Fossils are quite rare in the Durness Limestone, but the few that exist show that these rocks are about 530 million years old at the base and 500 million years at the top (i.e. Cambrian to Ordovician in age; see Table 1.1). As in many other places around the world, the Cambrian represents the time in Earth history when the first substantial numbers of animals with hard skeletons appeared, forms such as trilobites, brachiopods, gastropods and cephalopods. The fossils found in Northwest Scotland are more or less the same as those in eastern Canada and Greenland, indicating that Scotland was still part of Laurentia 600–500 million years ago. However, they are very different from fossils of the same age in Wales and the Welsh borders, so we have evidence that the northern and southern parts of the present British Isles were widely separated then.

In the northwest, the Durness Limestone is thin compared to the Torridonian sandstone, at 1000 m or so, and the underlying Cambrian quartzite is about 250 m thick. Where the Durness Limestone appears at the surface, typical limestone features occur, including swallow holes and disappearing streams at Durness,[*] extensive underground limestone cave systems, as at Smoo Cave (not all of which is yet fully explored) and Inchnadamph, and dry valleys at Traligill River above Inchnadamph on the lower slopes of Ben Mór

Figure 3.26 Pipe Rock, Skiag Bridge, Loch Assynt: the vertical pipes are fossil worm tubes in quartzite (former clean quartz sandstone of Cambrian age); the white pipes are trace fossils.

---

[*] The limestone caves at Inchnadamph once contained bones of bear, arctic fox, reindeer, lynx and lemmings – animals that lived in Scotland at the end of the most recent ice age, but which are now extinct.

Figure 3.27 Folding of Cambrian–Ordovician rocks in Assynt Thrust zone; Stronchrubie Cliffs, Inchnadamph, Loch Assynt, Sutherland.

Assynt. Above the 100 m-high Stronchrubie Cliffs is a limestone pavement, 300 m above sea level. Where well drained, the limestone yields fertile soils, which make a very visible impact on the landscape, as seen in the bright green pastures around Elphin (Fig. 3.12). All these features are to be seen in the Inverpolly National Nature Reserve. The Durness Limestone is important in another regard. Where it has been cut by granite, heat and fluids have metamorphosed the limestone into a colour-banded or streaky marble, conspicuous at Ledmore in Assynt and near Broadford in Skye. Quarries have been operating for the extraction of the attractive green and white speckled marble as a decorative stone (e.g. for fireplaces). Indeed, the Ledmore marble is actually exported to Italy, the home of marble.

The Moine Thrust zone
The western edge of the Caledonian mountain belt is marked in Scotland by the Moine Thrust zone (see Figs 3.1, 3.6, 3.8), a low-angle reverse fault running down the northwest coast from Loch Eriboll to Skye, then on southwards through the Sound of Iona. All the basement rocks (i.e. Lewisian, Torridonian, Cambrian–Ordovician) lie beneath and to the west of this great fault. It can be visited at the Knockan Cliff centre in Assynt, in the Inverpolly National Nature Reserve, where you can actually put your finger on the thrust

plane itself, which is marked by badly sheared and crushed cream Durness Limestone below, and older dark grey Moine schist above (see Fig. 3.6). Thrusting took place at about 400 million years ago, in a final episode of the Caledonian orogeny. Moine schists were transported many kilometres to the west and northwest over weaker rocks, such as the relatively soft Durness Limestone, which were ground down and recrystallized to create a sort of lubricated glide surface. In the Assynt district, it is possible to see several of these Caledonian thrusts, with the Sole Thrust being the lowest (at Loch Assynt), followed upwards by the Ben More, Glencoul and Moine thrusts (see Figs 3.6–3.8, 3.27). Between these thrusts, the underlying rocks have been folded into anticlines and broken by many small-scale reverse faults and minor thrusts, to create incredibly complex packets of deformed rocks. Some of this folding can be seen in the top half of the Stronchrubie Cliffs at Inchnadamph (Fig. 3.27). Erosion of the great pile of thrusts in the Assynt district, combined with some Caledonian intrusions (e.g. Loch Borralan and Loch Ailsh syenites about 430 million years old) has given rise to the so-called Assynt bulge (or culmination) in the thrust belt, now exposed on maps as the peculiar square eastward indentation (see Figs 3.1, 3.8, 3.27).

The Outer Hebrides

From the Butt of Lewis to Barra Head, almost the only rocks you will see are the ancient Lewisian rocks, the exception being around Stornoway, where thick New Red Sandstone (Permian and Triassic, 210 million years old, once thought to be Torridonian or Old Red Sandstone) pebble beds are found encircling Broad Bay, and Tertiary dykes cutting across from the Skye and Mull volcanoes. The Lewisian was named in 1819 by Dr John MacCulloch (famous for his discovery of a petrified tree in Mull) after the Isle of Lewis. In addition to the folded, banded gneisses that are seen on the mainland outcrop (Fig. 3.28), Harris is well known for its metamorphosed igneous complexes in the Uig Hills (Fig. 3.29) and the South Harris hills – Roineval (Fig. 3.30), for example, which is now the object of a controversial plan for a superquarry at Lingarabay near St Clement's chapel at Rodel (dating from 1528, the most important medieval monument in the Western Isles). The rock that is of interest here is an attractive white anorthosite, speckled with green pyroxene (Fig. 3.31). In places, associated with the igneous rocks are belts of metamorphosed sediments, such as the Langavat and Leverburgh belts. These are now slate, soft black graphite schist, garnet–mica schist, quartzite, marble and amphibolite, representing rocks that were once laid down on the surface – mud, shale, sandstone, limestone, lava – and were then deformed and metamorphosed with the gneisses. The metasedimentary belts trend northwest–southeast, parallel

Figure 3.28 Folded Lewisian gneiss, showing rotated basic blocks (dark) in migmatite, Benbecula, Outer Hebrides.

to the Langavat shear zone and to the Sound of Harris inlet.

The oldest Lewisian rocks in the Outer Hebrides are metamorphosed sediments and associated basic and ultrabasic igneous intrusions, and are probably about 3000 million years old. These rocks were then intruded by granitic and related igneous rocks deep within the crust, at 2900–2800 million years ago, then deformed and recrystallized soon after they were intruded. Today, it is these grey, creamy and black-and-white striped or banded gneisses, rich in quartz, feldspar, biotite and hornblende, that form the bulk of the Lewisian rocks in the Western Isles (including the Inner Hebridean islands of Coll, Tiree and Iona). At about 2400 million years and again to a lesser extent at 2000 million years ago, the gneisses were intruded by basic dykes (dolerite in composition). By analogy with

Figure 3.29 Uig Hills, South Harris, Outer Hebrides: rounded granite hills of the Harris Igneous Complex.

Figure 3.30 St Clement's, Rodel, South Harris, in the shadow of proposed site of Lingarabay superquarry (anorthosite).

the mainland, these may be related to the Scourie Dykes (Fig. 3.32), although here they are shown on maps as the Younger Basic Dykes, to distinguish them from the basic and ultrabasic pods and lenses that form part of the older gneiss complex that was intruded by the dykes. Although in broad terms the

Figure 3.31 Roineval, South Harris: anorthosite in a roadstone quarry at Rodel; dark streaks are diopside in white anorthosite rock.

Figure 3.32 Folded basic dyke (dark) in banded Lewisian gneiss, South Harris, Outer Hebrides.

Lewisian of the Outer Hebrides is comparable to that on the mainland, there are differences in composition indicating that the two areas formed part of the western margin of Laurentia, but may have been separated by some distance and were subsequently brought close together by fault movements. It has always been difficult to correlate the structures, rock types and sequences of events in the Outer Hebrides and the mainland, and it was not until the development of new analytical techniques at the end of the twentieth century that researchers were able to demonstrate significant differences and to conclude that these segments of the crust formed in different places. Lewisian rocks in the Outer Hebrides appear to have more in common with gneisses in eastern Greenland than with the Lewisian of the Scottish mainland. Despite more than a century of intensive research, the Lewisian has still not yielded up all its secrets.

Landforms in the 250 km-long chain of Outer Hebridean islands are dominated by the effects of glaciation and the present-day attacks by rain and the Atlantic waves. Apart from the granite hills of Uig on Lewis and the igneous complex of Harris, much of the land is under 120 m, and many places are below 30 m. The prominent triangular hills along the east coast of the Uists (Eaval, Hecla and Beinn Mhór) owe their existence to the proximity to the Outer Isles Thrust fault; erosion along the thrust line has produced an escarpment that is a prominent landscape feature. The coastline is highly indented in the Hebrides, with many sea lochs and long narrow islands orientated northwest–southeast as a consequence of the movement of ice from the mainland across the Minch. North Lewis is dominated by low, gently rounded hills, whereas south Lewis and Harris have mountains showing many features typical of glaciated upland terrain, including corries, rock troughs and U-shape valleys (see Ch. 8). Knock and lochan (knock, from Gaelic cnoc: a hillock) landscapes dominate the lower ground, and there are extensive areas of rocky wilderness, where almost nothing grows but a few prostrate dwarf junipers.

The west coasts of most of the islands have important machair lands, with windblown sand on top of gneiss. The best machair occurs in the Uists and Barra, and also on Tiree. Machair sand has a very high shell content, which distinguishes it from the white quartz-rich sands found in the golf course links of the east coast of Scotland. During the most recent ice age, low sea temperatures killed off many of the carbonate-shell marine animals, and the meltwaters at the end of the Ice Age transported debris onto the continental shelf. Onshore currents and wind subsequently re-worked this material and deposited it to form the beaches backed by dunes. The high carbonate content helps create fertile soils, which are free draining but lacking in bulky organic matter. The white sands and their wild plant colonies create one of the most impressive features of the Outer Hebrides. Machair is one of Europe's rarest landscapes and is unique to Northwest Scotland and the west of Ireland.

Basement rocks of the Inner Hebrides

It was mentioned above that Lewisian gneiss also occurs on Coll, Tiree and Iona. There are relatively small outcrops on Rum, southeastern Skye in the Sound of Sleat, within the Moine Thrust zone (and consequently affected by crushing and alteration), and northern Raasay, whereas the uninhabited island of Rona is made entirely of Lewisian rocks, and is a continuation of the Raasay outcrop. Lewisian gneiss is found as blocks (or xenoliths; Greek: foreign rocks) thrown up by the volcanic eruptions centred on the Western Red Hills volcanic centre, indicating that there is a basement of Lewisian gneiss beneath Skye. There is also a small area of Lewisian rocks on Rum,

and the Tertiary volcanic centres (see Ch. 7) were intruded into a rigid basement of Lewisian and Torridonian rocks. However, rocks in Colonsay and Islay that were once thought to belong to the Lewisian Complex are now known to be part of a quite different sequence (see below). Generally the Lewisian in the small islands is very similar to that in the Outer Hebrides and Northwest Highlands (especially the gneisses found in the Loch Torridon area): that is, banded gneisses 2800–2900 million years old with pegmatites and granite sheets 1750 million years old, and intrusions of dykes that are probably related to the Scourie Dykes are particularly common on Raasay and Rona. Patches of metasediments are found here and there, the most important being the famous green speckled marble found on Iona, and a pinkish marble from Tiree, both of which have been used as ornamental stones; the Iona marble can be found in Westminster Abbey. Other metasediments include banded ironstone (found on Iona just outside the marble quarry), quartzite and garnet–mica schist. Extensive trials were carried out on a magnetite body (an iron ore) in Tiree, but no mines were ever opened. Where Lewisian rocks occur in the Inner Hebrides, we see the familiar topography of knock and lochan, with a significant amount of bare rock in low rounded ice-smoothed hillocks. Most of the small islands have extensive machair formations around the coast, with fine white sand and dunes stabilized by marram grass (Fig. 3.33).

Torridonian rocks occur in northern Raasay, on Scalpay, the Sleat Peninsula of southeast Skye and on Rum. The Sleat Group of Skye was mentioned above (p. 56). The rocks are pebbly sandstone, grey siltstone and sandy shale, laid down by rivers flowing into a freshwater lake. Thin, dark grey or black slaty rocks are found on the east coast of Iona. They are nearly vertical low-grade metamorphosed sediments with a conglomerate at the base containing pebbles resembling Lewisian gneiss. Previously, it was thought that these rocks belonged to the Torridonian, but they are now referred to as the Iona Group and they may correlate with the Colonsay Group.

The Precambrian rocks of Colonsay and the Rhinns of Islay are rather special. Previously, they were assumed to be Lewisian gneiss, because of their outward similarities in the field to the rocks of the Outer

Figure 3.33 Kiloran Bay, Colonsay, Inner Hebrides: recent machair and sand dunes, backed against Precambrian Colonsay Group gneisses.

Hebrides and the northwest mainland; but more recent research has shown that they are very different, and that they form an igneous complex 1700–1900 million years old. In western Islay, the Rhinns Complex is made of syenite (a feldspar-rich coarse-grain igneous rock) intruded by gabbro; these igneous rocks were folded and metamorphosed to produce gneisses 1780 million years ago. This age is nearly the same as the date of the Laxford pegmatites and granites of the mainland, but the chemistry of the rocks is not the same. On Islay, the gneisses lie to the southeast of the Great Glen Fault, whereas Iona, with its proven Lewisian, lies on the opposite side, indicating that this fault may separate different types of crust. Some researchers have suggested that the Rhinns Complex may be related to the Annagh gneisses of Mayo in western Ireland, but those rocks seem too young.

On Colonsay and also the Rhinns of Islay, there are thick quartz-rich metamorphosed sedimentary rocks that seem to be unique in Scotland. They are referred to as the Colonsay Group, and amount to nearly 5000 m of marine sediments, now metamorphosed at greenschist facies (see p. 50). The junction between the 600-million-year-old Colonsay Group and the underlying Rhinns Complex is highly sheared. At present, there is no certainty about the origin or relationship of these rocks on Islay and Colonsay, nor is it known if they are the same as the Iona Group.

Loch Lomond looking north from Balloch (Old Red Sandstone): a chain of islands marks the Highland Boundary Fault; Arrochar Alps in background (Dalradian) (photograph by Pat & Angus Macdonald).

CHAPTER 4
The Caledonian mountains

Introduction

The Caledonian Mountains form a narrow belt that extends all the way from Svalbard in the Arctic, through eastern Greenland and western Norway, to Shetland, the Scottish Highlands and Southern Uplands, Wales and Ireland, and then on to the Appalachians in eastern North America (Fig. 4.1). In Scotland we see only a small section of this 7500 km-long mountain chain, now deeply eroded. Despite the fact that the segment of the Caledonian Mountains in Scotland is at most 200 km across, this narrow zone, like the Northwest Highlands (Ch. 3), has for over a century provided incredibly fertile ground for the growth and development of new ideas and techniques relating to the origin of mountain belts. Major advances in our understanding of how rocks deform and respond to changes in pressure and temperature inside the Earth were made in the Caledonian belt of

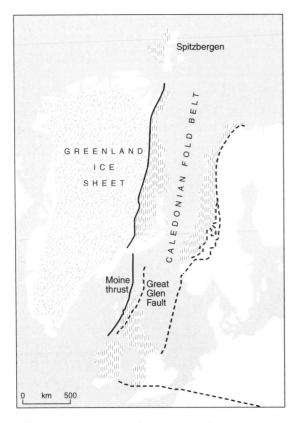

Figure 4.1 The extent of the Caledonian fold belt, prior to the opening of the North Atlantic Ocean.

the Highlands. Thrusting was first demonstrated here: Peach & Horne revealed the effects of the Moine Thrust in Northwest Scotland in 1907 (see Ch. 3); and, in 1893, George Barrow was the first to introduce the concept of regional metamorphic zones in the Southwest Highlands (see p. 50). In 1930, Sir Edward Bailey used sedimentary structures in metamorphic rocks for the first time, to deduce the correct way-up of folded rocks. And as early as 1897, C. T. Clough demonstrated that, in mountain chains, rocks could be folded and re-folded several times over: the concept of polyphase deformation, which is fundamental to our understanding of the evolution of all mountain chains. These concepts and techniques are today used routinely throughout the world. The Scottish segment of the Caledonian mountain belt is one of the most intensively studied mountain chains anywhere in the world. Despite this, it has remained full of enigmas and controversies, relating to when the sedimentary units were deposited, when and how often they have been subjected to major mountain-building phases, and where the various segments of crust came from that were brought together to complete the geological jigsaw. In 1860, at the very beginning of serious study of the Caledonian belt, a major controversy arose, dubbed the "Highlands controversy", between survey geologists and academics (see box on p. 69).

The rocks forming the Caledonian mountain belt account for the largest terrane in Scotland, and the growth of the belt was a highly complex set of events, starting with the break-up of a large supercontinent, Rodinia, 750–600 million years ago, and ending with the closure of the Iapetus Ocean 420 million years ago. This chapter follows the development of the Caledonian orogenic belt in terms of plate-tectonic theory, and shows how the various crustal segments or terranes were finally assembled together in the concluding stages of the orogeny, by 400 million years ago.

The Caledonian belt was formed when the Iapetus Ocean that separated Laurentia (to which Scotland was attached) from Avalonia (to which England and Wales were attached) closed 420 million years ago. Closure of the ocean resulted in continental collision: the ocean disappeared and the sediments that had formed on the ocean floor and shelf became folded and metamorphosed. Melting of the deep thick crust led to the formation of granites, which forced their

The Highlands controversy

In the 1860s, the Director of the Geological Survey was Sir Roderick Murchison (from Tarradale on the Black Isle), and the survey geologists considered themselves as professionals and that academic geologists were amateurs. When Murchison worked in the Northwest Highlands, he assumed that the rocks became younger from the Lewisian upwards, and that there was continuity between the strata; he had no idea then that thrust faults had been responsible for creating a much more complicated story. He assumed, wrongly, that the rocks above the Cambrian and Ordovician were Silurian sediments, but in fact they are Moine schist and quartzite, which are very much older. This notion was challenged by James Nicol, who was Professor of Geology at Aberdeen University, and therefore an "amateur". Nicol stated that older rocks lay above younger rocks as a result of fault movements. Murchison enlisted the help of Sir Archibald Geikie (1835–1924), Director of the Survey in Scotland, to discredit Nicol. Geikie was an accomplished author and communicator, and his first major text, on the geology and scenery of Scotland, was published in 1865, to popular acclaim. Geikie enthusiastically supported and broadcast Murchison's ideas on the structure of the Highlands, and in 1871 he became the first Professor of Geology at the University of Edinburgh, thanks to an endowment of Murchison's. However, this hampered progress for the Survey, a situation that was to be exacerbated still further in the Southern Uplands by the work of another amateur, Charles Lapworth, a schoolteacher from Galashiels, who mapped the graptolite shales in great detail. Lapworth proved that the same beds were repeated many times by folding, and that the

original Survey maps were wrong. In 1881 he was appointed to the chair in geology at Birmingham University on the strength of this work. Geikie was then forced to remap the Southern Uplands using Lapworth's tried and tested methods. The final coup for Lapworth was his interpretation, published in 1883, of the structure of the region between Loch Eriboll and Assynt as being the result of thrust faults that were responsible for transporting sheets of Moine schist westwards across the basement of Lewisian, Torridonian and Cambrian rocks. Lapworth saw that there were great similarities between the structure of the Scottish Highlands and that of the Alps, and he pointed out that Murchison's fundamental error was in assuming that the thrust planes were bedding planes. Geikie by then was Director General of the Geological Survey. He commissioned his colleagues Ben Peach and John Horne to carry out detailed mapping of the Northwest Highlands. After a single season in the field, Peach & Horne reported to Geikie that the "amateurs" had in fact got it right. Peach & Horne then went on to complete their mapping of the Northwest Highlands over the next two decades, eventually publishing their map and memoir, which is recognized around the world as a major classic in geological literature. Geikie was unwilling to acknowledge the work of the "amateurs". A great deal of public bitterness then ensued, which culminated in an official inquiry in 1900 (by the Wharton Committee) into the work of the Survey. Geikie retired in 1901 at the age of 65, and the Survey was reorganized. The history of this fascinating controversy has been published by David Oldroyd (see Bibliography, p. 232).

way up into higher levels, often along deep-seated faults and pre-existing weaknesses in the crust, and helped to elevate the mountain chain. Finally, faulting and westward thrusting brought together the various segments that now make up the belt (Fig. 4.2). We shall now consider the processes involved in mountain building, then explore how the processes have created the Caledonian belt.

How mountains are built

We begin our exploration of the Highlands by looking in general terms at the geological features common to those mountain chains that are formed by continental collision. The geological history of Scotland is inextricably linked with the growth of the Caledonian mountain belt. West of the Moine Thrust zone there are relatively small remnants of older rocks that formed in the root zone of Precambrian mountain chains: the Lewisian gneiss (1750–3300 million years old). In the Northern Highlands we find extensive outcrops of Moine schists (1100–800 million years old), whereas the Grampian Highlands consist of the much more varied and younger Dalradian schists (700–450 million

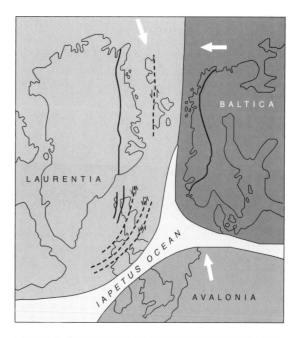

Figure 4.2 Continental collision that resulted in the formation of the Caledonian Mountains and the closure of the Iapetus Ocean.

years old) and the Southern Uplands contain the youngest rocks involved in the Caledonian events – the Ordovician and Silurian sediments, 490–420 million years old. These blocks of the crust were brought together 420–400 million years ago in the final stages of the Caledonian orogeny, a collision of three of the Earth's continental plates. This particular event gave Scotland its varied geology: the country has always been at the edge of a plate.

Mountain chains such as the Caledonian form by a sequence of events involving crustal stretching and eventual rupture, with the creation of new oceanic crust. Oceans grow outwards from the centre by seafloor spreading, as new basalt is intruded from the upper mantle into mid-ocean ridges. Elsewhere on the planet, older oceanic crust is destroyed by slabs being subducted down into the mantle, where it is recycled. Subduction zones are directed beneath deep ocean trenches at continent/ocean collision zones, or beneath island arcs at ocean/ocean collision zones. Since plates are in constant motion, eventually oceans will disappear completely and continents will collide to be welded together, forming much larger landmasses. During the Earth's 4600 million-year-long history, there were occasions when sequences of collision events led to the formation of gigantic supercontinents. As far as Scotland is concerned, the two most important of these were Rodinia, which formed by 1100 million years ago, and Pangaea, which formed by 300 million years ago and started to break up 200 million years ago with the opening of the present Atlantic Ocean, once again leaving Scotland at the edge of Eurasia.

All of this theory is brought together under the unifying concept or model of plate tectonics, which envisages the outer part of the Earth as being divided into rigid plates that are continually moving (Fig. 1.6, 1.7). Beneath the plates, about 150 km down, is a weak zone in the upper mantle called the asthenosphere formed of partially molten plastic material capable of movement. Hence, the plates of the rigid material (the lithosphere) move by gliding over the asthenosphere at speeds of up to 150 mm per year. It is believed that currents operate in the mantle, with new molten rock being injected into ocean ridges, causing the oceans to expand with the addition of new crust, in the form of basalt lava erupted at ridges. Since the Earth is not

expanding, old crust must be destroyed to make room for the new. Oceanic crust is thin, dense and made of basalt, whereas continental crust is much thicker, lighter and buoyant, and it appears to float above the dense layer. Only oceanic crust is destroyed by subduction, as slabs that are pulled down into the mantle. This is one of the main forces acting on plates that drive plate tectonics, the other being the push at ocean ridges. Stretching of the oceanic crust causes it to thin, with the result that the underlying peridotite mantle moves up closer to the surfaces causing some of it to melt as a response to the lower pressure. Once the crust splits apart, the basalt that results from partial melting of the mantle is extruded onto the ocean floor in the form of pillow lavas. These are blobs of magma that quickly form a skin, rather than flow freely across the ocean floor, because pressure of the overlying body of sea water prevents that from happening. Additional impulses of magma burst through the earlier-formed pillows and continue to pile up. This process is still going on today, especially around Hawaii and on the mid-ocean ridges. Fossil examples of pillow lavas are well exposed at Stonehaven, close to the Highland Boundary Fault and within the Highland Border Complex (Fig. 4.3).

Most of the activity we see today as volcanoes and

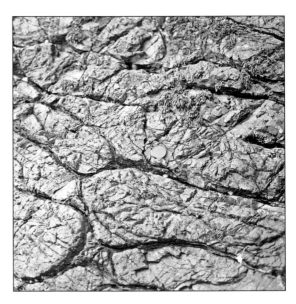

Figure 4.3 Stonehaven, Aberdeenshire: pillow lavas in the Highland Border Complex adjacent to the Highland Boundary Fault.

earthquakes takes place where tectonic plates meet (i.e. along their margins). The two main types of plate boundary that are relevant here are the constructive boundaries at mid-ocean ridges where new crust is being created, and destructive boundaries where plates collide. During collision, one plate is forced down (or subducts) beneath the other as a slab dipping on average at 45° into the mantle. Enormous forces are generated at subduction zones; earthquakes are abundant along and beneath deep ocean trenches, and melting of the descending slab produces magma that makes its way to the surface through volcanoes. The actual products of plate collision will depend on the nature of the crust of each plate. For example, when two ocean plates meet, one is forced under the other at the destructive margin, and a chain of volcanic islands – an island arc – is produced in the ocean above a subduction zone. This is the situation in Japan today. If an ocean plate such as the Pacific plate collides with a continent, South America, then the ocean plate is subducted beneath the continent. A mountain chain is formed at the continental edge, above a trench. Volcanoes and earthquakes are common in these zones, as in the Andes. When two continental plates move together (or converge), the intervening oceanic crust is subducted beneath them both. Eventually, collision will take place and, since neither continent can be subducted beneath the other, a wide zone of thickened crust emerges to produce a continental collisional fold-mountain chain. As the two continental margins are compressed together, sediments in basins around these edges are intensely folded and metamorphosed. Great nappe-like folds spill over from the collision zone onto each continent, and eventually the nappes are sheared off at the base with the formation of thrust faults that pile up slices of crust and help thicken the crust. Since the crust has been so hugely overthickened, the base at 100–150 km deep begins to melt and form granite magma that, being lighter than the surrounding folded rocks, makes its way to higher levels to be intruded as large molten masses of magma called plutons. The Himalayan mountain chain is the product of the collision of India with Asia. The line joining two continents in such a collision zone is known as a suture. In the case of the British Isles, Scotland and the north of Ireland meet England, Wales and southern Ireland along the

Iapetus Suture, running from the Shannon estuary to the Cheviot Hills, which marks the site of the previous Iapetus Ocean. This collision, which was completed 420–400 million years ago, was responsible for the creation of the Caledonian Mountains. The idealized sequence is best summarized in a series of diagrams (Fig. 4.4). It is mainly crustal thickening by thrust faulting and the intrusion of large amounts of granite that are responsible for creating elevated topography in continent/continent mountain chains. Once formed, the rocks at great height are immediately subjected to the powerful forces of erosion: water, wind, ice and gravity. Although young mountain belts themselves may appear very high by human standards, the crust beneath forms very thick root zones, down to 150 km. As material is stripped off at the surface, these buoyant mountain roots rise up to maintain the elevation of the chain, until eventually even the roots are worn down and the continental crust returns to normal or average thickness, which is 35–40 km.

In the Scottish part of the Caledonian mountain belt, it appears that several different crustal segments were involved in the collision between the continental

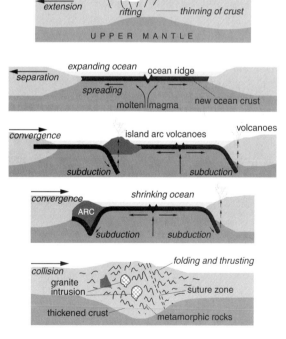

Figure 4.4 Sequence illustrating the opening and closing of an ocean and the collision that results in the formation of fold mountains.

How do rocks deform?

A characteristic feature of the rocks in the Highlands and Southern Uplands is that everywhere you see them they are folded. Folds normally form in bedded rocks by the action of compressive forces acting within the crust. Material is squeezed together and layers are bent into new shapes. The rocks are then permanently deformed; they cannot be unfolded but they may be re-folded. At depths greater than about 15 km in the crust, the temperature can be 250–300°C or more, which is enough to enable rocks to fold by slow plastic deformation in the ductile zone. Individual grains are distorted and may recrystallize to give the resulting rock a new texture. During compression, rocks will thicken; during tension they will thin and stretch, and possibly rupture eventually, if the rock strength is exceeded. In the case of rocks closer to the surface, or if force is applied suddenly, they respond in a brittle fashion and break by faulting and physical displacement. Here, individual grains are ruptured and blocks move apart. Faults also represent permanent deformation: it is impossible to fit faulted segments of crust together again to re-create the original

situation. The only type of temporary deformation is the relatively small amount of sagging of the crust that takes place when thick ice sheets accumulate during glacial periods. Once the ice melts, the crust eventually rebounds to its original level. This is termed isostatic recovery and is an example of elastic behaviour: when the load is removed, the crust completely regains its shape.

Folds have many different shapes or forms, depending on a complex interaction of rock chemistry, grain size, depth in the crust (and therefore pressure and temperature acting on the rock), the presence of fluids between grains, and the rate of deformation. The types of forces acting on rocks include compression, buckling, extension and shearing. Depending on whether the rocks are located in the brittle or plastic zone, the resulting deformation will be faults or folds. Various terms are used to describe the different parts of a fold, and fold shapes themselves (see Fig. 1.7). Faults similarly have a range of terms to describe the components and the types of fracture (see Fig. 1.9).

blocks of Laurentia, Baltica and Avalonia. As the Iapetus Ocean closed, subduction created island arcs. These arcs, together with sedimentary basins adjacent to the arcs and on the edges of the continents, were squeezed together and slices of ocean floor were detached and thrust into the complex fold belt as the continents collided, to produce a very complex structure. This was made all the more complicated because the three continents collided at different times, and obliquely (i.e. not head-on), and further because sedimentation was still taking place into the ever-narrowing Iapetus Ocean, so that the same events did not take place simultaneously everywhere along the Caledonian fold belt.

Earthquakes and faulting

Large earthquakes are the result of the rupture and sudden movement of crustal blocks along fault lines when strain that has built up over time is released in a few seconds, as soon as the rock strength has been exceeded. The adjacent blocks then stick together again, allowing strain to build up until another earthquake takes place. Most of the world's earthquakes occur along plate boundaries, principally subduction zones and continent/continent collision zones.

Scotland presently has relatively few earthquakes, and most of those that do occur are slight and cause little damage. They are all located along fault zones, principally the Great Glen Fault, the Highland Boundary Fault, the Southern Uplands Fault, the Moine

Thrust zone, the Loch Maree Fault, the Loch Tay Fault and the Ochils Fault (Fig. 4.5). Of these, the Great Glen Fault is the most seismically active fault in Britain, with over 60 tremors recorded in the past 200 years. The strongest ever shock recorded was in 1816, centred on Inverness, which caused damage to buildings, and was felt as far away as Glasgow, Edinburgh and Coldstream. The Caledonian Canal towpath was cracked by another earthquake in 1901. In 1964 a very strong tremor was felt around Glasgow, associated with a shatter zone known as the Paisley Ruck. Earthquakes occur along these faults as a result of re-arrangements of crust blocks on either side, caused by weathering and removal of debris (i.e. the faults are not experiencing any physical displacement at present). Small tremors also occur in the North Sea as a result of oil extraction, also around large dams and in association with the collapse of underground workings in coal-mining areas.

This chapter is subdivided into a discussion of the geology of the main blocks of the Caledonian Highlands, which are made up of quite different rock units and are separated now by steep boundary faults. These are:

- The Northern Highlands, between the Moine Thrust zone and the Great Glen Fault, mostly made of Moine rocks.
- The Grampian Highlands of northeastern Scotland, between the Great Glen Fault and the Highland Boundary Fault, consisting of an older basement

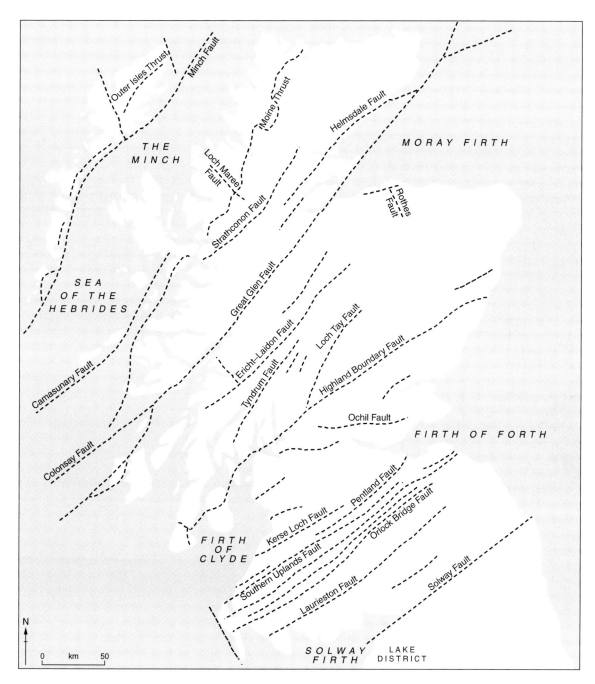

Figure 4.5 The main Caledonian faults in Scotland (northeast–southwest) and older northwest–southeast faults (based on BGS sources).

group in the Central Highlands (immediately south of the Great Glen Fault), overlain by Dalradian rocks.

- The Southwest Highlands, consisting of Dalradian schists and narrow outcrops of Highland Border Complex rocks along the Highland Boundary Fault zone.
- Shetland, which is an incredibly complex segment of the Caledonian belt, consisting of Lewisian, Moinian and Dalradian rocks, and in Unst and Fetlar a piece of oceanic crust and upper mantle rocks.
- The Southern Uplands, between the Southern Uplands Fault and the Solway Fault (which marks the site of the closure of the Iapetus Ocean), where we find mainly Ordovician and Silurian sediments, highly folded and very weakly metamorphosed – now mainly slates and greywackes (poorly sorted, coarse, dark, gritty and muddy sandstones with fragments of volcanic material).

The Northern Highlands

Broadly speaking, the rocks lying east of the Moine Thrust, and north and west of the Great Glen Fault, are metamorphic schists, intruded by granites and overlain around the Moray Firth coast by younger sediments, mostly the Devonian Caithness Flagstones, part of the Scottish Old Red Sandstone (see Ch. 5).

The metamorphic rocks of the Northern Highlands were originally deposited as a thick sequence of sandy sediments in a shallow sea 1000–870 million years ago. These are the Moine schists, named after the boggy headland in northern Sutherland known as A'Mhoine (Gaelic: peat). Fossils have never been found in the Moine rocks; at the time of their deposition, no animals with hard parts yet existed, and any soft-body remains would have been crushed in the coarse sands or destroyed during folding and metamorphism.

In the field, Moine rocks are remarkably uniform in appearance. They are silvery grey, well banded schists, rather coarse and granular, with abundant flakes of white mica (muscovite) or black mica (biotite) glistening on the flat surfaces. Growth of sheet-like mica crystals in parallel layers gives the rock their structure, which is known as schistosity. Since the rocks have been folded and metamorphosed

(Fig. 4.6), it cannot be assumed that these bands represent original bedding, unless original features can be found, such as grading from coarse to fine grains, or cross bedding, or mud cracks. Instead, repeated folding, flattening and thinning of the rocks, accompanied by recrystallization during metamorphism, produced the schistosity (which is also known as foliation). Such schists usually break readily into thin flat slabs along the aligned micas to produce a type of flagstone. One exception to this generalization is the migmatites in northern Sutherland, especially around Bettyhill and Strath Halladale, where granite veins and patches frequently invade the schists. In places, it can be easy to mistake the Moine rocks for Lewisian migmatites of similar appearance to the west of the Moine Thrust in northwest Sutherland (Ch. 3).

Figure 4.6 Folds in Moine schist: Monar Dam, Inverness-shire. The buckle folds are formed in a quartz-rich layer (white).

In the west of the region, there are many patches of Lewisian gneiss caught up in the folded Moine schists. These are referred to as inliers, since the Lewisian is older than the Moine and these patches lie entirely within the Moine schist outcrop. The Lewisian inliers are always found as slices interleaved with the oldest Moine rocks, suggesting that Moine sediments were deposited on a basement of Lewisian gneiss, which then became involved in folding and thrusting with Moine rocks during Caledonian deformation. Normally, the Lewisian is easily distinguished from the Moine in the west of the region between Loch Eriboll and the Kyle of Tongue, thanks mainly to the fact that the Lewisian rocks are at much higher metamorphic grades, and are more varied in type than the Moine schists. However, in eastern Sutherland, in the so-called migmatite zone where a great deal of partial melting has occurred, the rocks are streaked and veined with granite patches and sheets, and it becomes almost impossible to differentiate between the Lewisian and Moine, in the field at any rate.

The Moine rocks represent an enormously thick pile or suite of sediments deposited in a shallow, subsiding sea 1000–870 million years ago. This sea is assumed to have lain on the eastern margin of Laurentia and could have originated as the supercontinent was stretched and thinned, prior to rifting apart completely. Rifting created the Iapetus Ocean (about which more later). In total, about 12 km thickness of sand, silt and mud was laid down, having been brought into the basin from the south, as is shown by sedimentary structures still preserved in the rocks, mainly cross bedding. Zircon grains found among the detritus indicate that the source rocks being eroded to provide this abundant sediment were gneisses 1900–1100 million years old. However, almost nothing seems to have come from the Lewisian, which now lies to the west. This suggests that the sedimentary basins were distant from the Lewisian basement. Almost all the material was derived from the Rhinns Complex (Islay and Colonsay) or the Annagh Gneisses (Co. Mayo, western Ireland). Metamorphism of these sediments produced quartzite and mica schist.

The Moine is usually subdivided into three groups, named after localities in the West Highlands. From west to east (and oldest to youngest), these are the Morar, Glenfinnan and Loch Eil groups (Fig. 4.7). Yell in Shetland is made entirely of Moine schist, but it is not possible to say to which of the three groups in mainland Scotland these rocks belong. Similarly, the migmatites of eastern Sutherland cannot easily be tied in (or correlated) with these three groups. Small outcrops of Moine rocks are found in the Ross of Mull and Ardnamurchan. Rocks of the Morar Group are separated from the Glenfinnan Group by faults: the Sgurr Beag Thrust in the West Highlands and the Naver Thrust in eastern Sutherland (see Fig. 4.5).

In the south of the Sleat Peninsula in southeast Skye there is an interesting sequence of rocks called the Tarskavaig Moines, low-grade schists and quartzites, that have long been considered intermediate between the Torridonian and the Moine, and were thought to represent the metamorphic equivalents of Torridonian rocks. However, this view assumed that Torridonian rocks were deposited by rivers flowing across a land surface and that the Moine represented beach and continental-shelf sediments that were deposited immediately off shore at the same time. The

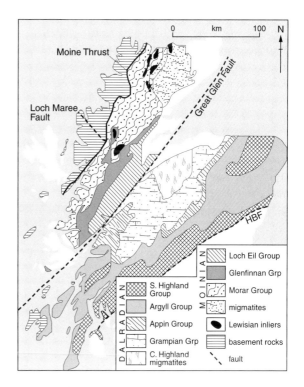

Figure 4.7 The distribution of Moinian and Dalradian rocks in the Caledonian Highlands.

Torridonian and Moine were then brought together along the Moine Thrust, the Moine having been folded in the middle of the Caledonian orogenic belt. But equating the Torridonian and Moine has always been controversial. First, the Torridonian contains Lewisian pebbles and cobbles, and was derived from an area to the west and northwest of the present outcrop. In the case of the Moine, there are no Lewisian fragments in the metasediments, and sedimentary-current indicators where found (especially cross bedding) consistently indicate that material was derived from the south and deposited in northeast-trending basins. Detailed chemistry of the Tarskavaig Moines does not support equivalence with the Torridonian, since the source materials are completely different.

What then about the original position of the Moine block? In spite of a century of investigations, which were especially intensive in the 1990s, it is fair to say that the controversy still rages. The main problem arises from an interpretation of the structural features in the Moine. Looking at Figure 4.7, it is apparent that the Moine is found within separate complex thrust slices that have been pushed together. The rocks went through two metamorphic events at 870–800 million years ago, when nappe folds formed (imagine pushing a table cloth across a polished surface – the cloth piles up and folds over), and again at 450 million years ago in the main phase of the Caledonian orogeny. This indicates that the Moine formed by crustal thickening during continental collision, and it is quite possible that this took place far away from its present location, much farther than the estimated 100–150 km movement on the Moine Thrust. So the Torridonian and Moine were deposited in widely separated sedimentary basins. It is also possible that the basement to the Moine, including the slices of high-grade gneisses, may not have been Lewisian gneiss. We shall return to this theme, of how various blocks in the Highlands were brought together in the final stages of the Caledonian orogeny, at the end of this chapter.

Running from southwest to northeast through the Northern Highlands, from Morvern to Caithness and Sutherland, are a series of migmatites and granitic gneisses, including those at Loch Shiel, Strath Halladale and Ardgour (more commonly known as the West Highland Granitic Gneiss). Within the Moinian, the migmatites appear as veins and patches of quartz and feldspar, injected parallel to the foliation in mica schists, and the overall grain size of the rocks is much coarser than in areas outside the migmatites. In addition, the rocks are usually cut by extensive veins and sheets of granite and pegmatite. Moine quartzites generally do not form migmatites, since virtually the only mineral component is quartz, whereas schists have a more varied composition and more readily act as host rocks. An age of 870 million years has been obtained for the West Highland Granitic Gneiss, using the lead/uranium method on zircon crystals (see p. 52). An even older age of 1100 million years was obtained using the rubidium/strontium method on whole rock samples. The cause of the event, which led to partial melting of Moine rocks and the formation of the migmatites, is a matter of controversy.

Reference was made on p. 57 to the Grenville mountain-building event 1200–1000 million years ago, which resulted from the collision of Laurentia and Baltica to produce (with Gondwana and Siberia) the supercontinent of Rodinia. It is possible that Baltica rotated 870 million years ago, to create the conditions (thickened crust, compression and high heat flow) that led to the formation of the migmatites by partial melting deep within the continental crust. This is a topic of continuing research, and there is as yet no conclusive result.

Other igneous rocks found in the Moine region include basic sheets and dykes intruded more or less parallel to the foliation of the schists, and metamorphosed from dolerite and gabbro to hornblende schist and amphibolite. These are well foliated rocks, containing black hornblende, biotite and white feldspar, and frequently with red garnet crystals. Pegmatites dated at 450 million years old cut these metamorphosed igneous rocks, which are often referred to as the Early Basics, meaning that they were folded and metamorphosed during Caledonian events. In addition to the metadolerites, there are also early granite intrusions cutting the Moine schists, the most important being Carn Chuinneag and Inchbae, between the dammed Loch Glascarnoch and Ben Wyvis (which is made of Glenfinnan Group Moine schists). The Inchbae intrusion is particularly spectacular, since the rock, the Inchbae augen gneiss, is highly distinctive and boulders can be found all the way east to the Moray Firth shore, having been carried there by moving ice. Augen gneiss formed from the flattening and recrystallization of a coarse granite, with large eye-shape crystals of feldspar surrounded by biotite mica. Much of Carn Chuinneag is covered in forest and is inaccessible, but the Inchbae rock is easily found on the A835 in the Blackwater River at Inchbae Lodge. Uranium/lead dating of these rocks has given an age of 580 million years, which dates the time of the main folding of the Moine schists. A remarkable and highly unusual feature of the Carn Chuinneag granite is the fact that a thermal-contact metamorphic aureole is still preserved around the body, related to heating of the country rocks during intrusion, in spite of the later regional metamorphism. The schists around the granite contain small deposits of iron and tin ores (magnetite and cassiterite respectively), although these have never been exploited commercially.

Finally, one notable igneous body in the Moinian is the Glen Urquhart Complex, near Drumnadrochit on the A831. This consists of a soft dark-green altered rock known as serpentinite, an ultrabasic igneous rock containing a rich variety of minerals in addition to serpentine (which was derived from the effect of hot water-rich fluids on olivine and pyroxene). Surrounding the serpentinite are bands of marble, skarn (limestone altered by hydrothermal fluids emanating from the serpentinite during intrusion), kyanite schist, coarse quartzite and gneiss. In the past, these rocks were thought to belong to the Lewisian, but this has not been proven.

The Grampian Highlands

The boundaries of the Grampian Highlands are two of Scotland's principal geological features that cross the entire country, both marked with striking changes in rock types and landforms, and both clearly expressed in the scenery – the Great Glen Fault and the Highland Boundary Fault. Geographically and geologically, this large and complex area is subdivided into the Northeast Highlands around Aberdeenshire, with mainly Dalradian rocks; the Central Highlands with Dalradian rocks of the Grampian Group and the Caledonian granites of the Moine and Dalradian mountains; the Southern Highlands along the Highland Border with Southern Highland Group Dalradian

Figure 4.8 Devonian lava flows, Glencoe Igneous Complex.

rocks; and the Southwest Highlands in Argyll and the isles, with Dalradian schists overlain by younger (Devonian) lavas of Lorne and intruded by the igneous complex of Glencoe (Figs 4.8–4.10). Apart from some Old Red Sandstone around Turriff and New Red Sandstone along the Moray Firth coast between Buckie and Inverness, this vast region consists almost entirely of metamorphic and igneous rocks belonging to the Caledonian orogeny. In contrast to the Northern Highlands, the Dalradian metamorphic rocks here are younger and much more varied than the Moine schists. Within the Grampian Highlands is the highest mountain in the British Isles, Ben Nevis (1344 m), which is part of a granite and lava complex, and the largest high plateau in Britain, the Cairngorm granite massif (over 1100 m high). And on the western seaboard is some of Scotland's most rugged and precipitous relief in Lochaber, Glencoe, Appin and Cowal, with a coastline deeply scarred and indented by firths, fjords and rocky sea lochs.

Dalradian rocks

The Grampian Highlands are dominated by huge thicknesses of repeatedly folded and metamorphosed Dalradian slate, phyllite, schist, quartzite, limestone and metavolcanic rocks. In 1891, Sir Archibald Geikie of the Geological Survey was the first to use the term "Dalradian" for the rocks south of the Great Glen Fault. The term is derived from the ancient kingdom of Dalriada, now mainly Argyllshire. Dalradian rocks

Table 4.1 Divisions of the Dalradian rocks in the Grampian Highlands

Group	Subgroup	Rock types	Interpretation
Southern Highland (over 4 km thick)	Not subdivided	Grits, slates, meta-volcanics (green beds)	Deep sea turbidite basin with volcanic rocks
Argyll (9 km thick)	Tayvallich	Pillow lavas (595 m.y. old) and limestone	Deep turbidite basin with volcanism – opening of the Iapetus Ocean
	Crinan	Coarse grits	Submarine fans
	Easdale	Slate (turbidites)	Faulting produces deep basins
	Islay	Port Askaig tillite (653 m.y. old), then dolomite and Jura Quartzite	Glaciation, then tidal shelf and rapid deposition, controlled by faults
Appin (4 km thick)	Blair Atholl	Slate and limestone	Continental shelf and basin filling
	Ballachulish	Limestone, slate, mica schist and quartzite	Shelf, with basin subsiding and filling
	Lochaber	Schist and quartzite	Shallow-water shelf
Grampian (at least 8 km thick)	Glen Spean	Quartzite	Shallow-water tidal marine shelf
	Corrieyairack	Quartz–mica schist	Deepwater marine turbidites (rifting)
	Glenshirra (Ord Ban)	Quartzite and conglomerate	Braided rivers and shallow seas
Basement Complex (Central Highland Migmatite Complex; over 1 km thick)	Dava succession	Migmatites and quartz-rich gneiss	Unknown source; possibly equivalent to Moinian; folded before Grampian Group was deposited
	Glen Banchor succession	Quartz–feldspar gneiss	

also occur extensively in Shetland (see p. 91) and in the north of Arran; they continue from the Argyll Peninsula to Donegal in Ireland. Age dating has shown the rocks to be late Precambrian to early Ordovician (750–480 million years old). Including the outcrops in Scotland, Shetland and Ireland, Dalradian rocks cover an area of 48 000 km², and have a total thickness of at least 25 km, although this figure is

Figure 4.9 Ben Nevis granite: view from Caledonian Canal.

ROS NEWLANDS

Figure 4.10 Buachaille Etive Mór ("the big shepherd of Etive"; 1022 m), Glencoe Igneous Complex.

cumulative and is not seen in any one locality. This great thickness is divided into four main groups, which are further subdivided into subgroups, all based on place names (Table 4.1).

The Dalradian rocks of the Grampian Highlands are noted world wide for the detailed investigation that has taken place in relation to folding and metamorphism within a mountain belt, and for the establishment of some fundamental principles involved in interpreting extremely complex geology.

Ben Nevis and Glencoe

Ben Nevis and Glencoe are perhaps synonymous with the geology of the Scottish highlands and the region is justly renowned for its magnificent scenery.

The Ben Nevis Complex consists of a granite body 7 km wide, within which sits a 600 m-thick plug of andesite lava 2 km wide. Lava and some explosive agglomerate form the upper part of the mountain, and reach to the summit at 1344 m. Granite and lava would not normally be juxtaposed in this way, since lava is erupted from surface volcanoes, whereas granite forms at depth, having been intruded into Dalradian schists. The explanation is to be found in the process of cauldron subsidence, whereby the roof of a magma chamber sank and fell down into the molten granitic magma beneath (Fig. 4.11). A second pulse of finer granite was intruded into the very coarse first or outer granite, and the block forming the original roof sank further, causing a ring fracture at the surface, which resulted in the collapse of the lavas into the chamber below. Erosion produced the present feature, with Glen Nevis being made of schist, and lava above granite on Ben Nevis. Carn Mór Dearg is made of the inner granite, and Aonach Mór of the outer granite. A swarm of northeast-trending andesite dykes (intermediate rocks; see Table 1.1), centred on Ben Nevis, cuts the outer granite. The dykes themselves are cut across by the inner granite, showing clearly that they were intruded between the two granite episodes. Heating of the surrounding Dalradian schists and limestones by the granite has produced a tough hard rock known as hornfels, which is brittle and it splinters readily when hit. The Ben Nevis Complex is Devonian in age (about 400 million years old), and was intruded at the end of the Caledonian orogeny.

An equally famous example of cauldron subsidence is that of Glencoe, which was long considered a classic, with a relatively simple piston-type structure. This is analogous to pushing a cork into the neck of a completely full bottle: liquid squirts up and over the top. This model, introduced by Sir Edward Bailey in 1909, was applied in many parts of the world to explain the formation of younger caldera structures, in which only the top part is presently visible. Glencoe, on the other hand, being Devonian in age, has suffered considerable erosion, and the very bottom of such a structure is now wonderfully exposed in three dimensions. Detailed remapping of the complex at the end of the twentieth century has thrown new light on Glencoe. Instead of a single, continuous ring fault (actually an ellipse), the structure is bounded by a series of eight straight faults that map out a zone trending northwest–southeast, within which are two main graben structures, also northwest–southeast, cut by smaller faults at right angles. Movement along these faults tapped magma from beneath, and the result was a series of violent explosions from volcanic vents, sited above fault intersections, that produced 500 m of andesite and rhyolite lavas, ash, welded tuff (ignimbrite, formed from burning clouds of ash and dust) and intrusive sills. The eruptions caused instability along the faults, leading to more volcanic activity at the surface. Erosion was continuous inside the structure, with rivers being channelled along the grabens (rift valleys). Pebbles of the nearby Moor of Rannoch granite (only slightly older) are found in conglomerates, indicating that the intrusion must have been rapidly uplifted to the surface. The Glencoe cauldron subsidence was of the order of 1400 m on the internal and boundary faults, and the overall structure was controlled by the interaction of Caledonian faults, including the Etive–Laggan and Great Glen systems, with older northwest–southeast weaknesses in the basement.

Landscape features in Glencoe are the result of glacial action on rocks of markedly different hardnesses. The very fine-grain lavas and sills are exceptionally tough and resistant to erosion (Fig. 4.9), in contrast to the adjacent Dalradian phyllites and schists, which have been eroded out into a broad U-shape valley (Fig. 8.6). The Three Sisters of Glencoe – Aonach Dubh, Gearr Aonach and Beinn Fhada – are made of rhyolite lava. Many of the rhyolites show fine flow banding that formed as the material was moving over the surface, rumpling and puckering on being pushed from behind. Rhyolite is an acidic igneous rock (i.e. very rich in silica) and such lavas tend to be erupted from highly explosive volcanic vents, in which fast-flowing and highly destructive clouds of extremely hot ash and molten lava droplets are produced. Further details can be found in the booklet and map of Glencoe in the series on Classic Areas of British Geology (see Bibliography, p. 232).

In terms of their geological environment, the Dalradian rocks represent sediments deposited on a continental shelf that lay on the edge of a basin that opened to the east, probably as the result of rifting of the Laurentian margin related to an early stage in the opening of the Iapetus Ocean. Sedimentary structures such as graded bedding, ripple marks and cross bedding are quite common, and these provide evidence for the environment of deposition as well as way-up evidence useful in determining the overall fold structures. The Grampian Group was laid down unconformably on top of a crystalline basement that included the Central Highland Migmatite Complex, which was deformed and metamorphosed 840 million years ago. These basement rocks could be equivalent to the Moine rocks found north of the Great Glen Fault. On Islay and Colonsay, the basement to the Dalradian consists of the Rhinns Complex (see p. 65). The Grampian Group occurs mainly in the Grampian Highlands (see Fig. 4.7); on Islay it is represented by the Bowmore Sandstone, and in Shetland it is absent. Previously, these rocks were referred to as the Younger Moines on the southeastern side of the Great Glen Fault. Deepwater sediments are overlain by river-borne and deltaic deposits derived from the west and south. Rocks of the overlying Appin Group indicate development of the main Dalradian basin, where sedimentation took place on a stable continental shelf. The Ballachulish Slate is an important member of the group; in the eighteenth and nineteenth

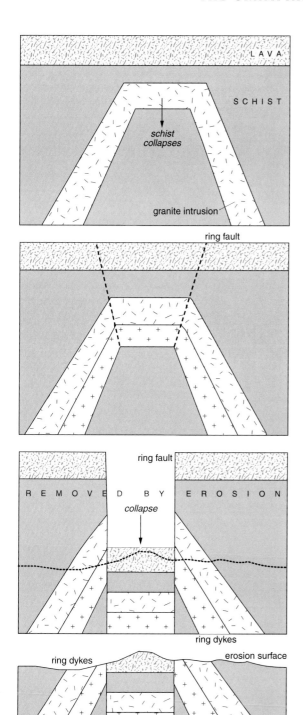

Figure 4.11 Sequence to demonstrate formation of cauldron subsidence by collapse around a ring fault.

centuries it was quarried extensively. The black slates (metamorphosed muds) are up to 400 m thick and have been used throughout Britain as roofing material. At the top of the Appin Group is the Lismore Limestone, which formed in a warm shallow sea. It is found in Islay, Lismore, Lorne and Appin, where it often forms broad northeast–southwest fertile valleys (Lismore means "the large garden" in Gaelic).

The Lismore Limestone is followed above by the Argyll Group, which begins with the Port Askaig Tillite, named after the type locality on Islay. This 750 m-thick unit represents sediments deposited by melting icebergs or from extensive ice sheets, and indicates one of several late Precambrian ice ages that affected the northern hemisphere 800–570 million years ago. The best outcrops are on the Garvellach islands, in the Firth of Lorne between Jura and Mull, where boulder beds are to be found containing large blocks of Rhinns Complex rocks. The tillite is an important marker horizon, since it is easily recognized and can be found as the Schiehallion Boulder Bed in Perthshire and the Claggan Boulder Bed in Connemara, western Ireland. Immediately on top of the glacial beds is a carbonate unit, the Bonahaven Dolomite, a type of limestone. Such dolomites are frequently found directly on top of glacial sediments, and for this reason they are usually referred to as cap carbonates; they indicate a sudden climatic change, during which carbonates were precipitated very rapidly on a shallow marine shelf in a warm arid climate. The Bonahaven Dolomite on Islay contains the best-preserved stromatolites in Britain. Such beds were formed as algal mounds in a shallow tidal sea affected by storms. The fact that this dolomite formed immediately on top of glacial deposits indicates that Scotland lay much closer to the Equator in the late Precambrian and that the glaciation may well have been world wide and have reached down to sea level even at the Equator; this is the so-called Snowball Earth hypothesis. The period from 800 to 570 million years ago is one of the longest and most extensive periods of glaciation in the entire history of the Earth, when the world was an icehouse. Jura Quartzite then follows in succession (Fig. 4.12), representing clean, shallow-water sands.

After the formation of the shelf limestones, the Dalradian basin rapidly became deeper as a result of

Figure 4.12 Jura from Islay, Inner Hebrides: Paps of Jura, Dalradian quartzite; a raised beach is clearly visible along the shore line.

movement on faults that bounded the basin, and fine-grain muds were deposited from rapidly flowing turbidity currents. Turbidity currents are dense mixtures of sediment (mud, silt and sand) and sea water, dislodged from the continental shelf by earthquakes, that flow rapidly down slope to the ocean floor, where the suspended sediment settles out; the resulting rock type is known as a turbidite. Metamorphism converted the muds into the Easdale Slates, which, like the older Ballachulish Slates, were extensively quarried, especially on the islands of Seil and Luing. On Kerrera, Easdale slates and interbedded limestones are spectacularly interfolded. Slates on Jura are of this same formation. Farther to the northeast, higher grades of metamorphism than on the Argyll seaboard have produced schists around Ben Lawers, Ben Eagach, Schiehallion and Killiecrankie, rather than slates. This part of the Argyll Group is important in terms of economic mineralization. Bedded baryte (barium sulphate) and metal ores (sulphides) occur in these rocks, north of Aberfeldy. Such deposits are termed stratabound, meaning that they formed during sedimentation and occur within beds. The baryte at the Foss mine (Aberfeldy, Perthshire) that is currently being worked lies at the top of the Ben Eagach schist. Baryte is used in the petroleum industry as heavy mud during drilling operations. Since 1984, 50 000 tonnes of baryte have been produced each year from Foss, and 10 000 tonnes annually since 1990 from the Ben Eagach quarry, the total since 1984 being 550 000 tonnes. At Duntanlich in Perthshire, the baryte reserves have been put at 7.5 million tonnes. Sulphides of lead and zinc are associated with the baryte.

Two important units at the top of the Argyll Group are the Loch Tay Limestone (Deeside Limestone in the northeast, Boyne Limestone on the Banffshire coast; Fig. 4.13), which is another marker horizon in the Dalradian, and the Tayvallich lava formation in the Southwest Highlands, some of which was erupted under water and shows typical pillow-lava structures. These are the most extensive volcanic rocks anywhere in the Dalradian. Many basic sills and dykes were intruded at the same time as the lavas were erupting onto the sea bed, the dykes probably representing vertical feeder pipes to the lavas. This period of volcanic activity resulted when the crust thinned, stretched and ruptured with the formation of the Iapetus Ocean.

Most of the Southern Highland Group is made up of slate, phyllite, mica schist, quartzite and grit – poorly sorted sediments that formed by turbidity currents

Figure 4.13 Folds in Dalradian impure limestone: Portsoy, Moray.

slumping into the Dalradian basin. Green beds near the base represent lavas and volcanic ash that were metamorphosed at low grade, the colour being attributable to the presence of the mineral chlorite. Rocks of this group form a broad band at the southern edge of the Grampian Highlands, from Argyllshire by way of Loch Lomond side to Aberdeenshire, against the Highland Boundary Fault (Fig. 4.5). Ben Lomond and the Arrochar Alps are made of Ben Ledi Grit belonging to this group. Another important unit is the Aberfoyle Slate, which provided roofing material for much of Edinburgh and Glasgow for over 200 years until the 1970s; Scotland now has no working slate quarries. In the Trossachs area, the Leny Limestone was once worked locally for agricultural purposes.

Microfossils found in the Macduff Slates on the Banffshire coast, near the top of the Southern Highland Group, indicate that these rocks belong to the Ordovician period (about 480 million years old). The Macduff Slates also contain a boulder bed with many limestone pebbles that appear to have fallen onto the muddy sea floor from a melting iceberg. These deposits are known as dropstones and they indicate another glaciation during the Ordovician.

Folding and metamorphism of the Dalradian rocks During the Caledonian orogeny, the Dalradian rocks were folded three or four times in what are referred to as episodes of deformation. The earliest major folds run northeast–southwest, and these were later re-folded to produce complex overfolds or nappes, the most important being the Tay Nappe. Adjacent to the Highland Boundary Fault, the Tay Nappe has been folded over to produce a steep belt, known as the Highland Border Downbend or the Highland Border Steep Belt, in which the Southern Highland Group Dalradian rocks are vertical or even overturned. To the north, the main part of the Tay Nappe is upside down and is referred to as the Flat Belt; here, the Dalradian schists are gently inclined or flat lying. Between the Tay Nappe and the Great Glen Fault, the older Dalradian rocks are folded into a complex series of upright anticlines and synclines, such as the Low Awe Syncline and the Islay Anticline in the western Grampians. During the formation of the nappe folds, the Dalradian rocks were at the peak of regional metamorphism, and when most of the recrystallization and growth of

metamorphic minerals took place (see the box on p. 50). In the Southwest Highlands and in Buchan, the Southern Highland Group rocks nearer the top of the Dalradian pile were formed at relatively low grades of metamorphism, and to the north and northeast the Grampian Group and Central Highlands Migmatite Complex reached much higher grades, with the Central Highlands Migmatite Complex indicating that partial melting must have occurred. These rocks were therefore deeper in the crust at the time. This picture is complicated in Aberdeenshire, where the intrusion of many basic and ultrabasic igneous bodies at 490 million years ago (Belhelvie, Huntly, Insch, Morven, Cabrach, Portsoy) resulted in much steeper thermal gradient and therefore higher metamorphic grades.

Natural resources in the Grampian Highlands
Apart from coal and oil, most of Scotland's economically important natural resources are to be found in the Highlands, and most of these are in the Caledonian rocks, particularly the upper parts of the Dalradian. The Moine schists and the Grampian Group (Lower Dalradian) are by contrast rather devoid of useful minerals. Deposits are found in three quite different geological settings, and understanding how such deposits form is important in mineral exploration. The settings are:

- stratabound, where the material was concentrated during deposition and is now found in beds
- igneous rocks, i.e. in lavas, basic and ultrabasic intrusions, and granites
- veins cutting through any other type of rock.

The great bulk of metal-ore deposits are to be found in the Dalradian, principally in the Argyll Group in the middle part of the Dalradian sequence. Economically important stratabound baryte deposits (barium sulphate) at Foss and Aberfeldy in the Ben Eagach schist (p. 91) are defined as being of world class, that is, this is one of the world's largest known concentrations of barium, and of its type it is probably unique. The deposit formed by the action of metal-rich salts pouring out from hydrothermal vents, located along fault lines, onto a muddy sea floor. Sea water was able to find its way down the crust along active faults, to be heated to temperatures above 200°C at depths of about 10 km and thus able to dissolve minerals (barium, silicon, iron, manganese, lead, zinc, sulphur) from the host rock, then circulated upwards in an underground plumbing system that made use of cracks and faults in the rocks. Once at the sea floor, the barium was dumped as a precipitate from solution, after mixing with sea water and reacting with the muds on the sea floor. The original source of the barium was from feldspars in the host rocks. Other related baryte deposits occur in the Ben Eagach schists at Braemar, Glenshee, Pitlochry and Loch Lyon (see Fig. 9.1).

Stratabound mineralization also occurs in the Ben Lawers schist, which is immediately above the Ben Eagach schist, in the form of copper pyrite (iron–copper sulphide). Deposits are found in the Tyndrum area (between Crianlarich and Bridge of Orchy) and in Knapdale, southwest of Loch Fyne, where there are nineteenth-century copper mines at Meall Mór, the copper occurring in quartzite, adjacent to metamorphosed basic lava. Also in the Loch Fyne area of Argyllshire, notably at Inverary (nineteenth-century workings here also) and McPhun's Cairn on opposite sides of the head of the loch, massive iron and copper sulphides with some nickel, lead and zinc are found as deposits within the metamorphic rocks (Ardrishaig Phyllite). These small deposits are hydrothermal in origin, related to the end stages of igneous activity, which resulted in the sulphides being deposited on the sea floor and then becoming incorporated within the marine sediments. Later, the ores were folded and metamorphosed together with the host mudrocks that became phyllites. A manganese-rich ironstone deposit at the Lecht may be stratabound, or may have developed in an explosive volcanic pipe that intruded the Dalradian after metamorphism and caused the surrounding rocks to be shattered and brecciated. The deposit was worked in the eighteenth century for iron ore and in the nineteenth century for iron and manganese. Manganese compounds were used in the dyeing and bleaching industries. The restored ore-mill buildings and spoil heaps are still preserved beside the mill stream on the A939 Lecht road, half way between Cockbridge and Tomintoul. Much of the ore resembles rust-coloured cindery blocks. Investigations in the 1980s showed high levels of manganese, barium and zinc in the country rocks.

The metal-rich mineralization around Tyndrum (at Auchtertyre) includes ores of copper, zinc, lead, nickel and some cobalt, as stratiform deposits in the Ben Challum Quartzite (Argyll Group), which are therefore younger than the Aberfeldy baryte deposits. Here the sulphides formed in small hydrothermal vents on the sea floor. The deposits were assessed by the Geological Survey in a 1980s drilling programme, but no new mines were opened.

Igneous rocks are abundant in the Dalradian of the Grampian Highlands, and many contain metal ores. However, there are no working mines at present, and none of the deposits is currently economic. Exploration in the late 1980s showed that there may be some future potential for copper, nickel, gold and silver in the region. Mineral deposits in igneous rocks vary greatly, depending on the host bodies. Many of the Caledonian basic and ultrabasic rocks have copper–nickel sulphides, iron ores, and metals belonging to the platinum group. Chromite (chromium–iron oxide) is found in small amounts in serpentinites in the Grampian Highlands (and more importantly on Unst, Shetland; see p. 92).

Ultrabasic intrusions occur within the Dalradian all the way from Loch Fyne in the west to the Moray Firth in the northeast, and many of these are in the Ben Lui schist (Crinan Subgroup of the Argyll Group in the Middle Dalradian; see Table 4.1), below the Tayvallich Volcanics. It is possible that the intrusions were related to the beginning of the widespread volcanic activity in the Southern Highland Group of the Dalradian, associated with thinning, stretching and eventual rupturing of the continental crust. Ultrabasic rocks are readily altered to serpentinite by the action of water-rich fluids. At Corrycharmaig in Glenlochay, Perthshire, the serpentinite body was previously worked for chromite, which occurs as pods in a serpentinite sill intruded into Dalradian garnet–mica–quartz schist. About 60 tonnes of ore was produced in the mid-1850s. Talc is also present and could potentially be extracted for industrial use as a filler in paints, textiles, paper and rubber products; it is not pure enough for cosmetic use.

Highly altered ultrabasic horizons in the Dalradian near Tyndrum (between Creag Bhocan and Ben Challum) contain iron–nickel sulphides, with chromium minerals and sulphides of cobalt and arsenic scattered (or disseminated) among the dolomite–talc–chlorite schists. Hydrothermal alteration of an original ultrabasic rock resulted in its being enriched in several metals. Such rocks are important in mineral exploration, as they frequently act as hosts for gold mineralization, and potentially also for valuable platinum-group metals.

The Younger Basic igneous rocks of Aberdeenshire represent the deep root zones of a volcanic arc that formed on the northern margin of the Iapetus Ocean during the Ordovician period (490 million years ago). The intrusions include layered ultrabasic to basic bodies – peridotite and gabbro – and some host copper–nickel–platinum mineralization at Littlemill near Rothiemay. One such body is the Huntly–Knock intrusion, which has rich concentrations of copper–nickel–iron sulphides. Gold is also present in small amounts. As the molten magma cooled, dense minerals, including the ores, settled out and accumulated in layers between the silicate layers as a result of sulphides being unable to mix with silicates. In the same area, the Insch intrusion has high copper–nickel values. None of these deposits has been exploited, partly because the metal contents are not high enough and the rock bodies are rather deeply buried beneath glacial deposits.

The final stages in the closing of the Iapetus Ocean were accompanied by widespread igneous activity, when many granite bodies in particular intruded the base of the thickened continental crust in the collision zone. Ore minerals can be found in some of these granites, but none of the occurrences has any economic importance at present. Deposits are often found in the roof zones of granite intrusions, where water-rich fluids and rare minerals accumulated and crystallized late in the cooling stages, and at the margins of the granites where chemical reactions took place between the host granites and surrounding Dalradian country rocks to form ore deposits. For example, in Glen Gairn west of Ballater, the granite contains veins and cavities filled with tin, tungsten and molybdenum minerals, with smaller amounts of lead, zinc, copper and some silver minerals in addition to fluorite (calcium fluoride) and topaz, indicating that fluorine-rich fluids had a role to play in the deposit. The Ballachulish granite, which lies very close to the Great Glen Fault, also contains copper sulphide and molybdenum mineralization in a network of interconnected veins and fractures, referred to as a stockwork, in the middle

of the granite. The Etive Complex nearby is another granite containing molybdenum–copper mineralization. In western Argyll, at Lagalochan near Kilmelford, a small volcanic pipe has high concentrations of gold and silver. Hydrothermal alteration of breccia took place after explosive intrusion shattered the wall rocks. Early copper–gold–molybdenum veins were cut by lead–zinc–gold–silver veins with antimony and arsenic. The pipe was related to the eruption of the Lorne Plateau Lavas. Copper-rich veins with lead, zinc and molybdenum occur in the Beinn nan Caorach granite sheet at Kilmelford. Gold and silver are found at Rhynie in Aberdeenshire in Devonian lavas and sediments, where hot springs adjacent to a volcanic vent deposited mineral-rich chert and instantly fossilized some of the earliest known land plants that were living around the springs (see box on p. 117).

The Tyndrum area of Argyllshire is important in terms of both its vein mineralization and the ultrabasic mineralization described earlier. Gold is particularly important, as at Cononish, where a gold-bearing quartz vein was discovered in 1984, but was developed only briefly in the 1990s as a potential mine; possible ore reserves are up to a million tonnes. Silver, copper, lead and zinc are also present. The gold at Cononish, which is the most important gold deposit so far found in Scotland, occurs as minute particles inside pyrite (iron sulphide or fool's gold) and galena (lead sulphide), which themselves occur in quartz veins. Concentrations in the ore are 10 parts per million (ppm, equal to grammes per tonne) gold and 43 ppm silver. A possible source for the mineralization could be buried Caledonian intrusions at depth. The main vein at the previously worked Tyndrum mines is a lead–zinc deposit younger (at 360 million years) than the gold vein (late Caledonian, over 380 million years old). A total of 20 000 tonnes of lead and zinc ore is still present in the Tyndrum veins and, according to the Geological Survey, if this were to be exploited the yield would be about 6000 tonnes of lead, 100 tonnes of zinc, 100 tonnes of copper, and 1 tonne of silver. Veins run northeast–southwest, parallel to the Tyndrum fault, and the host rocks are quartzites belonging to the Appin Group of the Dalradian. When the mines were operational, a total of 10 000 tonnes of lead ore and 200 tonnes of zinc ore (from dumps) was produced during 1741–1862 and 1916–25.

Other metal-ore deposits in the Dalradian, which have been worked in the past, include lead veins cutting the Ballygrant limestone at Mulreesh on Islay. Copper, iron and zinc are also present. Copper was worked in the eighteenth century at Kilsleven, east of Ballygrant, and silver has been discovered at Gartness. The Mulreesh mines operated intermittently for 200 years from 1680, producing 1400 tonnes lead and half a tonne of silver in the 20 years to 1880. The nineteenth-century industrial ruins are still present. At the end of the eighteenth century, manganese was mined from veins at Dùn Athad near the Mull of Oa, southernmost Islay. Metamorphosed basic igneous rocks at the very important archaeological site of Kilmartin in Argyllshire contain veins of copper–iron sulphides with precious metals. The Loch Tay limestone at Corrie Buie east of Loch Tay contains sulphide veins with silver, lead, zinc, copper, iron and rare native gold. Gold is also found in the Calliachar Burn near Aberfeldy, in the Southern Highland Group Dalradian. This was assessed by exploratory drilling in the 1980s. Copper was extracted in the nineteenth century at Tomnadashan on Loch Tay from a granitic-type body.

Although gold has never been produced from the scattered gold veins in the Grampian Highlands, a period of exploration in the 1980s in the Perthshire goldfield resulted from a sharp rise in the price of gold on the world markets. The metal was initially concentrated in Dalradian organic-rich turbidite sediments and later migrated upwards in fluids that filled cracks and fissures at higher levels in the crust. Because of the rich variety of metal-ore deposits – gold, silver, barium, chromium, copper, iron, manganese, molybdenum, lead, zinc, platinum, tin, tungsten, at over a hundred localities – the Middle Dalradian of Scotland has been described as a metallogenic province.

Other useful natural-rock materials in the Grampian Highlands include building stone – granite in particular – but the industry has now all but ceased, with only the Kemnay granite in Aberdeenshire and the Monadhliath granite currently being quarried. Many of the igneous rocks and metamorphic quartzites in the Dalradian are used locally as crushed stone to produce aggregate for roadstone and concrete. Dalradian limestone has been used for a long time in agriculture, as a source of lime, and for concrete aggregate. Quartzite and quartz veins are abundant in the Dalradian, but only the Appin Quartzite at Kentallen in Argyllshire was ever worked. Slate is widespread in the Grampian Highlands, but all the quarries have long been abandoned (Ballachulish, Aberfoyle, Glens of Foudland, Easdale) on grounds of high labour costs and inferior quality, cheaper Welsh slate having replaced the Scottish varieties with the arrival of the railway into Central Scotland. Decorative stones are few, an exception being the attractive Portsoy "marble" of Banffshire. This is in fact a multi-coloured, streaked and banded serpentinite, rich in talc, which formed from the deformation and alteration of an ultrabasic igneous rock. Portsoy marble was a fashionable decorative stone in the seventeenth century, used extensively in the Palace of Versailles, near Paris.

Landscapes of the Grampian Highlands

The geology of the Grampian Highlands is dominated by fold and fault structures that resulted from the Caledonian activity. Most of the major faults and fold axes trend northeast–southwest, including especially the Great Glen Fault and the Highland Boundary Fault, which delineate the region. Thus, the main lochs, rivers and coastal outlines and island shapes are dominated by this pattern, especially on the west. Rocks now at the surface – quartzite, slate, schist, limestone, granite – formed deep in the crust and were exposed by uplift and erosion over many hundreds of millions of years. Material was stripped off to produce sediment for the Old Red Sandstone, followed later by tropical weathering during the Carboniferous and subsequent periods, most notably in the Tertiary, culminating in sculpting during the most recent ice age, and deposition of moraines in valleys, and sands and gravels on lower ground as the ice melted. Originally cut by Tertiary rivers, valleys such as the Great Glen were excavated by the ice, their edges smoothed and their floors deepened.

It is the effects of glacial erosion and deposition that are probably visually the most dominant in the Highlands. The Scottish ice sheet was at its thickest around Rannoch Moor and Ben Nevis, and it moved outwards from this centre, as can be seen from the radial drainage pattern around the moor. All the mountain tops were under the ice, as is evident in the Cairngorms, with the plateau at 1200 m above sea level, U-shape valleys and corries (Fig. 4.14). The plateau features date originally from the Tertiary period, when extensive planation of the land surface took place. Rare remnants of an older landform still exist in the Cairngorms. These are the tors that formed when the granite was very deeply eroded along vertical and horizontal joints during intense rainfall and leaching in the Tertiary period, about 50 million years ago. Glacial erosion removed most of the overlying soft weathered rock, leaving only a few isolated tors behind in the Cairngorms (see Fig. 8.3) and on Bennachie in Aberdeenshire.

The Monadhliath Mountains to the northwest of the Spey Valley are ice-smoothed Dalradian schists, 600–900 m high, generally lacking corries and deep glacial troughs. Ice covered this area, but the main sheet was moving across from Rannoch Moor to the west. Surprisingly, the Moor of Rannoch granite itself now forms not a mountain but an extensive area of flat marshy ground with interconnected lochs and moraine hummocks, but with little bedrock exposed, except on some low rounded rocky knolls. This resulted from long periods in the past 50 million years when the basin was being eroded by the gradual retreat of slopes. The moor is surrounded by high steep-sided rounded hills in the distance. The U-shape

valley of Glen Coe at the Rannoch Moor entrance is a spectacular landform feature, unrivalled anywhere in Britain (see Ch. 8 opening). Glen Coe is a deep glacial trough, with steep slopes rising to 900 m summits amid the granites and exceptionally tough andesite lavas. It formed when ice carved the valley as it moved out of the Rannoch centre of ice accumulation.

The Spey, one of Scotland's major rivers, rises between Creag Meagaidh and the Monadhliath mountains, and flows to the Moray Firth, aligned with the Caledonian structures. It is now in a valley wider than the river itself would suggest, indicating that, like all the major river valleys, it is pre-glacial in age and was eroded in Tertiary times when the drainage pattern of Scotland was first established. During the Ice Age, glaciers moved out of Rannoch Moor into the Spey Valley and travelled northeast into the Moray Firth. The valley was then filled by an enormous thickness of sand and gravel when the ice melted, and these sediments are now being reworked by the modern river in its exceptionally wide floodplain.

Compared with the eastern Grampian Highlands, the mountains of the west and southwest are more impressive in their ruggedness and the presence of deeply indented sea lochs, which are glacially eroded troughs, now flooded. Erosion was more intense on the west coast, because of the presence of high mountains intercepting the snow-laden winds blowing in

Figure 4.14 Cairngorm granite, Loch Avon: a glacial trough.

off the frozen North Atlantic to add to the glaciers. The effect of the underlying geology on the landscape is nowhere better seen than at Loch Lomond. To the south, the loch is wide and shallow where it lies on the soft Old Red Sandstone of the Midland Valley, but suddenly, north of the Highland Boundary Fault, the loch becomes much deeper (to 180 m) and narrower, and Dalradian mountains rise up steeply to 900 m. The deep cleft was scoured out by the valley glacier as ice spilled southwards from higher regions to the north. Then to the north of Loch Lomond, the Arrochar Alps emerged, formed from the differential erosion of Dalradian quartzite, schist and granitic type rocks.

Clear evidence of the role played by glaciers in landscape formation is seen at its best in the area around Glen Roy, Lochaber. Here, the famous Parallel Roads of Glen Roy were discovered in 1840 by Sir Roderick Murchison of the Geological Survey, and Louis Agassiz, a Swiss scientist who proposed the idea of a former glacial event in Europe. According to Agassiz, the parallel roads were the shorelines of meltwater lakes that formed when ice dammed the valley exits and filled them with water from the melting glaciers above. In Glen Roy, the temporary lake was 16 km long and 180 m deep at its most extensive, and the parallel lines a few metres wide on the hillside represent levels cut into the valley sides by waves before the water level fell. The area is now a Site of Special Scientific Interest (SSSI; see also pp. 184–185).

Igneous rocks in the Caledonian Mountains

Igneous activity was both frequent and widespread throughout the Caledonian belt, from the Tayvallich lavas at the top of the Argyll Group (595 million years old), the Silurian to Devonian Newer Granites (420 and 410 million years old), and finally volcanic rocks of Lorne, Lochaber, Glen Coe and Ben Nevis of Devonian age (400 million years old). Igneous rocks are an important component of all old mountain chains, and some indication of the proportion of the Caledonian belt occupied by igneous rocks can be gained from Figure 4.15. Scotland's highest mountain areas – Ben Nevis and the Cairngorms – are made of igneous rocks.

In the Buchan area of northeast Scotland there is an important concentration of large flat-lying igneous

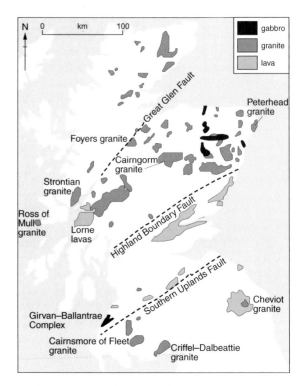

Figure 4.15 The distribution of Caledonian igneous rocks.

bodies that were intruded into Dalradian rocks during the peak of folding and metamorphism. These are layered ultrabasic (peridotite, mostly converted to serpentinite) and basic (gabbro) rocks, the layers having formed by the settling out of denser crystals, first olivine then pyroxene. The masses making up this complex are found at Belhelvie, Huntly, Insch, Morven, Cabrach, Portsoy and Haddo House, and there are many smaller ones. Several of these bodies have been quarried for building materials or roadstone. Dalradian rocks surrounding these intrusions were partially melted as a result of the additional heat produced by the intrusion into country rocks that were already at a high temperature during their metamorphism. The Younger Basics themselves were folded and disrupted during deformation, after they were intruded, and it has been suggested that one or more very large sheet-like basic bodies were intruded under the entire Buchan area, then broken up by folding, shearing and faulting.

As in other mountain chains around the world, the Caledonian belt is characterized by abundant granite intrusions (Fig. 4.15), which arose as a result of melting at the base of the continental crust, which had been thickened because of folding and the piling up of thrust sheets during continental collision. Granites in the Grampian Highlands occur in several groups, of different ages. Those in the southwest belong to the Argyll granites, including Etive, Rannoch Moor, Strath Ossian, Ballachulish and Ben Nevis. The Foyers granite, adjacent to the Great Glen Fault, is grouped with the Argyll granites, most of which are about 405 million years old.

By far the largest concentration of granites is in the northeast of the region, and here we find the Aberdeen, Kemnay and Strichen granites, mostly about 470 million years old. The Rubislaw quarry (once known as the deepest hole in Europe, at 122 m) provided most of the building stone for the city of Aberdeen (see Fig. 9.10), the piers supporting the Forth railway bridge (see Fig. 9.11), and many other locations around the world. A younger group, termed the Cairngorm suite, has ages mostly about 405 million years and includes Cairngorm itself, as well as Monadhliath, Lochnagar, Glen Gairn, Ballater, Ben Rinnes, Mount Battock (Kincardine), Hill of Fare, Bennachie and Peterhead (Fig. 4.16), many of which are composed of coarse-grain pink granites. These granites were intruded after the folding, metamorphism and uplift of the Caledonian mountain belt, and have been termed the Newer Granites. Several of the bodies, namely Cairngorm, Lochnagar, Ballater, Glen Gairn and Mount Battock, lie in an almost east–west zone, which may point to the possible existence of some older long-lived crustal weakness that was used by the granites as they intruded the upper crust from near the boundary between the lower crust and the mantle. During the mid-1980s, these granites were investigated as possible sources of geothermal energy, because they have high heat flow, but the actual values were not sufficient to allow exploitation of these rocks at such isolated locations.

Also of Devonian age are the volcanic rocks of Lorne and Lochaber (the Lorne Plateau Lavas), found around Oban and as small outliers on Kerrera and southeast Mull. They are 800 m thick and consist mainly of basalts and andesites, with a few rare rhyolites. The lavas in the Ben Nevis and Glencoe complexes are part of this sequence, which used to be much more extensive. The lavas are frequently seen to contain gas bubbles that were later filled with secondary minerals, often white calcite. Age dates for the lavas are very similar to those for the Etive granites, to which they are related, that is, 400 million years old.

The Highland Border Complex

At ten different locations along the Highland border line, there are narrow outcrops of weakly metamorphosed igneous and sedimentary rocks bounded by vertical faults. These are fragments of a group referred to as the Highland Border Complex, and they range in age from Cambrian (515 million years old) to Ordovician (445 million years old). The main outcrops are at Stonehaven, the North Esk valley near Edzell, Keltie Water near Callander, Aberfoyle, Balmaha, Innellan,

Figure 4.16 Bullers o'Buchan, Peterhead granite (Caledonian): marine erosion along joints, producing steep cliffs, sea stacks, arches and bridges.

Loch Fad on Bute and Glen Sannox on Arran. Igneous rocks are mostly serpentinized (i.e. original ultrabasic peridotite was altered by water-rich fluids), although lavas are recognizable, most notably at Stonehaven (see Fig. 4.3), where the pillow structures indicate eruption onto the ocean floor. Sedimentary rocks are mostly quartzites, with black shales and limestones. Mudstone and chert (deepwater sediments) are commonly found with the lava, and frequently the chert (silica) is the bright-red jasper variety and could be a chemical precipitate derived from hot springs.

Since its first discovery by George Barrow in 1901, the Highland Border Complex has been the source of debate and controversy that is still not fully resolved. Part of the reason is that the rocks occur in narrow, poorly exposed, highly deformed slices, widely separated and bounded by faults and thrusts. Secondly, many of the metasediments resemble Dalradian schists, and the complex was thought to be the youngest part of the Dalradian. However, the sequence of folds is not the same. In the 1980s, the age of the complex was confirmed by the discovery in the limestones of very rare Ordovician fossils, some 470–460 million years old. An important feature of the Highland Border Complex is that the sediments contain no traces of the underlying (i.e. older) Dalradian rocks, which by that time would have been exposed in the Highlands and were being eroded. The Highland Border Complex could therefore not have been anywhere near the Dalradian at the time, because more than 20 km of rock was removed from the Highlands, and Dalradian boulders would be expected to be present in the Highland Border Complex rocks. It is now believed that the complex consists of two main units. The lower (and older) unit (serpentine and pillow lava) represents a fragment of the ocean floor that was sliced up and thrust over the Dalradian. Such slivers of oceanic crust are called ophiolites, fragments of which are also found at Ballantrae and in Shetland (Unst and Fetlar, see p. 91). Younger sedimentary rocks occur unconformably above the older ophiolite, forming the upper unit. The entire Highland Border Complex was then broken up and slid into its present position sideways along the Highland Boundary Fault over a long period of time (from 440 to 400 million years ago) from some distant and as yet unknown location, but probably part of an ocean basin that existed to the south of Laurentia during the Ordovician period (495–443 million years ago), at about 30° south of the Equator.

Geology of Shetland

The islands of Shetland are the northernmost remnants of the Caledonian mountain chain in Britain, and so Shetland presents a link between the Scottish and Scandinavian segments of the Caledonian orogenic belt, and therefore occupies a unique position in attempts to correlate the geology of these regions. The pronounced shape of the archipelago is a reflection of the north–south fold axes and faults that developed during Caledonian deformation. This trend was exploited during the Ice Age, and local glaciers on Shetland accentuated the underlying differences in resistance to erosion of the bedrock to produce the deep inlets and rugged cliffs, now subjected to further battering and shaping by the sea and wind.

For its size, Shetland has enormously complex and interesting geology, and contains representatives of Lewisian, Moine and Dalradian metamorphic rocks that can be more or less matched up with their equivalents on the mainland of Scotland. In addition, there are Devonian lavas and sediments (the Old Red Sandstone), and several large granite intrusions (Fig. 4.17). But what stands out as Shetland's most unique feature is the rock complex of Unst and Fetlar, the northernmost isles, where a fragment of a former Iapetus Ocean floor was caught up in the Caledonian folding and thrusting.

Caledonian folding and thrusting

One of the major features in Shetland geology is the Walls Boundary Fault, which separates west from east Shetland; running due north–south, it is a continuation of the Great Glen Fault on the Scottish mainland. All the structures in Shetland have this north–south trend, whereas in Scotland and Scandinavia the trend or strike is more northeast–southwest. The geology on either side of the Walls Fault is markedly different; the fault probably moved by at least 170 km in a sinistral (leftward) sense, like the Great Glen Fault.

In west Shetland, the oldest rocks are Lewisian gneisses, which belong to the continental crust that formed Laurentia between 3300 and 1750 million

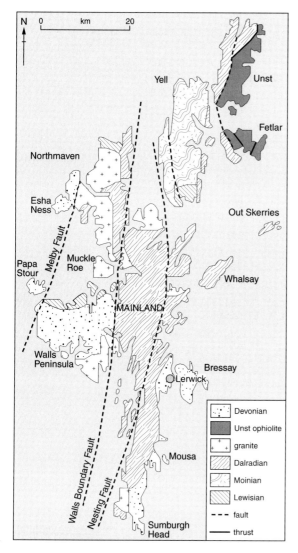

Figure 4.17 Shetland (based on BGS sources).

mica schist, marble and hornblende gneiss, the metamorphic equivalents of sandstone, mudstone, limestone and lava respectively. These metamorphic rocks are likely to have formed during the Grenville orogeny that affected Laurentia 1200–1000 million years ago, and are similar to rocks found in the Northern Highlands (see p. 57).

Moine schist is found east of the Walls Boundary Fault, the main outcrop being on Yell, but it has proved difficult to correlate the rocks accurately and in detail with Moine rocks on the Scottish mainland. The rocks on Yell are mostly coarse quartzite, quartz–feldspar gneiss and mica schist. As in northern Scotland, the Moine rocks most probably formed as shallow-water sediments 1000–800 million years ago, and were deformed and metamorphosed during the Caledonian orogeny 600–400 million years ago.

The eastern part of Shetland Mainland and parts of Unst and Fetlar consist of Dalradian rocks – quartzite, limestone, mica schist and metamorphosed lavas – 10–12 km thick in total. Appin, Argyll and Southern Highland Group rocks are represented. The Grampian Group at the base of the Dalradian appears to be missing from Shetland, possibly because the boundary between the Moine and the Dalradian is a shear zone (i.e. a narrow belt of intense movement). An extremely coarse-grain rock, the Valayre gneiss, occurs at the contact. Migmatites, granite and pegmatite intrude

Figure 4.18 North Roe, Shetland: Caledonian dyke cutting sheared Lewisian gneiss.

years ago. The Lewisian of Shetland consists of banded gneisses, with the dark and pale layers of hornblende and feldspar appearing very thin and streaked out as a result of further flattening and shearing during the later Caledonian orogeny. Outcrops are sparse and occur mainly in North Roe, north of the Ronas Hill granite (Fig. 4.18). These gneisses may have had an igneous origin (i.e. lavas, tuffs or sills). Other high-grade gneisses and schists in north Mainland and around St Magnus Bay and on Foula originated as sedimentary and volcanic rocks: quartzite, the east Mainland Dalradian, as well as granite plutons about 400 million years old. Of the meta-sediments, the limestones are probably the most important. Erosion of these thick bands has given rise to a series of well marked narrow valleys, such as the one running from Weisdale to Dales Voe.

In Unst and Fetlar, the rocks are mainly dense, dark ultrabasic and basic igneous rocks – banded peridotite and gabbro – cut by a swarm of basic dykes. These rocks are part of the ocean floor and they were thrust over the Dalradian basement during the collision that formed the Caledonian Mountains about 500 million years ago. Originally, the dolerite dykes would have acted as feeder channels for ocean-floor lavas that are now missing. Such fragments of oceanic crust caught up in mountain belts are referred to as ophiolites. The Shetland ophiolite is 8 km thick, but was severely deformed and fragmented during overthrusting, and parts of the complete sequence have been sheared off. The thrust zone itself is marked by serpentinite and talc, minerals that formed by the action of water-rich hot fluids on the ultrabasic rocks. When the ultrabasic igneous rocks were initially forming in the upper mantle and base of the oceanic crust, dense minerals settled out first to form layers, including the ore mineral chromite, which has been mined on Unst, as have the metamorphic serpentine and talc. Platinum-group elements have been discovered in association with the chromite-rich rocks and the serpentinite of Unst and Fetlar, and relatively high gold levels have been found in the ultrabasic rocks. The Skaw granite in northeast Unst is about 495 million years old, and was deformed by thrust movements that pushed the granite into its present position against the Dalradian schists of Saxa Vord.

Above the metamorphic basement rocks of Shetland lie the Devonian sedimentary rocks, laid down 400–380 million years ago by rivers flowing off the Shetland and Orkney hills, as well as the Scottish and probably the Norwegian mountains, into a lake-filled depression in the crust known as Lake Orcadie or the Orcadian Basin. These rocks are the same Devonian age as the Old Red Sandstone of Orkney, Caithness and Central Scotland, but being lake deposits they are generally of very fine-grain and grey flagstones. Devonian rocks are found on Shetland Mainland, as well as on Foula, Bressay, Mousa and Fair Isle. As with the

Figure 4.19 Walls Peninsula, west Shetland: Mu Ness; folded Old Red Sandstone sedimentary rocks.

metamorphic basement rocks, the Devonian sedimentary rocks are quite different on either side of the Walls Boundary Fault. The greatest thickness of sandstone is found on the Walls Peninsula, where volcanic rocks also occur. These rocks have been spectacularly folded, as is best seen at Mu Ness on the west coast (Fig. 4.19). Folding was related to the intrusion of the Sandsting granite complex 360 million years ago, which baked and hardened the rocks by up to 2 km away from the contact. Volcanic agglomerate and lava are also present on Papa Stour and Esha Ness, where rhyolite and andesite alternate with ash beds and ignimbrites; welded pumice and glassy shards erupted from a violently explosive volcano and were transported as a superheated ash cloud. A few of the sandstone beds contain fossil fish, such as at Melby, which are the same age as the famous Achanarras fish beds of Caithness. Fish beds are also found in east Shetland at Exnaboe, north of Sumburgh Head.

Granite is widespread in Shetland Mainland, on both sides of the Walls Boundary Fault, those intrusions to the east being older than the Old Red Sandstone sedimentary rocks and those to the west being younger. The Northmaven granite is one of the largest Caledonian granite intrusions in Shetland; it makes up Ronas Hill, which is the highest part of the island chain. Many of the granites are sheared and have closely spaced vertical joints, which are picked out by the sea to form spectacular cliff scenery (Fig. 4.20).

Figure 4.20 Western Shetland, Erne Stack, Muckle Roe: cliffs and sea stacks in Caledonian granite; drowned landscape being altered by marine erosion; skerries and islets are part of the granite outcrop.

Mineral resources of Shetland

The economic geology of Shetland in the past was dominated by the extraction of metal-ore minerals, mainly copper, lead, zinc and chromium. Copper and iron mines were worked from about 1790 until 1920 at Sand Lodge mine on the south Mainland, where some 12 000 tonnes of ore were extracted. On Fair Isle, 15 tonnes of copper ore was removed from Copper Geo on the northwest coast, from a vein cutting through the Old Red Sandstone. The chromite deposits on Unst and Fetlar have been referred to above in the account of the ophiolite complex. Opencast quarries on Unst produced 50 000 tonnes of ore between 1820 and 1944. Extensive exploration took place in the 1950s for chromite and again in the 1980s for platinum, but no new extraction began, mainly because the occurrences are scattered and relatively small. However, a great deal of research was carried out on the platinum ores and for this reason the Unst ophiolite has gained worldwide importance. At Clothister Hill on the west side of Sullom Voe, iron was mined in a magnetite body (magnetic iron oxide), which yielded 10 000 tonnes between 1954 and 1957. A further 20 000 tonnes of ore is estimated to remain. The ore was used not to produce iron metal, but as a heavy mud in coal flotation. The only important non-metallic deposit is talc, which occurs along the edges of serpentinite at the base of the ophiolite in Unst and Fetlar. Talc was quarried at Queyhouse quarry for about 40 years from 1945, yielding up to 10 000 tonnes annually for export, to be used in making roofing felt. On Shetland Mainland, talc (also known as soapstone or steatite) is found in a serpentinized Dalradian igneous body at Cunningsburgh, south of Lerwick. The ancient Catpund Steatite Quarries are found here, with rudimentary carved pots still remaining. The soapstone was exported all over Orkney and Shetland and to Scandinavia. Semi-precious stones occur in the highly varied schists on North Roe. Some of the felsite dykes of this area (see Fig. 4.18) were quarried to produce stone knives, probably in Neolithic times. Finally, the local Old Red Sandstone flagstones have been used locally for building purposes, including in Lerwick. The most impressive example is the 2000-year-old Mousa Broch (Fig. 4.21), which is still intact, resting on foundations of the stone from which it is built, flagstones of Devonian age. The igneous and metamorphic rocks of Shetland have also been used locally as roadstone.

Figure 4.21 Mousa, Shetland: iron age broch (Atlantic round-house), made of and standing on Devonian flagstones.

Landscapes of Shetland

The modern landscapes of Shetland have developed from the interaction of rivers, ice and sea on rocks of varying hardness. Ice from Scandinavia moved across Shetland some 100 000 years ago, to be followed by a local icecap on the central ridge 25 000 years ago. The glaciation of Shetland was first described by Peach & Horne, famous for their pioneering work in the Northwest Highlands and the Southern Uplands of Scotland. Scandinavian ice coming from the northeast brought boulders across to Shetland, including the Dalsetter stone in southeast Mainland, which was derived from its only known occurrence in southern Norway. Ice sheets deepened many of the channels between islands and peninsulas by scouring. Deep basins such as St Magnus Bay also owe their origin to glacial excavation. The few upland areas in Shetland are granite, including Ronas Hill, which is more resistant than the Dalradian schists, and these hills have gentle rounded contours. Hill ridges, valleys, islands and promontories are strongly aligned north–south, reflecting the fault and fold pattern and the alternating hard and soft quartzite, schist, limestone and granite outcrops. Once the ice melted 10 000 years ago, the over-deepened depressions were drowned by the rapidly rising sea. Today, Shetland coastal landforms are still being shaped in a most dramatic way, with the strongly jointed granites and volcanic rocks forming precipitous cliffs, particularly in the Walls Peninsula and Esha Ness, both of which form headlands projecting out to the west, in contrast to the north–south elongation of the main island mass. The sea also has the ability to build up landforms, as well as destroy tough rocks, and the sand spits, ayres and tombolos of the coast have built up from the movement by sea currents of sand, gravel, pebbles and shingle. Some of these spits of land join small offshore islands to larger neighbouring islands (Fig. 4.22).

Figure 4.22 Shetland: three ayres or spits of sand joining headlands on opposite sides of voe or drowned inlet.

The Southern Uplands

Southern Scotland was made famous in the annals of geology by the original work of James Hutton, the father of modern geology, as he is often known, who visited the region in the late 1780s in search of evidence for his theory of the Earth. Proof for his theory that the Earth was immeasurably old and that rocks formed by processes that are still active today was discovered by Hutton at Jedburgh and Siccar Point. The locality at Siccar Point (Fig. 4.23) is world famous and in the history of geology is a uniquely important site; photographs of the rocks exposed there appear in countless textbooks around the globe. The region was also made famous by the pioneering work of Charles Lapworth, a teacher from Galashiels and later Professor of Geology at the University of Birmingham, who studied the graptolite fossils in shales near Moffat and published the details in 1878. His scheme (the correct time sequence of graptolite zones) remains unchanged, and to this day Lapworth's fossil zones are used throughout the world (see p. 97). Building on Lapworth's discoveries, Peach & Horne re-examined the Southern Uplands and published a key memoir in 1899, which was rivalled only by their other monumental work on the Northwest Highlands, published in 1907. These two remarkable geologists did field work in the Southern Uplands in winter because the Highlands were inaccessible at that time of year. A considerable amount of research was done from the mid-1800s onwards into the mineral deposits at Leadhills, although the coalfields of Sanquhar and Thornhill were not described in detail until 1936. Significant advances in the study of the folding and faulting in the Southern Uplands were made from the 1980s onwards, but it is true to say that there are still problems to be solved and controversies to be resolved in the region with regard to the details of the plate-tectonic evolution of southern Scotland.

The Southern Uplands region, covering 10 000 km², is one of rolling hills, mainly covered in glacial deposits, with heather moors and conifer forests. Beneath the superficial cover lie rocks that were folded during the Caledonian mountain-building events (Fig. 4.24). Most of these rocks were laid down on the ocean floor from 470 to 420 million years ago in the Ordovician and Silurian periods. The region lies between the

Figure 4.23 Siccar Point, Berwickshire: Hutton's unconformity; Old Red Sandstone conglomerate (fossil scree) above vertical Silurian shales. A camera bag (arrowed) lies on the unconformity to the right.

Southern Uplands Fault, which runs northeast to southwest, from Dunbar to Ballantrae, and the Solway Fault, which coincides with the Iapetus Suture, a structure that joins northern and southern Britain (see Fig. 4.2). Within this 75 km-wide block of country, the older rocks run northeast–southwest, parallel to which are many faults, the most important being the Stinchar Valley, Orlock Bridge and Laurieston faults. These faults divide the region into many thin slices, most with quite distinct rock sequences.

Compared to the Highlands, the Caledonian rocks of the Southern Uplands have been only weakly metamorphosed. The great majority of the older rocks are coarse greywackes with smaller amounts of finer siltstone, mudstone, shale, volcanic ash and submarine lavas, and chert, a rock made from the remains of silica-rich plankton and sponges. Metamorphism of the fine-grain rocks has produced slate, some of which was used locally to roof farm buildings. Greywacke, a grey, black, dark-green or deep-purplish hard rock has been used also as a building material and is especially noticeable in the drystone walls ("dykes") in farmland. The term greywacke is derived from the German Grauwacke (wacke, German: mining term for stony, or large stone, or lump of rock) and it signifies a coarse-grain poorly sorted angular type of grit or sandstone containing sharp fragments of quartz, feldspar, dark silicates and igneous or metamorphic rock fragments held together in a dark mud or clay matrix. Low-grade metamorphism has recrystallized the cement to produce tough hard rock that is

Figure 4.24 St Abb's Head, Borders, Berwickshire: folded Silurian greywackes.

referred to locally by the quarrying term of whinstone, but this is not to be confused with the whinstone in the Midland Valley, which refers to igneous dolerite. Because of the glassy crystal and rock fragments, greywacke may seem to be an igneous rock, but it is in fact a sedimentary rock laid down in deep water.

Fast-flowing currents transported sediment off the edge of the continental shelf down the continental slope to the deep abyssal plain of the ocean floor. Mud, sand and water mixed together into a dense slurry, and the currents were probably triggered off by earthquakes. These turbid flows – turbidity currents – gathered speed on the way down the slope, mixing the components together. On reaching the flat ocean floor, the current lost energy and suddenly slowed down; coarse mixed sediment was deposited first to form thick sandstone and greywacke layers, on top of which the fine sand settled to form siltstone, followed gradually by the much slower deposition of clay and mud to form shale and mudstone at the top of the sequence. The pattern repeated itself many times over to produce a series of graded beds, from coarse at the bottom to fine at the top. Given the complex nature of the folding during the Caledonian orogeny, this grading relationship has turned out to be extremely important in unravelling the structure of the Southern Uplands. Sediments deposited from

these currents of sand and mud are known as turbidites. Modern turbidity currents have been observed in the Atlantic and they are known to flow very fast, at speeds of over 60 km per hour, with tremendous erosive power so that they are able to excavate channels or canyons into the sea bed on their way down slope. Erosional features at the base of the beds are also very useful for showing the correct way-up of the sediments. Lines and streaks in the beds are lineations caused by stones being dragged along, and these show the current direction (Fig. 4.25) and hence can indicate where the sediment came from. Large volumes of sediment can be transported in turbidity

Figure 4.25 Innerleithen, Peebles, Borders: grooves on underside of Silurian slate beds, caused by scouring in turbidity flow.

currents for many hundreds of kilometres. Material will often excavate a submarine canyon at the edge of the continental shelf, and the deposits will spread out at the base of the slope to form a submarine fan. These sediments were deposited slowly and continuously, at rates of a few metres per million years, over a period of 50 million years, suggesting that the fine muds were probably laid down on the deep Iapetus Ocean floor, far away from the effects of continental margins.

Fossils in the Southern Uplands rocks tend to be restricted to the fine sediments, where dead animal remains would have been preserved by slow accumulation of mud far below the oxygen-rich upper waters. On the other hand, greywackes are too coarse and were deposited in too violent an environment to allow anything to be preserved, let alone fossilized. Their most important and most abundant fossils are the graptolites, a group of marine animals, now extinct, that floated in the early oceans as colonies of tiny polyps attached by a thread-like structure (see Fig. 1.12). Charles Lapworth studied these in considerable detail at Dob's Linn near Moffat, a site of international geological importance. He established some 30 different graptolite zones in a sequence of black shales spanning 30 million years. Because the graptolites evolved very rapidly, with new species appearing every million years, the graptolite zones have been used to determine the relative ages of the sediments with great accuracy. When the zones are tied in to isotope dates, then absolute ages can be quoted with a high degree of certainty for these rocks. Dob's Linn is the location where in 1985 the Ordovician/Silurian boundary was defined, which gives it global significance. A plaque commemorating Lapworth's work was erected in 1930 at Birkhill Cottage (close to the Grey Mare's Tail, near Moffat), where he lived.

Interbedded with the graptolite shales are at least 130 thin layers of a white volcanic ash, termed bentonite, a clay formed from the disintegration of the ash. It is possible that the explosive eruptions that produced the ash may have led to the extinction of some of the graptolite species. There is also an unfossiliferous (or barren) mudstone that was deposited during the late Ordovician ice age about 450 million years ago, which may have extended to near the Equator in the so-called Icehouse Earth. Evidence for this event can also be seen at Macduff on the Banffshire coast (p. 83), where there are dropstones that fell from icebergs onto seafloor sediments.

The Southern Uplands greywackes are divided into three main zones or blocks, separated by faults that run northeast–southwest, parallel to the strike of the beds (Fig. 4.5). These are the northern, central and southern belts, a division introduced by Peach & Horne in 1899 and still used today. The northern belt has Ordovician beds (470–460 million years old), the central belt contains early Silurian sediments (460–430 million years old), and the Southern belt has mid-Silurian rocks (430–420 million years old), that is, the rocks become progressively younger towards the south. In all three belts, the rocks dip very steeply to the southeast, and they may be vertical or even overturned by folding. Intense compression has produced a series of tight anticlines and synclines in which the limbs of the folds are nearly parallel. In the oldest parts of the sequence, volcanic lavas are present with compositions indicating that they formed in the Iapetus Ocean as volcanic island arcs. The many northeast–southwest trending faults in the Southern Uplands (see Fig. 4.5) all have their downthrow side on the southeast. Within each of the 30 or so fault blocks, the oldest beds are on the southeast side, but the blocks have progressively older bases going from southeast to northwest, so that in an overall sense the Southern Uplands appear to trend younger towards the southeast.

By examining the composition of the grains in the greywackes and the sedimentary structures that indicate current directions, it becomes evident that the sediments being deposited on the ancient ocean floor were derived from a continental landmass that lay somewhere to the north, and from a volcanic island arc situated to the southwest. When all the field evidence is considered together, the Southern Uplands block or terrane is interpreted as having formed at the Laurentian margin of the Iapetus Ocean when the ocean floor was destroyed by subduction northwards beneath the plate carrying the continent. Sediments lying above the subduction zone were folded, sliced by faults and thrusts, and stacked up in a pile against Laurentia. Finally, this pile of greywackes was thrust towards the south when Scotland and England collided. It is possible that the Iapetus Ocean was several thousand kilometres wide and, if the present

75 km-wide belt of the Southern Uplands could be unfolded and the effects of faulting removed, then the sediments would stretch out to at least 1500 km. The term given to this pile of sedimentary rocks, scraped off the ocean floor in a bulldozer fashion as England and Scotland collided, is an accretionary prism. Accretion signifies that the greywackes were folded and weakly metamorphosed as they became physically attached to the northern continental crust in the Caledonian orogeny. Overall, the Southern Uplands block ended up being thrust over continental basement as the ocean closed and was finally destroyed at the end of the Silurian, 418 million years ago. This was the subduction event that gave rise to the volcanic arc at the active northern margin of the Iapetus Ocean in the area now occupied by the Midland Valley of Scotland (see Fig. 4.2). During collision, the sediments were scraped off the ocean floor while oceanic crust was being destroyed by being forced down the subduction zone. Sediment piles were transported along faults with a reverse thrust movement, and successive thrusts pushed from underneath, causing the overlying material to be folded and pushed up into a much steeper orientation. The site of the former subduction zone is the Iapetus Suture, which marks the collision between Laurentia (including Scotland) and Avalonia (including England). Located on the suture is the 400-million-year-old Cheviot granite and volcanic complex, and to the south is the Lake District Igneous Complex, which emerged at the surface above the oceanic crust that was being subducted southwards beneath Avalonia.

Based on the same evidence, an alternative model for the evolution of the Southern Uplands has been put forward. The second interpretation uses the composition of the grains in the greywacke sediments and current indicators to infer two source directions, a continent to the north and northeast, and a volcanic island arc to the south and southwest. This would suggest that, at least early in the history of the Southern Uplands, there was a volcanic arc between the Iapetus Ocean and the pile of thrust sheets, with a sedimentary basin lying in between (a structure known as a back-arc basin). Eventually, continental collision resulted in the disappearance of the basin and the volcanic arc beneath the greywacke pile. These opposing views on the development of the Southern Uplands have not yet been reconciled. In particular, there are difficulties relating to how the arc could have been overridden and have disappeared, and why it has not been exposed by uplift and erosion.

During the final stages of continental collision, granites were intruded immediately north of the Iapetus Suture zone, about 400 million years ago. These are the Loch Doon, Cairnsmore of Carsphairn, Cairnsmore of Fleet and the Criffel–Dalbeattie granites; the first three may be surface exposures of a much larger buried intrusion known as a batholith (see Fig. 4.15).

The Ballantrae Complex

In Southwest Scotland, just north of the Southern Uplands Fault, lies the world-renowned Ballantrae Complex, which has been intensively studied for over a century. It is regarded as a classic area in terms of the geological field relations of its many and varied complexly interfolded rocks, many of which have distinctive and highly unusual compositions. Although strictly in the Midland Valley, it has played such an important role in our understanding of the closing stages of the Iapetus Ocean that it is best discussed in the context of Southern Uplands geology. The rocks on the coast between Ballantrae and Girvan have long been known to differ markedly from their neighbours. In 1851, Sir Roderick Murchison of the Geological Survey observed that the serpentinite resembled rocks from the Lizard Complex in Cornwall. The Ballantrae Complex consists of fragments of the ancient Iapetus Ocean floor – pillow lavas, gabbro, serpentinized peridotite, chert and black shales – and it formed in the early Ordovician period, 500–475 million years ago. This fragment of oceanic crust was then folded, faulted and pushed up by thrusting and carried onto the continental margin when the Iapetus was closing and Scotland and England were colliding together towards the end of the Caledonian orogeny. During this intense tectonic activity, rocks that formed at quite different depths in the oceanic crust and upper mantle were juxtaposed by compressive forces, so that we now have a complex mixture of igneous, sedimentary and metamorphic rocks side by side in a very small area. Oceanic crust that is thrust over and interleaved with folded continental crust in a mountain belt is ophiolite, and the process of forceful tectonic intrusion is known as obduction.

Rocks formed on the ocean floor are deepwater sediments: black mudstones (with graptolite fossils), chert, which formed from the remains of silica-rich plankton (radiolaria), and conglomerate, which probably slumped down as jumbled rock fragments onto the ocean floor from the continental slope. Lapworth's graptolite zones (p. 97) are crucial in determining the age of these rocks. Basalt lavas and ash represent volcanic outpourings, the pillow structures indicating deepwater eruption directly onto the ocean floor. Upper mantle rocks that originally underlay the ocean floor are now represented by serpentinite, which is altered ultrabasic peridotite, and by gabbro, a coarse-grain basic igneous rock that in places is layered or banded. It is possible that the upper-mantle rocks formed at depths of about 40 km, while the oceanic crust above was about 7 km thick, so the original peridotite may have been thrust upwards (or obducted) by as much as 45–50 km. Obduction of the Iapetus Oceanic crust began at about 480 million years ago, and by 470 million years ago the ophiolite complex was finally welded onto the Midland Valley. During this intense process, the base of the ophiolite metamorphosed the surrounding rocks because the mantle rocks were much hotter. Also, as the oceanic crust was being overthrust at a destructive plate margin, it was simultaneously eroded at higher levels, and great blocks of lava, chert, mudstone, shale, pillow lava, serpentinite and gabbro were incorporated into the complex. Some of these blocks are 575 million years old and may represent deep-level metamorphism of oceanic crust that once underlay the Ballantrae Complex.

Lying on top of the deformed Ballantrae Complex are sediments ranging in age from mid-Ordovician (470 million years old) to early Silurian (430 million years old). These conglomerates, sandstones, and thin limestones were deposited unconformably on top of the igneous basement, which by then formed the floor of a shallow shelf sea. The conglomerates contain many boulders, cobbles and pebbles of the underlying igneous rocks; they were transported for rather short distances from the north and northwest by rivers that flowed into basins at the edge of the Midland Valley. The basins were bounded by faults that moved during sediment deposition, so that water levels were constantly varying in a cyclical manner. Shallow-water conglomerates are overlain by deepwater sandstones

and mudstones with graptolites, then overlain by shallow sediments again, including limestones with a rich and varied fossil content: corals, brachiopods, gastropods, echinoderms and trilobites. The fossils are similar to those found elsewhere in Laurentia and are quite distinct from those of the same age in Wales, which then formed part of Avalonia. Scotland at the time was situated on the edge of Laurentia and lay at about 20° south of the Equator. Avalonia meanwhile was farther to the east and south, with the Iapetus Ocean separating the two, but gradually closing as the continents were heading for collision by 420 million years ago.

The Girvan–Ballantrae area helps to piece together the evidence for the closure of the Iapetus Ocean by subduction, followed by continent/continent collision in the final stages of the Caledonian mountain-building episode. Oceanic crust was being destroyed at this time in a major subduction zone that lay to the south, with a volcanic arc to the north, at the margin of a large continent, Laurentia.

By late Silurian to early Devonian time (420–400 million years ago), the closure of the Iapetus Ocean had led to the formation of a high mountain chain. Scotland was then in the interior of a large continent, some 20° south of the Equator. Erosion of the high new mountains produced copious amounts of sediment that were removed from the bare slopes by slumping and flash floods, and were carried down by intermittent torrential streams into valleys etched into the floor of an inland desert. Such sediments are typically coarse conglomerates and red sandstones, the Old Red Sandstone of the Devonian period. One such ancient valley is Lauderdale in the Borders, which has been excavated by more recent erosion to produce a modern valley on the same site. The final stages of continental collision led to crustal thickening and igneous activity. Products of this are seen in the volcanic rocks at St Abb's Head, the granite and lavas of the Cheviot Hills, which lie astride the Iapetus Suture, and the granites of Galloway. There is a gap in the record for the middle Devonian, indicating that further earth movements resulted in uplift and erosion of much of the sediment laid down in early Devonian times. Sedimentation resumed in late Devonian times, when the Upper Old Red Sandstone formed in extensive basins. These younger sediments rest on top of

Figure 4.27   Scott's View: Eildon Hills, Melrose, Borders: a laccolith intrusion; the three hills were named Trimontium by the Romans.

the eroded remnants of Silurian greywackes in the Southern Uplands. Nowhere is this clearer than at Siccar Point (see Fig. 4.23), the locality made famous by James Hutton in 1788. He was the first person to recognize the significance of the unconformity: that it represents a very long period of time during which the older rocks were uplifted and eroded before finally being inundated and buried by later sediments. John Playfair, the mathematician who accompanied Hutton on that historic field trip and who wrote Hutton's biography, famously remarked that "the mind seemed to grow giddy by looking so far into the abyss of time". These same folded Silurian greywackes are well exposed on the coast at St Abb's Head, immediately north of the Devonian volcanic rocks (see Fig. 4.24), and were to inspire Sir James Hall, who was a contemporary of and strongly influenced by Hutton, to perform the first laboratory experiments on folding of rocks in the early 1790s, as well as his experiments on igneous and metamorphic rocks (see box on p. 10).

Across the Southern Uplands are many northwest-trending normal faults, such as at the Rhinns of Galloway, Loch Ryan, Dumfries, Lochmaben, Thornhill and Sanquhar. These faults are perpendicular to the general northeast strike of the Caledonian folds and faults, and they may have been active in the early Carboniferous period 350 million years ago and controlled deposition in sedimentary basins into the Permian period (250 million years ago). The early part of the Carboniferous is marked by volcanic activity, with lava flows around Kelso (the Kelso Traps) and many volcanic plugs that now dot the landscape of the Borders, for example Rubers Law, Smailholm and the Eildon Hills, a mushroom-shape intrusion called a laccolith (Fig. 4.27). Sediment deposition during the Carboniferous was dominated by rivers, estuaries and

shallow seas, eventually giving way to swampy forests when Scotland was on the Equator some 310 million years ago, at the time when the Coal Measures of Thornhill, Canonbie and Sanquhar formed.

The climate changed at the end of the Carboniferous from tropical to arid as Scotland drifted northwards from the Equator. The red desert sandstones of the Permian (290–250 million years ago) have well preserved sand dunes and flash-flood deposits – coarse breccias and alluvial fans; these are present in the Loch Ryan, Thornhill, Dumfries, Lochmaben and Moffat basins. It is possible that movement on the northwest-trending faults caused frequent uplift of the granite mountains and consequent erosion and rapid deposition onto the desert floor. These rocks are often referred to as the New Red Sandstone.

Mineral deposits in the Southern Uplands
The granites of the Southern Uplands have mineralization associated with them, in particular gold, silver, copper, lead and zinc veins. Uranium has been found around the Criffel granite at Blackstockarton Moor in the form of 60 or more carbonate veins with pitchblende (uranium oxide), copper, bismuth and iron ores. Deposits close to the Southern Uplands Fault include Stobshiel, with gold, lead, zinc and copper veins, the Fountainhead antimony mine at Hare Hill (which also contains gold) and the Fore Burn deposit near Straiton in Ayrshire, which has 50 g per tonne of gold, as well as a wide range of other metal ores, including copper, silver, lead and zinc. Mines at Glendinning between Langholm and Eskdale Muir, which last operated in 1920, produced 200 tonnes of antimony; gold, silver, arsenic, copper, lead, zinc and barium are also present in veins cutting sulphide-rich greywacke sediments. Glenhead Burn, where the

Loch Doon granite cuts greywackes, has gold mineralization in quartz veins containing high levels of arsenic; some lead–zinc minerals are also present.

The largest and best-known mining district in the Southern Uplands is in the Lowther Hills at Leadhills and Wanlockhead, the highest villages in Scotland. Mining took place in this orefield for many centuries and the deposits of lead and gold may have been known to the Romans. Lead in the form of galena or lead sulphide has been mined in the district for over 700 years. Monks from Newbattle Abbey were given a charter in 1293 to allow them to mine for lead. Zinc is found in association with the lead and has been mined as zinc blende (sphalerite, zinc sulphide) from 1880. Prior to the seventeenth century, the main product of the mines was silver, which is found with galena. Exploitation of the ores continued until the 1930s and intermittently afterwards until 1958, and the dumps were worked over in the 1960s. Considerable reserves remain in this, Scotland's dominant source of lead. The ore occurs in veins of shattered and brecciated greywacke, but not often in the interbedded shales. Associated non-ore or gangue (German: mineral vein) minerals are quartz and calcite, which are discarded. There are more than 70 veins and they trend in several directions but predominantly lie within the main northeast–southwest ore zone. The age of mineralization is later than the host greywackes and has been dated at 360 million years old, which is Carboniferous and the same age as the widespread lead–zinc mineralization in Ireland. Lead was at one time smelted in the Leadhills–Wanlockhead area and resulted in significant lead poisoning, a hazard that appears to have been accepted as a necessary evil. The landscape around the smelters shows the scars in the form of old tips, abandoned machinery and poisoned vegetation. In the extension of the Iapetus Suture zone into Ireland, there is evidence of considerable lead–zinc mineralization associated with this major structure in the crust, and exploration possibilities exist in Southwest Scotland also.

When the mines were in active use, they produced 300 000 tonnes of lead metal, 10 000 tonnes of zinc ore and 25 tonnes of silver. A remarkable feature of the Leadhills deposits is the sheer variety of minerals: well over 70 different species have been found, more than for any other locality in Scotland. The list includes many rare and unique types: carbonates, sulphides, sulphates and oxides of lead, zinc, copper, nickel, iron, barium, cobalt, as well as native gold. For this reason the district has been declared a Site of Special Scientific Interest (SSSI). Leadhills is the type locality for rare lead minerals such as leadhillite, lanarkite, scotlandite and caledonite. These are described as secondary minerals, in comparison to the primary or main minerals such as galena, zinc blende (or sphalerite, zinc sulphide), baryte, calcite, pyrite, quartz, and so on. Most of the gold that was found in the area was extracted from stream sediments as placer gold, much of this in the sixteenth century. The term "placer" (Spanish: sandbank) is used in ore geology to describe a deposit of loose sediment containing heavy ore minerals that were washed in by streams flowing over the surface and sorted by gravity in the stream bed. Coins were minted from Leadhills gold in the reign of James V and Mary Queen of Scots, as well as being used for royal regalia.

There are mineralized springs and wells in the Southern Uplands, most notably St Ronan's Well at Innerleithen, Hartfell Spa at Hart Fell, and Moffat Well near the town of Moffat. These are natural waters emanating from joints and faults in the greywackes, and mostly contain sulphurous salts and chalybeate, a mixture of dissolved iron salts. They were reputed to have healing properties, but this is not proven.

Building stones in the south of Scotland are varied and abundant, and have been used extensively on a mostly local scale. The granites of the southwest, particularly the Criffel granite near Dalbeattie, have been quarried since the early 1820s for use as a building stone, an ornamental stone and for crushed aggregate. Dalbeattie itself is built of this granite, and the rock was exported to Glasgow and to Liverpool for use in the construction of the docks. Basic igneous rocks of Carboniferous age (355 million years old) form many of the hills in the northeast of the region, and these have been extracted from many quarries for use as roadstone. The rubble core of some of the Borders abbeys has been derived from volcanic plugs and the Eildon Hills laccolith intrusion. Old quarries from where this stone was derived still form a prominent feature at Melrose, and the viewpoint to Scott's View at Dryburgh is excavated in this same rock type (Fig. 4.27). Sandstone and greywacke have been used for buildings, greywacke forming the main structure and softer sandstone, which is more easily shaped (or "dressed"), being used around windows and doors (see Fig. 9.12). Sandstone predominates as the external stone of the abbeys. Pale cream Lower Carboniferous sandstone was used as a building stone in many towns, particularly Kelso. In the southwest near Dumfries and Annan, New Red Sandstone (Permian and Triassic, 260–210 million years old) dune-bedded red desert sandstone is used locally and was transported to Central Scotland in the nineteenth century. Limestone of Carboniferous age was once worked for agricultural lime, and the Carham Stone at Kelso was also used as a building material. Few quarries now operate in southern Scotland, except those at Galashiels, where thick greywacke is being taken out for roadstone, and in the southwest there are the sandstone quarries at Locharbriggs (Dumfries, Permian), Newcastleton (Carboniferous), Gatelawbridge (Thornhill, Permian), Corncockle (Lochmaben near Lockerbie, Permian) and Corsehill (Annan, Triassic). Slates interbedded with greywacke in the Borders region were once used for roofing material locally within southern Scotland. Quarries at Stobo and Innerleithen provided much of the slate for the town of Peebles.

Landscapes of southern Scotland

The shape of the land in southern Scotland is strongly influenced by the underlying geology, even though, in much of the interior part of the country, bare rocky outcrops are scarce, except in the southwest where the granite hills of Galloway have something of a Highland appearance. In the case of the Loch Doon granite, it is actually the surrounding metamorphic aureole or heat-altered zone where the highest summits are found. Indeed, the highest point in the Southern Uplands is the Merrick (843 m), at the edge of the Loch Doon granite in the north Galloway Hills. Dominating the landforms in most of the region are the northeast–southwest ridges and valleys defining the Caledonian trend. In particular, the Moorfoot and Lammermuir Hills run closely parallel to the Southern Uplands Fault, to form a highly distinctive low straight ridge visible from Edinburgh and Fife. The Cheviot Hills form the southern boundary and mark the political border. These hills too have a Caledonian trend, running southwestwards in the direction of the Solway

Firth, and sitting astride the Iapetus Suture. Volcanic outcrops of Devonian age – andesite and rhyolite lavas, together with explosive tuffs and agglomerate – form the uplands east of Jedburgh, and the dome-shape Cheviot (815 m), just inside England, is a Caledonian granite mass that intruded the Cheviot volcanics of Devonian age (418 million years old). A second major structural feature, particularly evident in the southwest, is the set of northwest–southeast faults that run perpendicular to the Southern Uplands Fault and the Solway Fault. This set of parallel faults defines the Rhinns of Galloway coastline, Loch Ryan–Luce Bay, Wigtown Bay, the valley of Glenkins from Castle Douglas to Dalmellington, Nithsdale and Annandale, valleys now used as major transport routes but which date back millions of years. These same valleys, together with Lauderdale to the northeast, were the locations of sedimentary basins in which Old Red Sandstone and New Red Sandstone were deposited 410 and 250 million years ago respectively. Erosion has removed most of the soft sedimentary rocks in these valleys. Smooth rounded hills dominate much of the upland landscape of the Borders, Dumfries and Galloway, having been carved by streams and ice from the Ordovician and Silurian greywackes, shales and mudstones, to form a plateau 450–600 m high, dissected by rivers. Summits tend to be at uniform heights, and jagged peaks, cliffs and rocky outcrops are mostly absent. Higher hills occasionally protrude above this plateau, in the Tweedsmuir and Moffat Hills between Moffat and Peebles, and the Lowther Hills south and east of Leadhills, which in places are over 750 m high. These hills tend to be made of tougher and more resistant quartzites and thick beds of grit, whereas the thinner greywackes and shales have been weathered into gentler rolling hills. Exceptions to the general pattern are to be found in the 360-million-year-old Lower Carboniferous igneous intrusions and lava flows: the Birrenswark lavas north of the Solway, the Glencartholm volcanic beds near Newcastleton, and the Kelso Traps farther north. In the case of the 160 m-thick Kelso Traps, weathering of the horizontal flows has created a stepped landscape around the town ("trap" comes from the Swedish "trappa" for "staircase"). Many isolated volcanic necks, also Carboniferous in age, intrude the greywackes, examples being Rubers Law,

Figure 4.28 Smailholm Tower, Borders: outcrop of Carboniferous dolerite intrusion, standing above sedimentary rocks, and smoothed by ice; corners of the tower are Old Red Sandstone.

Dunion Hill, Smailholm, Black Hill and Penielheugh, from where there is an unrivalled panorama of the entire geological landscape of the Borders. Glacial erosion of these dolerite intrusions has created crag-and-tail features (Fig. 4.28). The triple hills of the Eildons near Melrose (Fig. 4.27) are certainly the most famous of the Borders volcanic outcrops. All three of the Eildon summits are part of the same mushroom-shape intrusion.

Landforms in the southwest of the Southern Uplands differ in some important respects from those in the northeast as a result of the bedrock geology. The dividing line can be taken as the watershed between the Teviot and Tweed systems, in which streams mostly flow northeast into the North Sea, and the rivers to the west of the Tweedsmuir Hills, which flow southeast off the high granite and greywackes into the Solway Firth along the northwest–southeast faults. The Moffat Water follows the line of a northeast–southwest fault, which has been excavated by streams and ice to form an impressive, very straight U-shape valley (see Fig. 8.7). Elevation of the land surface after the ice sheets melted has left valleys hanging high above the main valley floor, none more stunning than the Grey Mare's Tail, with its waterfall tumbling out of the corrie occupied by Loch Skene (see Fig. 8.11). Terraces in many of the wider rivers are another result of the pulsed uplift of the land after the ice had melted. St Mary's Loch and the Talla reservoir occupy rock

basins, hollows scoured out of the bedrock, then over-deepened by ice moving along pre-existing valleys.

The various Galloway granites have different landscape features. The Criffel and Cairnsmore of Fleet bodies form massive hills with rounded shapes and, where glaciated, knock and lochan topography and craggy faces predominate. Cairnsmore of Carsphairn is also a high granite hill, reaching 797 m, with corries on the rugged northern and eastern sides. Large blocks of rock are found at the summit, carried there by the ice from elsewhere and are termed erratics because they are of rock unrelated to the rocks in the vicinity. In the case of the Loch Doon granite, the igneous rock itself tends to form ground at around 300 m, but the metamorphic aureole or heat-affected zone around the intrusion (i.e. baked and toughened greywacke called hornfels) is responsible for much higher ground including the Merrick (842 m) in the west and the Rhinns[*] of Kells, with Corserine at 814 m on the eastern flank. The highest hill within the central granite ridge is Mullwharchar, at 692 m. At the Loch Doon dam, there are good examples of ice-plucked rocks called roches moutonnées: bare rocky outcrops with smoothed whalebacks and craggy faces where the moving ice pulled off loose rock. Corries in the granites of the Galloway Hills tell of the erosive power of ice during the most recent glacial period. The Devil's Beeftub in Dumfriesshire is an example of a rough corrie shape having been excavated in the greywackes, but it is much less perfectly formed than those found in granite uplands (Fig. 4.29). During the last glaciation, the thickest accumulations of ice were on the Galloway granite hills, from where the ice moved south towards the Solway, and on the Moffat and Tweedsmuir Hills, with ice-flow directions to the east into the North Sea. The bedrock responded in different ways to the effects of the moving ice, which itself had different properties, depending on whether it was frozen to the rock or moved over a thin film of water separating the glacier from the bedrock. In the case of the granites of the wetter and warmer southwest, corries are quite common, but not so in the greywackes of the high Borders hills, where the ice was frozen directly onto the rock outcrops and was therefore

---

[*] "Rhinns" (also spelled Rinns) is Gaelic for "headland" or "rounded hill".

Figure 4.29 The Devil's Beeftub, near Moffat, Borders region: a deep corrie excavated by ice in Silurian shale and greywacke, it is the source of the River Annan. The summits in the background are at about 800 m.

unable to scour deeply. Ice also moved down the Firth of Clyde and scraped marine deposits off the sea bed; shell deposits from this event are found on the Rhinns of Galloway.

The lowlands in the southwest are mostly located on younger sedimentary rocks that overlie the grey-wackes. These are the Carboniferous rocks of the Thornhill and Sanquhar basins, and the New Red Sandstone at the head of the Solway Firth, the Dum-fries and Lochmaben basins, the upper part of the Annan valley, and the low ground between Loch Ryan and Luce Bay. New Red Sandstone is also present in the Thornhill basin, above Carboniferous rocks. The coastal areas include cliffs where folded greywackes are present, such as Whithorn. Softer sedimentary rocks have mostly been extensively eroded and form low coastal features, particularly the broad mudflats, salt marshes and dunes, several of which are Sites of Special Scientific Interest, including the Solway Estuary, the Cree Estuary and Luce Bay.

In the northeastern part of the region, there are also impressive cliffs in folded greywackes (see Fig. 4.24) and in Devonian volcanic rocks at St Abb's Head. Carboniferous sedimentary rocks to the south, mostly limestone, form the low marshy ground of the Merse[*] between Duns and Berwick. This area is renowned for the glacial landforms created by meltwaters when the ice sheets were melting at the end of the most recent ice age. These include the Bedshiel kames near Green-law, where sand and gravel ridges were formed at the edge of the melting ice, or in tunnels beneath it (see Fig. 8.12, p. 182).

---

[*] The Merse is the low ground on the Scotland/England border, underlain by Carboniferous limestone, which was formerly marsh land and is now a low-lying fertile plain, around the Tweed, bounded by the Pentland, Lammermuir, Muirfoot and Cheviot hills.

Plate tectonics and drifting continents

In this section, we explore how the Caledonian Mountains were formed by the collision of continents and the closure and disappearance by subduction of an ocean, the Iapetus, which had previously separated these continents. From the previous sections in this chapter, it is evident that the Caledonian fold belt in Scotland is made up of segments of the crust, trending northeast–southwest, separated by large-scale faults (see Fig. 4.2). Between the boundary faults, the geology of each crustal segment is self-contained and unique, with distinctly different rock units forming discrete segments, usually referred to as terranes: the Hebrides, Northern Highlands, Grampian Highlands, Midland Valley and Southern Uplands, the Hebrides terrane forming the ancient basement to the Caledonian fold belt. Structures within the Hebrides terrane (the Outer Hebrides and the Northwest Highlands) are older than those in the Caledonian belt, but have been mostly unaffected by Caledonian movements, although faults such as the Outer Isles Thrust (see Fig. 3.1) may have been re-activated at the time of formation of the Moine Thrust, which in Scotland marks the northwest margin of the Caledonian belt. South of the Iapetus Suture, the Caledonian belt in England and Wales forms part of the crust of the continent of Avalonia, and the terrane forming the southern margin of what was the Iapetus Ocean is known as the Leinster–Lakesman terrane (after Leinster in Ireland and the Isle of Man–Lake District in England).

The Caledonian mountain belt was created by the collision of three continental masses – Laurentia, Baltica and Avalonia – that moved together and closed the Iapetus Ocean. But to establish the complete history of the Caledonian mountain belt, we need to go back in time to the period when the Iapetus Ocean itself did not exist. About 750 million years ago, the supercontinent of Rodinia began to break up and the Pacific Ocean was created. Further rifting at 600 million years ago led to the formation of the Iapetus Ocean. In Scotland, this rifting event is marked by the eruption of the Tayvallich lavas, at 595 million years ago, in the Dalradian outcrops of Argyllshire. These particular lavas have the chemical characteristics of basalts erupted at mid-ocean ridges, suggesting that new oceanic crust began to form at the time of rifting

and continental break-up. Basaltic dykes on nearby Jura may represent channels that fed the lavas erupting onto the newly forming ocean floor. As the Iapetus Ocean grew wider by seafloor spreading, deposition of the sediments continued on the Laurentian margin of the ocean to form the Southern Highland Group Dalradian, then younger sediments. Rifting of the supercontinent and the creation of new ocean floor led to Baltica separating from eastern Greenland while Laurentia (North America with Scotland and northwest Ireland at the eastern edge) separated from Gondwanaland (the southern continents, including what was to become England, Wales and southern Ireland; see Fig. 4.2). The new continents were separated by mid-ocean spreading ridges. These plate movements were all taking place south of the Equator.

By the middle to late Cambrian, 510 million years ago, the Iapetus had reached its maximum width, (about 5000 km), and seafloor spreading stopped. Mid-ocean ridges were replaced by chains of volcanic island arcs (but not in the same places), and by subduction zones around the margins of Iapetus, rather similar to the situation in the present-day Pacific Ocean. One such volcanic arc was situated between the Grampian Highlands and the Southern Uplands, in the area that later became the Midland Valley. The Iapetus Ocean then began to shrink by subduction around its edges as continental positions changed, and Laurentia, Baltica and Avalonia started their respective journeys towards one another, sweeping up the island arcs and forcing them to collide with the margins of the continents. At this time, 480 million years ago, the Grampian Mountains began to emerge in the north, and the Girvan–Ballantrae Complex formed at a time of widespread volcanic activity associated with the northward subduction, beneath the Southern Uplands and future Midland Valley, of oceanic crust into the mantle. On the south side of the Iapetus, a southerly dipping subduction zone beneath Gondwanaland and Avalonia gave rise to a volcanic arc in Wales 490 million years ago, and the Lake District volcanic arc 460 million years ago. By 450 million years ago, the Lake District volcanic arc had stopped erupting. The ocean narrowed progressively and, by 450 million years ago, Laurentia, Baltica and Avalonia were at collision point, with Scotland caught up between all three. First, Baltica collided obliquely with

Laurentia and slid sideways, then Laurentia was thrust over the edge of Avalonia. Subduction and volcanic activity continued until about 420 million years ago, by which time the Iapetus Ocean had closed completely, and Scotland and England had collided along what is now the Iapetus Suture. The line of the suture runs from the Shannon estuary and central Ireland, then onwards immediately north of the Isle of Man, through the Solway Firth and then northeastwards to an area just south of the Cheviot Hills (see Fig. 2.1). Geophysical studies indicate that the suture dips down to the north, beneath the Southern Uplands. The narrowing and eventual closure of the Iapetus Ocean can be understood from the fossils contained in the Ordovician and Silurian rocks. Initially, 570 million years ago, graptolites predominated. Gradually, more shallow-water shelf animals are found, such as trilobites and brachiopods, followed by inshore forms such as ostracods, and eventually freshwater fish appeared 430 million years ago, indicating that the open ocean had now disappeared, to be replaced by inland seas, which dried up finally when continental collision was completed by 420 million years ago.

Between 480 and 420 million years ago, the various crustal segments of Scotland, which now form the terranes, had slid into place along large-scale fault lines trending northeast–southwest, by sinistral (left-directed) slip movement (in response to the collision between Baltica and Laurentia) amounting to several hundreds of kilometres (see Fig. 4.2). In the case of the Southern Uplands, sediments derived from Dalradian and other rocks to the north continued to be deposited above a subduction zone as Avalonia was converging obliquely on Laurentia, and slices of ocean-floor sediment were scraped off and folded and overthrust to the northwest, to form the complex accretionary prism that became attached to the emerging Midland Valley. Collision of the continents created greatly thickened crust, and partial melting of the folded and metamorphosed sediments at the base of the continental crust 400 million years ago produced granite that made its way up into the crust, helping to raise the level of the mountain chain by buoyancy effect. Scotland was then part of another supercontinent during the Devonian period – the so-called Old Red Sandstone continent of Laurussia (named after Laurentia and Russia), and vast amounts of sediment, possibly

totalling 25 km vertical thickness, were stripped off the bare rocky mountains that had been rapidly uplifted. We cannot tell for certain how high these young Caledonian Mountains were, but they may have been of the same order as the present-day Alps, about 5000 m high. The Iapetus Ocean therefore existed for some 180 million years, from the end of the Precambrian at 600 million years ago to the middle of the Silurian period 420 million years ago. This time interval is roughly the same as the lifetime of modern oceans, where the oldest oceanic crust is about 180 million years at a point farthest away from mid-ocean ridges. A major result of the collision event was the creation of additional continental crust in the form of the 7500 km-long Caledonian mountain belt at the junction of Laurentia, Baltica and Avalonia: North America, Greenland, Scandinavia, Britain and northern Europe were now part of a much larger landmass.

Caledonian faults in Scotland

We have already seen how the shape of Scotland has been strongly influenced by the events of the Caledonian orogeny. One of the most obvious features is the northeast–southwest orientation of the main faults, most of which resulted from the sinistral strike-slip movement of crustal fragments during continental collision. These are the Great Glen Fault, the Highland Boundary Fault, the Southern Uplands Fault and the Iapetus Suture (see Fig. 4.5). In the far Northwest Highlands we find the Moine Thrust, which represents the edge of the Caledonian mountain belt. The entire mountain chain was physically pushed westwards over the ancient Lewisian basement by the collision of Laurentia and Baltica, some 425 million years ago. In the case of the strike-slip faults, these were initiated slightly later, in response to the oblique collision of three plates, which brought together fragments that had formed in different places at the edge of Laurentia. Calculating the amount of movement on these faults is extremely difficult. Previously it had been thought that the Great Glen Fault had moved to the southwest, based on the supposition that the Foyers granite in the Grampian Highlands near Inverness and the Strontian granite in the Northern Highlands west of Fort William had once been a single intrusion split in two by the fault (Fig. 4.30). However, we now know that these granites have different compositions and are not of the same age. Since no rock masses can be matched across the Great Glen Fault, its sinistral (leftward) movement must be greater than the present length, which is more than 200 km. Geophysical evidence indicates that the Great Glen Fault and the Highland Boundary Fault reach depths of 40 km (i.e. they penetrate the entire crust). Also, it appears that the Great Glen Fault becomes shallower to the southwest and eventually splits up into smaller faults beyond the Firth of Lorne. On the other hand, many features of the rock compositions on either side of this fault – in the Northern and Grampian Highlands – are similar, and this has led some geologists to propose that the amount of movement is a few hundred kilometres at most. Several of the faults were responsible for the location of Caledonian granites, which appear to have made their way up to higher levels in the crust along these lines of weakness, which opened up spaces during fault movement. Later, many of them were re-activated at the end of the Caledonian orogeny, but this time as normal rather than strike-slip faults. After the compressive events that gave rise to fold mountains, the crust rebounded in a phase of relaxation, so that tensions were set up and the crust was stretched. Hence, the Midland Valley formed as a fault-bounded trough or basin (a rift valley or graben) and sank down to accommodate great piles of sediment being transported by rivers from the Highlands, Scandinavia and the Southern Uplands. Still later, movements took place on the Great Glen Fault and the Helmsdale Fault during the Jurassic (from 160 million years ago), such that the Moray Firth Basin formed in part by downfaulting of the southeast side of the fault. Faulting kept pace with sedimentation, thereby allowing enormous thicknesses of oil-bearing strata to form. The northwestern coastline of the Moray Firth, the Black Isle, Tarbet Ness peninsula, Caithness and the shape of the Shetland archipelago are defined by the Great Glen Fault and its extension, the Walls Boundary Fault in Shetland. Finally, the Caledonian faults played a major role in the siting of the Tertiary volcanic centres on the west coast at 60–55 million years ago. Intrusions and lava flows are to be found in Skye, Rum, Mull, Ardnamurchan, St Kilda, Arran and Ailsa Craig, which are situated at the intersections of these faults with a north–south line of weakness in the crust. Ice moved along the fault lines, causing the

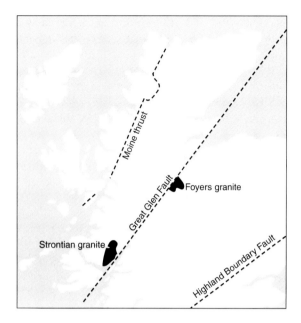

Figure 4.30 The Strontian and Foyers granites.

pre-existing river valleys to widen and deepen, so that these are now main transport routes. It is therefore no exaggeration to say that Scotland was shaped in every sense by the events of the Caledonian orogeny.

Holyrood Park, Edinburgh: Salisbury Crags sill intruded into the Edinburgh volcano (photograph by Roger Jones).

CHAPTER 5
Lowland Scotland – after the mountains

Introduction

The Caledonian Mountains of Scotland, Ireland, Wales and Scandinavia arose 410 million years ago as a response to the collision of three continents to form a much larger supercontinent (see Fig. 4.2). At the time, Scotland lay in the interior of this continent, about 20° south of the Equator. An ocean existed some way to the south, in the area now occupied by the extreme southwest of Ireland and southwest England. Most of Scotland was mountainous, with several valleys and lowlands that were to become sedimentary basins: areas of crust that received the deposits washed down from the Scottish and Scandinavian mountains. Rapid downcutting of the young high mountains resulted in the root of the thickened crust rebounding to compensate for what was removed by gravity, torrential flood waters, and wind. Plant life during the Devonian period (418–362 million years ago) was very sparse on land, being restricted initially to lichens, ferns, reeds and mosses. With no tree or grass cover, the bare stony mountainous terrain was subjected to rapid erosion. Because Scotland was far from the effects of wet winds, rainfall would have been intermittent and probably torrential, so that large amounts of scree and boulders were washed down off the mountain tops and spread out across valley floors as alluvial fans during flash floods. A thickness of 20–25 km of material was removed and redistributed in this way, filling up the valleys with boulder conglomerates and sandstones. The latitude of 20° south is in the zone where desert conditions exist, and the continental deposits formed in such an environment are frequently stained red with hæmatite (iron oxide). For this reason, the rocks that accumulated in the sedimentary basins, which emerged after the Caledonian orogeny, are called the Old Red Sandstone. The Devonian period of geological time was named after Devon (by Sir Roderick Murchison, from Tarradale in Easter Ross, in 1839), which at the time was at the edge of a young ocean, while the rest of the British Isles area was a continental landmass.

The terms Old Red Sandstone and Devonian are often confused, but they are not interchangeable. The sedimentary rocks (conglomerates and sandstones) that make up the Old Red Sandstone sequences were deposited under arid and semi-arid conditions, which

began and ended at different times in different places, from near the end of the Silurian period until the early part of the Carboniferous period.[*]

At present, the main outcrops of Old Red Sandstone sedimentary rocks are found in the Midland Valley and in Caithness, Orkney and Shetland. Smaller amounts can also be seen in the Southern Uplands and at a few locations in the Highlands. Because the rocks formed in the Caledonian orogeny are intensely folded, Old Red Sandstone deposits generally lie unconformably on top of the older rocks. One of the best-known examples world wide of an unconformity is the famous section at Siccar Point, known as Hutton's unconformity (see Fig. 4.23). The sequence of events giving rise to an unconformity is illustrated in Figure 1.12.

During the late Silurian and early Devonian, the Midland Valley began to develop as a basin because of crustal tension that caused the Highland Boundary Fault and Southern Uplands Fault to act as a pair of parallel normal faults, allowing the crust in between the faults to sink and form a rift valley or graben (German: furrow or trench). Within this valley, northeast–southwest-trending basins developed as sediment accumulated. From northeast to southwest these are the Crawton, Strathmore and Lanark basins. As sediment was brought into the basins, faulting continued to allow the base to sink and accept more and more material. Old Red Sandstone rocks are found north of the Highland Boundary Fault and south of the Southern Uplands Fault, so the two main faults did not themselves actually control the deposition. Gravels were transported by large rivers flowing off the Highlands, Southern Uplands and the Scandinavian mountains to the northeast, and they were deposited in the basins that were developing along the faults bounding the Midland Valley. Scotland then was closer to Scandinavia, because the North Sea Basin had not yet opened up. Volcanoes existed in the region of the Ochil and Sidlaw Hills, and many of the

---

[*] Rock successions are subdivided into Lower, Middle and Upper, whereas time periods are subdivided into Early, Mid- and Late. Further subdivisions of time are used, and are based on internationally agreed type sections, usually of marine sedimentary rocks containing characteristic fossils. For example, the Late Devonian is divided into the Frasnian and above it the Famennian, named after Frasnes and Famenne in Belgium, where the European type sections were defined, in 1862 and 1885 respectively.

Figure 5.1 Aberfoyle: Old Red Sandstone boulder conglomerate at the Highland Boundary Fault.

pebble beds there contain volcanic rock fragments. The earliest deposits are of very coarse grain, for example the boulder conglomerates at Aberfoyle right up against the Highland Boundary Fault (Fig. 5.1). Later sediments, and those laid down farther away from the Highland Boundary Fault, are finer-grain sandstones. About 9 km of lower Old Red Sandstone was deposited in the Midland Valley basins, so that much of Central Scotland would likely have been covered in these sediments. Decrease in grain size as rocks become younger upwards indicates that the source area was becoming less rugged, that is, the mountains were being worn away and fast-flowing torrents carrying large boulders gave way to slower rivers that transported sand and small pebbles.

When the Midland Valley was receiving boulder beds and sandstones onto a continental desert floor, the region around the Moray Firth – Moray, Inverness, Caithness, Orkney, Shetland – was a large low-lying area surrounded by mountains, which gradually

formed an inland shallow lake, sometimes referred to as Lake Orcadie, or the Orcadian Basin. Here the sands are finer than in Central Scotland, and in Orkney and Caithness at least they form tough, well bedded grey and buff flagstones. This basin formed by subsidence of the crust around the Great Glen Fault, and it was filled with sediment flowing mainly off the Northern and Grampian Highlands that lay to the west and south, built of Moine and Dalradian metamorphic rocks (Fig. 5.2).

Ordovician and Silurian rocks

The oldest rocks in the Midland Valley are Ordovician and Silurian sandstones, mudstones and limestones laid down 460–430 million years ago in shallow tropical seas bordering the Iapetus Ocean. Rocks of Ordovician age at Girvan were described in Chapter 4 (p. 98), and here we look at the series of inliers found in the south of the Midland Valley running north-

Figure 5.2 Sketch map of the Devonian Lake Orcadie.

east–southwest adjacent to the Southern Uplands Fault: the Pentland Hills, Carmichael, Lesmahagow and the Hagshaw Hills. An inlier is an area of older rock completely surrounded by younger formations. During the Silurian, life was abundant in the sea, and fossil brachiopods, molluscs, trilobites, starfish and crinoids (sea lilies) can be found, and sometimes also primitive fish and water scorpions. There are large collections of these fossils in the National Museum in Edinburgh. In addition to these marine sediments, lavas are abundant, and this string of inliers may represent a volcanic arc and an associated sedimentary basin behind the arc, which formed along the edge of the Laurentian continent when the Iapetus Ocean was closing as Avalonia approached from the southeast. Generally, the rocks in these inliers are steep to vertical and they strike northeast–southwest and in most cases are overlain by younger rocks, usually gently dipping Old Red Sandstone, above an unconformity.

At Lesmahagow and the nearby Hagshaw Hills, the Silurian rocks are exceptionally important for the unique fossils they contain. Some of the world's oldest complete fish come from these localities, as well as arthropods (the earliest-known water scorpions, named eurypterids). Most of the early fish were small (5 cm long), armour-plated, jawless creatures that lived in shallow muddy lagoon waters, feeding as scavengers or parasites, rather like the modern-day lamprey. Algae, plants and primitive shrimp-like fossils are also found. The Lesmahagow inlier was the site of various Camp Siluria fossil-collecting expeditions on the banks of the Logan Water, from the late 1880s to 1973. The expeditions were organized by the Glasgow Geological Society, and many unusual fossils were collected. All these important localities are protected as Sites of Special Scientific Interest. Fossils from these sites are to be found in museums around the world, their importance being first recognized in 1855 by Sir Roderick Murchison, director of the Geological Survey, who wrote the first treatise on the Silurian system, having established the name, and who also named the Devonian system.

The Pentland Hills are a block of older rock in the Midland Valley, upfaulted from deeper levels. Much of the material is igneous in origin and more resistant to erosion, which is the main reason that the hills form such a prominent feature around Edinburgh. Old

sedimentary rocks are found in the North Esk, Bavelaw and Loganlea inliers. Shallow-water shelly fossils are particularly abundant in the North Esk outcrops. These include a rich variety of corals, crinoids, bivalves, brachiopods, gastropods, nautiloids, worms, bryozoans, ostracods, starfish, echinoids, fish and trilobites, as well as algae. Some of these limestone beds in these rocks are made entirely of fossil shells. Sediments that formed in deeper waters have none of these shelly animals, but do have free-swimming cephalopods (related to squids) and floating graptolites, which after death sank into the fine muds on the bottom. One of the most extensive water-scorpion faunas (eurypterids) in the world is found here. Clay bands among the siltstones and mudstones represent sudden deposition of volcanic ash carried by the wind, from distant explosive eruptions on islands in the volcanic arc at the edge of the Iapetus Ocean. Evidence for basement rocks in the Midland Valley comes from nodules found in volcanic rocks of Carboniferous age in north Ayrshire, East Lothian and Fife (see p. 127). In addition, there are pebbles of igneous rocks and metamorphic quartzite in some of the Silurian conglomerates in these inliers, indicating that a basement of continental crust existed, which has now disappeared, partly as a result of being thrust beneath the Southern Uplands and partly through being covered by younger rocks. Sandstones in some of the earliest Devonian rocks also contain minerals such as garnet and staurolite, which were derived from metamorphic rocks that no longer exist in the region.

Devonian rocks

Old Red Sandstone in the Midland Valley

Devonian sandstones and conglomerates, and andesitic and basaltic lavas, were laid down in a series of basins that existed in the Midland Valley, at the base of mountains in the Highlands and Southern Uplands 418–362 million years ago. The Devonian is divided into three parts, with only the lower and upper being found in the Midland Valley. Most of the sediments are continental and formed in desert conditions where the cement between sand grains was rich in iron oxide (hæmatite), giving the rocks an overall distinctive red colour. Hence they are usually referred to as "Old Red

Sandstone", but since desert conditions began late in the Silurian (from 420 million years ago) and continued into earliest Carboniferous times (until 350 million years ago), Old Red Sandstone is not exactly the same as Devonian in extent, but rather represents a set of environmental conditions called facies. Time boundaries at the top and bottom of the Devonian are gradational in places. On the other hand, the Devonian rocks of southwest England, southwest Ireland and continental Europe are marine sediments. Because of the conditions of deposition in Central Scotland – fast-flowing rivers and arid valley floors – the Old Red Sandstone has relatively few fossils, save for some fish from temporary shallow lakes, rivers and oases, and some early primitive plants and spores. During the Devonian, Scotland was 20° south of the Equator.

Close to the northern and southern margins of the Midland Valley, Old Red Sandstone sedimentary rocks are thick and very coarse, being typically boulder beds and conglomerates with rounded pebbles of quartzite and granite (Fig. 5.1, 5.3). Towards the centre of the Midland Valley the grain size decreases, and conglomerates give way to sandstones and still-finer siltstones. Volcanic rocks are very abundant, in the

Figure 5.3 Old Red Sandstone cliffs at Dunnottar Castle, Stonehaven, Aberdeenshire.

Ochil and Sidlaw Hills in particular, and also in the southwest at the Carrick Hills south of Ayr. Erosion of these volcanoes contributed to pebbles of volcanic rocks found in Old Red Sandstone conglomerates; most of these, which are andesitic and basaltic in composition, have been dated at 420–410 million years old, including those in the Pentlands. However, those at Crawton south of Stonehaven are rather older, at 430–420 million years old. Being more resistant to erosion than the surrounding sediments, these volcanic outcrops form prominent landscape features in the Midland Valley. In the case of the Ochils and the Pentlands, this is in part related to the existence of fault scarps running along the margin of the hills (Fig. 5.4). In Ayrshire and Angus, the lavas have yielded many examples of agates (Scotch pebbles): colour-banded quartz deposited in gas cavities that formed when the lavas were being erupted, but were subsequently filled when silica-rich solutions percolated through the lavas. Erosion has removed the pebbles and they are now found in the overlying soils and glacial deposits. Associated with the lavas in the Ochil Hills are mineral veins, principally silver, cobalt, iron, copper, baryte and some lead and zinc. The veins are related to the Ochil Fault and tend to run nearly north–south at right angles to the main fault. Mines operated in the eighteenth century at Silver Burn, near Alva, from which both silver and cobalt were extracted. Mineralization in the lavas dates from the late Carboniferous (300–280 million years ago), when a set of east–west fractures affected rocks in the Midland Valley. Hydrothermal waters circulating in the cracks dissolved the metals and re-deposited them in mineral veins. About 45 tonnes of silver was produced in the early 1700s. The cobalt extracted from the same mines in the middle of that century was used as a pottery glaze at Prestonpans, just east of Edinburgh.

The Devonian lavas amount to a maximum thickness of 2000 m, with individual flows being a few metres thick. Actual volcanic vents are very rare, and it appears that the lavas were erupted from fissures. Some of the eruptions took place under water, within temporary lakes. Alternations of conglomerates, lavas and volcanic debris (rocks formed by mudflows rushing in torrents from the volcanic hills) indicate rapid erosion of the volcanic pile and redistribution of lava pebbles in the Midland Valley by river systems.

Figure 5.4 Ochil Hills, from Stirling Castle: Devonian lavas and Ochil Fault; flood plain and meanders of River Forth.

Devonian volcanic rocks in the vicinity of Edinburgh include those of the Pentland Hills, Braid Hills and Blackford Hill, which, together with the younger Carboniferous volcanic hills of Arthur's Seat, Calton Hill, Craiglockhart and Corstorphine Hill, contribute to the city's fame as being built on seven hills. These rocks are basalt and andesite, amounting to 1800 m thick in the Pentlands. Small volcanic vents are found at Swanston and the Braid Hills. In addition to lava flows, there are also beds of tuff representing volcanic ash from explosive eruptions.

The Carrick Hills in Ayrshire are made of 400 m of basalt and andesite lava. Farther south, there are 600 m of lavas in the hills around Dalmellington in a belt close to and parallel to the Southern Uplands Fault. Southeast of Kilmarnock, the Burn Anne area is known for its jasper and agate within the lavas; these have been quarried and collected as decorative stones. These outcrops are more or less in line with the Devonian lavas of Ayrshire and they form part of the volcanic ridge in the centre of the Midland Valley that includes the Ochil Hills (720 m high) and the Sidlaw Hills (450 m high) and stretches towards Montrose and Crawton in the northeast. The ridge, which runs parallel to the Highland Boundary Fault and Southern Uplands Fault, separates the two areas of Old Red Sandstone deposition, the Strathmore basin and the Lanark basin, and may have formed a topographic feature in early Devonian times, during the eruption of the lavas.

As well as producing surface lava flows, igneous activity in the Devonian also resulted in material being intruded beneath the surface, in the form of sills or laccoliths. Dundee Law (174 m) is a prominent conical hill; it is a volcanic plug and was the site of an Iron

Age hill fort. There are many others in the Sidlaw Hills. Tinto Hill (707 m) southeast of Lanark, which forms a prominent landscape feature in the south of the Midland Valley, consists of a fine tough resistant red rock called felsite; it has been dated at 412 million years old. Other felsite bodies include Lucklaw Hill at the northeast end of the Ochils, now being quarried for roadstone and clearly visible from many parts of Fife. Black Hill in the Pentlands is also a felsite dome. The rock is rich in feldspar and quartz, and the original granitic magma cooled so rapidly after injection that it formed a volcanic glass. Over time, the glass devitrified (i.e. lost its glassy texture) and the red coloration is a result of the high feldspar content. Felsite is a hard brittle rock that breaks down into small flat flakes and angular shards; when weathered, these tend to produce hills with smooth rounded shapes.

Large Caledonian granitic intrusions are very rare in the Midland Valley, in comparison with the Grampian Highlands. One body is the Distinkhorn granite in east Ayrshire, southeast of Darvel, and the nearby Hart Hill, which is more basic. This intrusion cuts Old Red Sandstone sedimentary rocks and, like Tinto, is the same age as the lavas (412 million years old).

Devonian igneous activity related to the closure of the Iapetus and the ending of subduction, having been caused by partial melting of the crust and upper mantle above a descending slab. Igneous activity ceased in the early Devonian, 410 million years ago: continental collision and suturing had then taken place.

Volcanic rocks of Devonian age are also present in the Southern Uplands, particularly the St Abb's Head vent, filled with red ash and boulders (i.e. volcanic agglomerate). The Cheviot lavas and granite intrusion of 396 million years ago have already been described (p. 98). Some of the volcanic material is tuff and agglomerate, suggesting explosive eruptions. The granite intruded the lavas shortly after surface volcanic activity had ceased. Old Red Sandstone sedimentary rocks in the Southern Uplands are represented by valleys filled with conglomerates, a particular example being Lauderdale in the Borders.

The Midland Valley has no rocks of mid-Devonian age, so there is no record of events during the time-span from 390 to 370 million years ago. If any sediments were deposited during that 20-million-year episode, they were eroded and removed shortly afterwards. Possibly there were movements on the Highland Boundary Fault then, and the Midland Valley could have been elevated enough for the Lower Old Red Sandstone to be eroded in part. Sedimentation resumed in late Devonian times, during which the Upper Old Red Sandstone was formed as a series of sandstones and shales laid down by large rivers meandering across a plain. These rivers rose in the Highlands and Southern Uplands and carried material to the northeast, towards what is now the North Sea. Evidence from offshore drilling indicates the presence of marine deposits, so that the sea may have extended to the north during the late Devonian. Current directions in the Lower Old Red Sandstone are completely opposite from those in Upper Old Red Sandstone sediments, showing that there must have been crustal movements in the period 390–370 million years ago related to collision and uplift. During this time, the main northeast–southwest folds formed, the Strathmore syncline and the Sidlaw anticline being recognized as two prime examples of folds by Sir Charles Lyell in his 1838 textbook of geology. The Sidlaw anticline is a symmetrical fold, whereas the Strathmore syncline is highly asymmetrical, the northwest limb being nearly vertical as a result of movements on the adjacent Highland Boundary Fault.

Wherever the Upper Old Red Sandstone is seen in the Midland Valley, the sediments (mostly bright red sandstones) are unconformable on the rocks beneath. Lavas are absent. At the top, the Old Red Sandstone gradually passes up into Carboniferous sedimentary rocks, with sandstones and fossil soils eventually giving way to muddy shallow marine deposits, indicating that the sea had encroached into the Midland

Figure 5.5 The unconformity between the Upper and Lower Old Red Sandstone at Arbroath Pier. The sandstone beds at the base dip to the right; the conglomerate beds above are horizontal.

Valley and drowned the low-lying areas. The same unconformity between Upper Old Red Sandstone and all older rocks is the one recognized by James Hutton, not only at Siccar Point (p. 95) but also at Jedburgh and Lochranza on Arran. An excellent example of the Lower/Upper Old Red Sandstone unconformity can be seen on the shore at Arbroath (Fig. 5.5). Upper Old Red Sandstone rocks are thickest in the west, i.e. close to the original sediment source, especially in the Firth of Clyde area, where they are over 2000 m thick and very well exposed in the raised-beach cliffs at Weymss Bay and on Arran. Exposures are also found north of the Clyde around Helensburgh–Loch Lomond and from Dumbarton to Stirling, also north of the Tay and in Kinross and northwest Fife; thickness decreases towards the northeast. The Fife outcrops include the very important fossil fish locality at Dura Den south of Cupar, where some ten distinctive species of armour-plated fish were discovered in 1915 and completely removed by the British Museum. Fossils are rare in the Old Red Sandstone; these specimens indicate an age of about 370 million years for the sandstones and shales that were deposited possibly in a shallow temporary lake that dried out and killed the fish, judging by the mudcracks and many fish on the same bedding plane. In the southern part of the Midland Valley the main Upper Old Red Sandstone outcrops are in the Pentland Hills, where there are 300 m of pale red sandstones, increasing to 600 m nearer Edinburgh. Upper Old Red Sandstone sediments are in most cases much finer than Lower, most probably because the Lower

Old Red Sandstone was eroded to provide sedimentary material for the Upper Old Red Sandstone rivers. These are described as mature sandstones, meaning that they are more consistently rich in quartz, having been deposited by large slow-moving rivers that meandered across broad flat floodplains. The maturity of sediment increases towards the top of the Upper Old Red Sandstone and gradually cornstones are developed at the boundary with the Carboniferous (362–360 million years ago). Cornstones, also known as caliche or calcrete, are types of fossil soil deposits formed in arid conditions on a flat low-lying land surface that was subject to intense evaporation. Some of the thickest and best examples of cornstones are found at Milton Ness, north of Montrose. The deposits are rich in calcium carbonate (limestone) that formed a hard pan at the surface, which in places was dissolved by water during brief humid intervals, to form cavities and collapse structures and sometimes riddled with tubes that mark petrified tree roots. The Midland Valley at the time lay at 5–10° south of the Equator, and the few rivers flowing into Central Scotland brought very little sediment from the now-denuded uplands. Fossil sand dunes also point to an arid climate at the end of the Devonian period.

Old Red Sandstone north of the Highland Boundary Fault
Beyond the Midland Valley, the greatest extent of Old Red Sandstone rocks is around the Moray Firth, Caithness and Orkney (Figs 5.6), forming the Orcadian Basin. In contrast to those in the Midland Valley, the majority of these rocks belong to the Middle Old Red Sandstone and were formed by rivers flowing into a broad, low-lying area that was surrounded by mountains and periodically flooded by shallow lakes that came and went as rainfall patterns changed. During periods of high lake-water level, there were probably connections with the sea to the southeast, during which fish entered the area. Movements on faults may in part have controlled deposition, so that the floor of the basin sank to allow more sediment to accumulate. Fine well bedded buff siltstones and sandstones were deposited in the basin, and these have been exploited as the flagstones that make up many a pavement in Edinburgh and Glasgow. Living in these lakes were primitive fish, whose anatomy and evolution were studied by Hugh Miller, the Cromarty stonemason,

Figure 5.6 Old Red Sandstone cliffs at New Aberdour, Moray coast.

who published the first account of the Old Red Sandstone in 1841.

In addition to the rocks of the Orcadian Basin, there are important Old Red Sandstone outliers in the Grampian Highlands, principally at Tomintoul, Rhynie and Turriff, where the local building stone bears witness to the surrounding geology. Lavas, sandstones and conglomerates are found in Argyllshire, resting on Dalradian basement, at Kerrera and around Oban, and in Kintyre the sandstones contain fossil fish, millipedes and plants that prove an age of 418 million years. In the Ailnack Gorge near Tomintoul, the Lower Old Red Sandstone forms a coarse conglomerate lying on top of Dalradian schists. Subsequent erosion at the end of the Ice Age has cut a very deep hidden gorge as the pebbles in the conglomerate were removed and converted the meltwater stream into an effective rock saw.

In the Orcadian Basin, most of the Old Red Sandstone is represented by the Caithness Flags, which are Middle Old Red Sandstone, and are well exposed in the Reay Cliffs at Duncansby Head (Fig. 5.7). Outcrops occur on both sides of the Moray Firth, from

The Rhynie Chert

At Rhynie, the Lower Old Red Sandstone forms a 20 km-long narrow north–south outcrop bounded on the west by a fault. Within the sandstone is the now world-famous unique Rhynie Chert locality containing the remains of many primitive plants, perfectly preserved in three dimensions. These plants grew around hot volcanic springs at the surface, which occasionally erupted silica-rich waters that engulfed the plants and instantaneously petrified them. The plants could not rot down, and now their internal structure can be studied in immense detail. Chert beds are interspersed with sandstone and shale, and the sediments were deposited on an alluvial plain occupied by marshes and temporary lakes. Chert is a type of non-crystalline silica that forms when hot hydrothermal solutions (200–300°C possibly) cool quickly. Dissolved silica is then deposited at normal temperatures on the surface (20°C, say). Dating of the chert gives a result of about 400 million years, making this deposit one of the earliest known examples of hot springs at the surface. In addition, the chert has traces of precious metals, particularly, gold, silver, antimony, mercury, tungsten and molybdenum. The deposit is described as sinter (layered crusts around hot springs) and there are various veins and fractured or brecciated sections with infilled cavities that suggest

geyser activity, similar to the features seen today in Yellowstone Park, Wyoming. It is not known whether the Rhynie plants were specially adapted to living in a hot-spring environment or whether they were typical land plants that simply were suddenly overwhelmed by boiling water. The plants have mostly been preserved in their growth position as roots and stems, and include representatives of many different types, including algae, bacteria, lichens and marsh reeds. Some of the plants resemble primitive living plants, particularly the lycopods or club mosses. The Rhynie deposit is exceptionally important because such early land plants are very rare and usually poorly preserved. Small animals living with the plants were also fossilized in the hot springs. These are mostly crustaceans and arthropods – shrimps, spider-like organisms and insects – which may have fed on the plants, judging by the evidence of damage on the stems of *Rhynia*, the freshwater reed and commonest plant. Thus, we have here a well preserved ancient land-based animal–plant ecosystem, one of the oldest in the world, whereas previously nearly all life had been confined to the sea. Further details of this fascinating locality can be found on the Rhynie website, hosted by the University of Aberdeen: http://www.abdn.ac.uk/rhynie/.

Nairn through the Black Isle, east Sutherland, Caithness, Orkney and Shetland, as well as off shore in the Moray Firth Basin and central North Sea. Although most of the rocks are Middle Old Red Sandstone, there are a few areas of Lower Old Red Sandstone conglomerates, lying unconformably on Moine rocks (Fig. 5.8). These include Ben Alisky, Morven and Brora in the north, and the spectacular example at Loch Duntelchaig south of Inverness. At Reay (Red Point) on the north coast, the Middle Old Red Sandstone rests directly on the Moine schists and granitic rocks of the basement, and the junction is marked by a fine example of an unconformity where angular fragments of

the underlying granite can be seen at the base of the Old Red Sandstone. This would have represented fossil scree on the land surface. Conglomerate beds are often more resistant than the sandstones and, in the case of Knock Farril near Dingwall, the steeply dipping conglomerate forms a prominent landmark, and 2500 years ago was used as the site of a vitrified fort (Creag Phadraig). It is thought that vitrified forts were constructed by setting alight large piles of timber so that the heat fused the outer surface of the stones and welded them together to create a highly resistant wall. Dr John MacCulloch (1773–1835) attempted the first scientific study of these monuments.

On the northwest shore of Loch Ness, Mealfuarvonie has some of the highest outcrops of Old Red Sandstone in the country. This strip of land has been thrust up by faults associated with the Great Glen Fault. Similar thrusts have contributed to the step-like features of the Old Red Sandstone on Struie Hill, near Edderton. Some of the shales within the Old Red Sandstone are rich in organic matter, including the spa beds at Strathpeffer. These are the source of the sulphurous waters that were exploited in the nineteenth century; the pumps and wells have been restored to working order and the waters can be sampled once more.

At Helmsdale in eastern Sutherland, the Lower Old Red Sandstone lies directly on top of the 420-million-year-old Helmsdale granite. Because the

Figure 5.7 Devonian flagstones at Reay Cliffs, Caithness.

Figure 5.8 Maiden Pap, an isolated peak of Old Red Sandstone conglomerate resting on Moine basement rocks at Dunbeath, Caithness.

sandstones contain fossil plants and spores dated at about 400 million years old, there must have been considerable erosion during this 20-million-year period to bring the granite to the surface. A rather similar situation exists along the Great Glen Fault, where Old Red Sandstone conglomerate rests on top of the Foyers granite, which was intruded at a depth of more than 7 km, so that much of overburden must have been removed in the course of 20 million years.

Southern outcrops of Old Red Sandstone are mainly river deposits, whereas in Orkney and Caithness the rocks were laid down in a shallow lake. Fossil fish beds are common here, the best known being at Achanarras, south of Thurso. These beds are about 380 million years old and are correlated with the fish beds found in the Black Isle (Ethie and Cromarty), the Tarbat Ness peninsula (Edderton), Orkney (Sandwick) and Shetland (Melby). Fish living in Lake Orcadie probably inhabited the shallower waters around the edge. When they died, their bloated remains were carried away from the shallows by currents, and on becoming waterlogged they sank into deeper waters, where the lack of oxygen allowed the carcasses to be preserved in fine mud. On occasion, the fish beds are densely packed with fossils, suggesting that high temperatures and evaporation caused the lake to dry up; high salinity and insufficient oxygen would have killed off the fish in mass-mortality events; algal blooms in shallow water could also have contributed to this. Mudcracks on the surface of many of the beds suggest too that the lake dried up periodically. The Caithness Flags are mostly of fine-grain,

delicately laminated grey or buff siltstones, frequently with carbonate cement. Sedimentation was slow, regular and cyclical, reflecting repeated changes in climate from wetter to drier, which would have controlled water levels in the lake. Cycles change from impure limestone, to flagstone (fine siltstone), then ripple-marked mudstone, and finally mudstone showing mudcracks at the surface. These cycles were repeated many times through the 3000 m of Middle Old Red Sandstone in Caithness.

Climatic cycles are related to the spin of the Earth on its axis and to its elliptical orbit around the Sun, together with sunspot activity, all of which control the amount of energy reaching the Earth's surface. We will return to this theme in Chapter 8, when dealing with the causes of ice ages.

At least 15 different types of fish are found at Achanarras (south of Thurso), which is an important Site of Special Scientific Interest, owing to their excellent preservation. There are representatives of predators, scavengers and omnivorous fish. They may have migrated into Lake Orcadie from the sea when water levels were high, then became adapted to changing conditions as water levels fell. One of the most abundant fish is *Dipterus* (a lungfish), representatives of which are still found today, and this appears to have been able to tolerate widely varying conditions, including low oxygen levels and high salinity as the lake dried up. The largest predator is *Glyptolepis*, over a metre long. Most of the other fish were 20–50 cm long. The fish beds themselves are very dark, because they are rich in organic matter, much of it from algae.

It is possible that some of the fish beds may have acted as source rocks for oil in the North Sea, thus making the Devonian the oldest source rock in Scotland.

Flagstones have been quarried in Caithness for centuries and used as building stones, roofing material, pavements and field boundaries, this last feature giving the Caithness landscape a highly distinctive character. The material is hard, strong, durable and easily worked, and it produces slabs of constant thickness. Many famous archaeological monuments in the northlands are made of flagstones, of special note being Mousa Broch, Shetland (see Fig. 4.21) and Maes Howe and Skara Brae on Orkney (see Fig. 9.3). Famous coastal landforms – sea stacks, pyramids, caves, hollows, geos – result from wave erosion of evenly spaced joints in the finely bedded sandstones. Two good examples are Duncansby Head stacks near John o'Groats (see Fig. 5.7), and the Old Man of Hoy on Orkney (Fig. 5.9), which stands on a lava flow and is now in a precarious state and has had to be pinned down by bolts and cement. It is on the brink of collapse because of marine erosion, but has been artificially preserved since it is such an important symbol of the Orkney landscape.

In Orkney, the Old Red Sandstone was deposited on a basement of granite and gneiss of unknown age, but probably related to the Moine rocks of the Northern Highlands. This rock was used as a local building stone and it can be seen in the old quarries in the higher ground near Stromness. On nearby Graemsay Island, the conglomerate at the base of the Old Red Sandstone contains rough angular blocks of the underlying metamorphic rock, and so represents fossil scree that was washed in by sudden floods. This situation is almost identical to that at Red Point near Reay in Caithness. By far the most extensive rocks in Orkney are the Stromness Flags and the Rousay Flags. These Middle Old Red Sandstone sediments have upwards of 50 cycles (regular repetitions of siltstone and sandstone), 2–10 m thick, of dark siltstone, laminated siltstone with mudcracks, followed by coarser rippled sandstones. The Sandwick Fish Bed is unusual in being 60 m thick and it represents a time when the lake had greatly expanded in volume; it is a prolific source of fossil fish. An important locality is the Cruaday quarry in the west of Mainland, from which many perfectly preserved fish have been collected since the nineteenth century, being so abundant here that they form a carbonized layer. This site is internationally important in the evolution of fish in the Devonian, a period when the primitive, jawless, armour-plated fish gave way to the ancestors of modern types.

Upper Old Red Sandstone rocks in Orkney are seen only on Hoy, where an unconformity with the Middle Old Red Sandstone suggests that erosion of 3000 m of Middle Old Red Sandstone rocks had occurred during a period of earth movements. The earliest rocks are volcanic tuffs (ash carried by the wind), followed by basalt lava then 1000 m of tough well cemented red sandstone, laid down by rivers.

Mineral deposits in Orkney are scarce, save for uranium ore in the flagstones. Reconnaissance work carried out in the 1970s did not result in further exploration or exploitation, because of local resistance within Orkney. The Caithness flags are equally rich in uranium. Attempts were made in Orkney to exploit the scattered deposits of lead, copper, iron and manganese, mostly in the late eighteenth century, but to no avail. Flagstones and sandstone have been used locally for building stones and as road metal, although the Orkney flagstones were not exported, as the quality is not as high as that of the Caithness flags.

FIONA MCGIBBON

Figure 5.9 Hoy, Orkney: the Old Man of Hoy, a sea stack of Old Red Sandstone resting on lava flow.

Scotland in the Carboniferous

By the start of the Carboniferous about 360 million years ago, the Caledonian Mountains had been worn down very effectively by the forces of erosion, although the Highlands and Southern Uplands still formed high ground as they do today. Central Scotland was flooded by shallow tropical seas in which corals flourished, and limestone formed as a result. Subsequently, rivers drowned the coral reefs with sediment and eventually the rivers themselves silted up and allowed dense forests to become established in the swamps and marshes around sea level. Rapid evolution and colonization by land plants in the Carboniferous was responsible for the formation of coal deposits, on which the economic development of Scotland in the Industrial Revolution depended.

Deposition of marine sediments therefore represents a change from the terrestrial river and lake deposits of the Devonian, and a change in climate. Scotland continued to drift northwards and crossed the Equator near the start of the Carboniferous. In general terms, sedimentary rocks formed in the 70 million years of the Carboniferous consist of limestone, then sandstone, and finally coal, reflecting changing environments from shallow tropical seas to river deltas then coastal swamps. However, within this broad scheme there are many repetitions of these three main rock types in each division, in a regular cyclical arrangement that can be explained in terms of a combination of climatic change, plate tectonics, and movements on faults, which affected sea level locally and globally. The Carboniferous period represents the first time in Earth history when land plants made a major contribution to climate and to the formation of rocks. Most of the world's high-quality coals are of

Carboniferous age; older coals are exceptionally rare.

In the Midland Valley, Carboniferous volcanic rocks are widespread and form important landscape features, such as the lavas in the Campsie Fells, Kilpatrick Hills, Renfrewshire Heights and Gargunnock Hills, which make a horseshoe-shape outcrop around Glasgow, covering a total area of some 3000 km^2. Equally significant are the volcanic plugs – eroded remnants of volcanoes that poured out these lavas – that now form prominent landmarks, such as Dumbarton Rock, Dumgoyne and Dumfoyne in the west, Arthur's Seat, North Berwick Law and the Lomond Hills in the east (Figs 5.10, 5.11). The volcanic rocks, together with the widespread sandstone, limestone and coal deposits, mean that Carboniferous rocks underlie most of the Midland Valley. They cover the Old Red Sandstone and Devonian rocks, except near the Highland Boundary Fault and Southern Uplands Fault.

The transition from Devonian to Carboniferous 360 million years ago was a gradual process, so that in places red sandstones are found at the base of the Carboniferous succession. Earliest Carboniferous rocks are sandstones, shales and carbonate beds, exceptionally well exposed in Campsie Glen. These are the Ballagan Beds, an alternating sequence of dolomite (calcium–magnesium carbonate), sandstone, cornstone (see p. 116) and gypsum (calcium sulphate). Such a rock sequence indicates deposition in a shallow lagoon that formed on the landward side of coastal sandbars. High temperatures caused the water to evaporate, leaving behind deposits of dolomite and gypsum. Mud cracks and cubic impressions found in the sandstones are the moulds of salt crystals, further indications that sea water was being evaporated. This succession at the base of the Carboniferous is sometimes referred to as the cementstones.

Figure 5.10   Edinburgh, Arthur's Seat: the Lion's Head and Lion's Haunch volcanic vents.

A volcanic episode suddenly ended this lagoonal period, with the outpouring of nearly 1000 m of basalt lavas in the west, referred to as the Clyde Plateau Basalts. These were erupted from fissures and vents, and the near-horizontal lavas give rise to a distinctive stepped topography, known as "trap topography". Between each lava flow there is usually a bed of red fossil soil ("red bole", iron-rich laterite or clay soil) that resulted from subtropical weathering of the lava tops, before the next eruption of ash or lava. Since the bole is softer than the massive basalt, it weathers preferentially to produce the stepped landscape on the cliff faces of the Campsie Fells and the Kilpatrick Hills. The Garleton Hills in East Lothian are made of 500 m of basalt of the same age (340 million years old) as the lavas in the west, as are the lava flows of Arthur's Seat and Craiglockhart Hill in Edinburgh. The Clyde Plateau lavas thin out rapidly to the east, and east of the lava hills sedimentation continued, so that eventually about 2000 m of sandstone was deposited by slow-moving rivers in what is now Fife. Occasional thin limestones indicate that the sea flooded the area from time to time.

In West Lothian, the barrier of the lava hills created conditions where oil shales formed in a restricted enclosed lagoon or lake. Much organic matter (mostly plants and algae) accumulated in the stagnant waters, and the lack of oxygen in the muddy lake floor allowed oil shales to form. Oil shales are abundant around Bathgate, and the shale-mining and oil-processing industry developed in the middle of the nineteenth century (see box on p. 211). The early Carboniferous volcanic rocks of the Clyde Plateau lavas in the west, the Garleton Hills in East Lothian and Arthur's Seat in Edinburgh are overlain by sedimentary rocks – sandstone, mudstone, shale, limestone and coal. In the Paisley area, coal of this age was worked as long ago as the early 1600s. Sandstone from this sequence has been widely used in Edinburgh in the construction of the eighteenth-century New Town, the most famous quarries being Craigleith, Ravelston and Hailes. Much of this sandstone was also exported to London and the continent.

There are important limestone horizons in this sedimentary succession, including those at Dunbar in East Lothian. Some of the beds have been worked for agricultural lime for over 200 years, and today the Blue Circle Cement works at Dunbar produces high-quality material in a huge opencast quarry. Many now derelict lime kilns dot the landscape. Formerly, these were used for manufacturing agricultural-quality lime, using locally derived limestone and coal. Fossils are generally abundant, and many of the limestones are in fact coral reefs, preserved in three dimensions. Crinoids, brachiopods and bryozoans are present among the many-branched colonial corals.

Carboniferous limestones were deposited in warm shallow tropical seas with abundant sources of food in well oxygenated waters, virtually clear of mud and silty debris. In such conditions, the animals living in the sea, such as corals and shellfish, secreted calcium carbonate dissolved in sea water into their skeletons, which after death formed part of the rock. Limestone is therefore an organic sedimentary rock, in which large amounts of carbon dioxide are locked up for millions of years until eventually the rock is exposed and weathered at the surface. Limestones therefore play a very important part in the carbon cycle, which in turn

Figure 5.11 Dumbarton Rock and Castle, River Clyde: a Carboniferous volcanic plug, with columnar jointing on the right.

has a fundamental role in the global climatic system.

Contemporaneous with the limestones, volcanic activity produced lava flows in the east of the Midland Valley, especially around Bathgate and Bo'ness and in Kinghorn in Fife. Basalt lavas and ash beds are interleaved with limestones. The Bathgate Hills have 500 m of basalt, which gives the hills their rugged appearance. Locally, the limestones were once extensively worked for agricultural use. One of the quarries at East Kirkton is an important Site of Special Scientific Interest, because it contains rare and unusual fossils, some extremely well preserved, including amphibians, millipedes, scorpions and eurypterids. Plant remains are abundant, whereas fish are rare, and it is thought that these rocks may have formed on land, in shallow pools and freshwater lakes close to hot volcanic springs, with dense forests around. This is the location of the world-famous oldest ancestor of the reptiles, *Westlothiana lizziae* (nicknamed Lizzie), discovered in 1988 and the scene of intense international research in the early 1990s. Lizzie can be seen in the National Museum in Edinburgh.

Above these Carboniferous limestones are sandstones and coal seams, on which the Scottish coalmining industry has been based. The rocks were deposited in a cyclical fashion by rivers whose deltas became silted up, to be replaced by stagnant swamps where dense vegetation flourished. Fossil soils, usually referred to as seat earth, packed with roots are common. In addition, there are beds of shale containing abundant ironstone nodules, which were formerly of economic importance for the iron and steel industry in Scotland at the start of the industrial revolution. The material in the ironstone bands is iron carbonate (siderite), together with carbon from plants that played a useful part in iron-ore smelting. Coal seams are many, and those over a metre thick were considered economic; very few are thicker than 2 m. Of the once-many opencast collieries and deep mines, rather few now exist in Central Scotland. Small-scale opencast workings are begun, then abandoned frequently (after 3–4 years), depending on their viability. Lack of care and maintenance of these poorly regulated sites has led to environmental problems related to water pollution, waste dumping, and underground collapse and landscape degradation. Damage to surface structures caused by mining subsidence is not uncommon

in the Scottish coalfields. For example, in 1885 the town hall at Bo'ness collapsed into a hole when the workings 15 m below the surface caved in.

Fossil soils associated with the coal forests have yielded another material that was formerly of economic importance – fireclay. Deep tropical weathering resulting from a warm humid climate caused minerals to be washed down from the surface, leaving a muddy residue rich in alumina as the mineral kaolinite. This is similar to the process that produces bauxite in modern subtropical regions. In the iron and steel industry, fireclay is used to line furnaces, since it resists very high temperatures (up to 1500°C). In many cases, the fireclay contains fossilized roots attached to tree trunks in the overlying beds. Often the tree fragments are so abundant that they probably represented logjams in the Carboniferous rivers of 320 million years ago, when trees were washed down from inland forests onto the deltas and became waterlogged. A magnificent remnant of a 325-million-year-old Carboniferous forest can be seen in Glasgow's Victoria Park. In 1887, during excavations in a roadstone quarry, several tree trunks were found in their growing position within shale and sandstone beds, showing their shallow branching roots (Fig. 5.12). One of the fallen trees is over 7 m long, and it seems likely that the forest was overcome by a river bringing down sand onto a marshy delta. These trees (lycopods) are now extinct, their only surviving relatives being the much smaller clubmosses, which grow in very damp conditions. They were spore-bearing trees with simple, dense leaves arranged spirally on the trunk; they first appeared in the Devonian and became very abundant in the Carboniferous (Fig. 5.13). This remarkable locality is a Site of Special Scientific Interest.

During the early part of the Upper Carboniferous, 325–315 million years ago, several thick dolerite sills were intruded into the sandstones, a particularly notable example being Salisbury Crags in Edinburgh's Holyrood Park (see the photograph on pp. 108–109 and Fig. 5.15). These sills have provided much whinstone, used mainly in road construction, and are therefore another valuable economic resource in the Carboniferous rocks of the Midland Valley.

Following the deposition of the deltaic sandstones and coals, there was a return to marine conditions, with the formation of an upper limestone group. This

Figure 5.12 Fossil Grove, Victoria Park, Glasgow: part of a Carboniferous forest, showing tree trunks in their growing positions, with roots penetrating shale beds (IPR/35-32C British Geological Survey, © NERC, all rights reserved).

again is cyclical in nature, with regular and systematic changes from limestone to sandstone, mudstone, seat earth, then coal, repeated over and over again, although limestone is not necessarily present in every cycle. Limestone bands are very important in correlat-ing the rocks, since they contain the same fossils throughout the region. The marine limestone bands indicate drowning by the sea and therefore a general rise in sea level. Sandstones and coals were normally deposited in separate basins controlled by fault move-

Figure 5.15 Edinburgh, Holyrood Park: Salisbury Crags dolerite sill intrusion; sandstone above and below.

Figure 5.13 Part of *Lepidodendron* plant stem, a Carboniferous tree, showing scale features.

Figure 5.14 Edinburgh, Holyrood Park: Hutton's section: base of dolerite sill (top) intruding sandstone beds (bottom right).

ments, so that thickness varies greatly from place to place, and sequences of rocks were not identical in the different basins. Identifying the relevant marine bands (limestones and limey mudstones, which are often more fossiliferous than limestones) from their distinctive fossil contents is therefore an effective way of obtaining ages of the rocks above and below the limestones in different basins. Such distinctive units are then used as reference points and are known as marker horizons or index limestones.

The Coal Measures

In late Carboniferous times (315–305 million years ago), the economically important Coal Measures were formed, when widespread and dense tropical forests became established in Central Scotland. Some 20 seams in the Coal Measures have been extensively worked in the Scottish coalfield. As elsewhere in the

Carboniferous, the sedimentary rocks are cyclical in nature: sandstone, siltstone, mudstone, seat earth and coal, with occasional ironstone bands and very rare limestones that represent marine incursions onto a predominantly river-delta environment. Sedimentary rocks of the Coal Measures are thick, and subsidence of the region kept pace with sedimentation, so that most of the rocks were deposited in shallow water, the surface being almost at sea level. Freshwater mussels (and plant remains, of course) are the commonest fossils in the Coal Measures. Rocks belonging to the Coal Measures are found in three main basins in the Midland Valley, corresponding to the coalfields: Ayrshire, Central and East Fife–Midlothian (connected beneath the Firth of Forth). Total thickness of sedimentary rocks in the basins averages about 500 m, although the small Douglas Valley basin (southwest of Lanark) has the greatest thickness, 700 m.

The youngest Carboniferous rocks in the Midland Valley are 300 million years old: grey sandstones that merge upwards into redbeds as a result of iron-oxide staining (hæmatite), indicative of a transition to the arid conditions that characterize the Permian period. Scotland by now had crossed the Equator and lay at 5–10°N, leaving behind the equatorial rainforest belt that characterized the Carboniferous.

The Hercynian orogeny

For most of the Carboniferous, the southern continental landmass of Gondwana – the consolidated mass of Australia, Africa, Antarctica, South America – drifted north to collide eventually with the Old Red Sandstone continent of Laurussia – North America, Greenland, Scandinavia, Europe and Russia – to create the new supercontinent of Pangaea. This continental collision created a new east–west mountain belt across southern North America and central Europe – the Hercynian chain. The orogeny that created this belt is known by various local names: Hercynian (Hertz Mountains), Variscan (from Variscia, the Roman name for Saxony), Armorican (Armorica in Brittany) and Alleghenian (Allegheny Mountains, part of the Appalachians). In Scotland, the effects were not strongly felt, as it was far from the collision zone itself. Such effects as are discernible are mainly some folding of Carboniferous rocks, many east–west faults, and a considerable number of dolerite dykes with this same trend. The dykes were intruded along fractures that opened up in response to the release of the north–south Hercynian compression.

By 300 million years ago, Scotland again found itself in the middle of a large continental landmass, but now 5–10° north of the Equator in the tradewinds belt. The Carboniferous seas had long since receded and Scotland was a dry desert. Cross bedding in fossil sand dunes indicates that the prevailing wind blew from the east. Most of the country was an upland region, save for parts of the Midland Valley and Dumfries. This period is termed the Permian and was named by the Scottish geologist Sir Roderick Murchison, after Perm province and town in the Urals of Russia. Desert conditions persisted into the succeeding period, the Triassic (250–205 million years ago), and in Scotland the Permian and Triassic rocks are usually grouped together as the New Red Sandstone, covering 100 million years of Earth history. Rocks at the top of the Carboniferous succession are in many places stained bright red as a result of oxidation effects from the overlying New Red Sandstone deposits.

In summary, the Midland Valley evolved as a rift-valley basin from 365 million years ago, following the uplift and erosion that marked the end of the Caledonian orogeny in the early Devonian. Crustal thinning during the Carboniferous allowed a basin to form, which initially was flooded by a shallow sea and later by terrestrial sediments. At the start of the Carboniferous, volcanic lava poured out over the surface; this activity continued intermittently because of crustal thinning and partial melting of the upper mantle rocks. Later igneous events were dominated by sill intrusion and east–west basic dykes, at right angles to the Hercynian orogenic north–south compression.

Carboniferous volcanic activity

Much of the high ground in the Midland Valley is associated with the lavas and volcanic plugs and sills of Devonian (400 million years) and Carboniferous (340–330 million years) age. Volcanic activity that had begun at the end of the Devonian continued throughout the Carboniferous across the Midland Valley, with many volcanic episodes punctuating the deposition of sediments in the shallow sea early in the Carboniferous, and the rivers and deltas at later stages. Activity continued into the early Permian (300–250 million years ago), so that we have evidence for igneous rocks – lavas, tuffs, plugs, dykes and sills – forming across a timespan of about 100 million years. Volcanic rocks like these are typical of rift valleys, including the Rhine Graben in Germany and the East African Rift Valley in Kenya and Ethiopia. The magmas that produced the igneous rocks were derived by partial melting of the peridotite that forms the upper mantle, which yielded different varieties of basalt as surface lavas, or dolerite in intrusive sills and dykes. Melting of the mantle occurred in response to thinning of the crust, caused by stretching following the compression that had produced the Caledonian Mountains. Crustal thinning caused the underlying mantle to bulge up nearer the surface into a lower-pressure environment, so that the peridotite of the mantle began to melt. Basalt magma is the product of partial melting of peridotite, and large volumes were generated rapidly, in several separate events. The Clyde Plateau Basalt lavas were erupted 340 million years ago over an area of at least 3000 km^2 to a thickness of more than 1000 m over most of the Midland Valley. These lavas now form the horseshoe-shape outcrop around Glasgow, but they also occur farther east, beneath the Upper Carboniferous sedimentary rocks. In the east, volcanic rocks at Arthur's Seat and Calton Hill in Edinburgh (see Figs 5.10, 5.14, 5.17), and the Garleton Hills of East Lothian, are of the same age. The Craiglockhart lavas (Edinburgh) and Kelso Traps (in the Borders) may be slightly older.

Volcanic vents are plentiful in the Clyde Plateau and several of them are located within a north-east–southwest zone 3 km wide and 30 km long, from Fintry to Dumbarton. Several of these vents form well known landmarks, including Dumgoyne Hill and Dumbarton Rock (see Fig. 5.11). Another similar zone occurs along the south side of the Kilsyth Hills. This trend follows the pattern established in the Caledonian orogeny, and the volcanic zones appear to have formed along some inherited zone of weakness in the crust. The vents represent the deeply eroded remains of individual volcanic necks, from which lavas, ash, cinders and bombs were erupted. A very large caldera structure has been mapped around the Meikle Bin and Dungoil area of the Campsies, southeast of Fintry. This has a circular ring fault 2 km in diameter centred on the Waterhead locality. The central part of the caldera collapsed by 100 m into the underlying magma chamber. Cinder cones built up around the volcano, and massive lavas poured out from the central vent to cover the surrounding landscape. Many of these lavas are rich in secondary minerals, particularly zeolites (hydrated sodium–aluminium silicates), which filled cavities originally formed by gas bubbles in the lava. The pipes, which brought the lavas to the surface, are in many cases funnel or trumpet shape, tapering down at depth. They were forced into the sedimentary rocks by a process of volcanic drilling under pressure,

Figure 5.17 Edinburgh, Holyrood Park: Samson's Ribs, columnar jointing in a dolerite intrusion, part of Arthur's Seat volcano.

from the upper mantle. Dungoil is one of the largest such neck intrusions now exposed in the Campsie Fells. About 1 km beneath the surface lies an igneous intrusion representing a solidified magma chamber, which fed the central volcano; it is 8 km in diameter and has a volume of 300 km^3. The caldera feature has been identified only by careful mapping of the ground, and there is no landscape feature similar to what can be seen in modern calderas, such as the one around Mount Teide on Tenerife in the Canary Islands.

Some of the volcanic rocks, particularly those in the east, are thought to have been erupted into lagoons or a shallow sea, so that water would have entered the volcanic necks and caused massive violent explosions driven by steam. As a result, much of the magma became broken into fragments and was carried up by volcanic gases and steam to form ash cones. An analogy might be the island of Surtsey in Iceland, which erupted from the Atlantic sea floor in 1963. Ring-shape structures frequently dip towards the centre of these volcanic necks, indicating that ash layers collapsed into the volcanic funnels. In east Fife, there are about 100 volcanic vents, including many around the coast, a feature that has contributed to the triangular shape of the Fife peninsula, because these igneous rocks are more resistant than the Carboniferous sedimentary rocks into which they were intruded. Fife in the Carboniferous was probably one of the most volcanically active areas that Europe has ever seen. As in the Campsies in the west, several of the Fife necks are strung out along northeast–southwest faults, including the Ardross Fault, from Elie Ness to Fife Ness. The Elie

Figure 5.16 Edinburgh, Arthur's Seat volcano: irregular blocks of agglomerate in Lion's Head vent.

neck is particularly interesting: the bedded tuff (black volcanic ash containing crystal fragments of pyroxene and hornblende) contains many small tangerine pyrope garnet crystals, the so-called Elie rubies. Garnet forms at high pressure in the Earth's mantle, and these volcanic pipes brought deep-seated fragments from the top of the mantle and base of the crust to the surface during eruption, thereby giving us some insight into the composition of deeper layers that are not amenable to sampling by drilling (the deepest drill hole in the world in Russia reached only 12 km). Pressure conditions needed to form the garnet exist at depths of more than 150 km. Pyrope garnet is a magnesium–aluminium silicate rich in titanium, whereas true ruby is aluminium oxide, a member of the corundum family of minerals. The Partan Craig vent at North Berwick (East Lothian) contains material from the very base of the crust: quartz–garnet–feldspar granulite and pyroxene granulite. These high-pressure metamorphic rocks form part of the buried basement of the Midland Valley. From the mineral content, it is possible to estimate the formation temperature as about 850°C and pressure equivalent to depths of some 35 km. Such blocks of foreign rock carried up by volcanic eruptions are xenoliths (Greek: foreign rock).

Vents in Fife were active for a long period during the Carboniferous, and continued up to the early Permian, about 300 million years ago. Prominent landmarks include the Lomond Hills and the Rock and Spindle near St Andrews (Fig. 5.18). Intrusive sills and east–west dykes constitute another important feature of Carboniferous igneous activity in the region. As far as sills are concerned, the Midland Valley Sill is the most prominent, covering an area of 1600 km² in the east of Scotland. Outcrops of the 150 m-thick body form important landmarks at Stirling Castle and North Queensferry. The northern piers of the Forth Rail Bridge have their foundations on this dolerite (see Fig. 9.11). Well formed columnar cooling joints are everywhere visible. The sill was intruded into very thick sedimentary rocks towards the end of the Carboniferous, about 300 million years ago, and appears to have flowed down dip into a major syncline, so as to fill the core of the fold, rather like water pouring into a saucer. In Fife, the twin volcanic necks of East Lomond and West Lomond intrude the horizontal Midland Valley Sill, creating a distinctive landscape

Figure 5.18 St Andrews, Fife: Rock and Spindle: radiating cooling joints in a cylindrical dolerite intrusion cutting volcanic ash.

feature, visible from Edinburgh, 30 km away. The Carboniferous sills and dykes in the Midland Valley have been an important source of raw material for use as roadstone and concrete aggregate. About 10 million tonnes per year is quarried, and future reserves are considerable. Once, the dolerite (or whinstone, to use the quarrying term) was shaped by hand to make cobbles for roads, but now the material is blasted and crushed by machine into angular chips. Dolerite is a particularly useful roadstone, because it bonds well with asphalt to form a strong skid-resistant surface. The internal random needle-shape feldspar crystals are often intimately intergrown with larger pyroxene crystals to form a robust three-dimensional structure devoid of mineral orientations that could cause inherent fractures to form. Transport costs are low, since much of the demand is local, within the Central Belt.

Eigg: marine erosion has created this natural arch in Jurassic sandstones; the headland in the middle distance is Blàr Mór, north of Bay of Laig, made of Tertiary basalt lava flows.

CHAPTER 6
The North Sea and the Inner Hebrides

Introduction

Since the deposition of the Carboniferous sediments in lagoons, shallow seas, river deltas and swamps, most of the Scottish mainland has remained a land area for the past 250 million years. The collision between the northern and southern continents led to the formation of the Hercynian (or Variscan) orogeny and to the creation of the supercontinent of Pangaea, consisting of all the continents in one giant structure. Pangaea was arranged almost symmetrically as a north–south landmass across the Equator, but did not stretch as far as the poles. Glaciers in the southern hemisphere towards the end of the Carboniferous period began to melt, and the poles became free of ice. On the eastern side of Pangaea was an embayment of the worldwide ocean or Panthalassa (Greek: the whole sea). This broad inlet was called the Tethys Sea, and its final remnant today is the Mediterranean, which now has no connection with the Pacific Ocean.

The creation of Pangaea had profound effects on life, and the reasons are not hard to find. First, the shelf seas, which contain many important habitats for life, were reduced in number and area, once all the separate continents had collided. Secondly, the sharp reduction in plate-tectonic activity meant that sea levels were lowered because there were very few mid-ocean ridges rising up as broad structures from the ocean floor. Finally, the configuration of Pangaea and the absence of polar icecaps would have had an impact on global climate, because ocean currents and wind patterns would have been very different from those of today.

During the Permian period, Scotland lay just north of the Equator, in the middle of a large continent, and the climate was hot and dry. By the start of the Triassic period, 250 million years ago, the country lay at 15° north, and by the end of the period, 40 million years later, continental drift brought Scotland to a latitude of 30° north.

New Red Sandstone

Because the sea was never present in Scotland during the Permian and Triassic, and rainfall was low and sporadic, fossils are generally lacking in the rocks formed in that time interval, most fossils being found in marine sedimentary rocks. The few deposits are mostly bright red dune-bedded sandstones, laid down in valleys at the foot of the Caledonian Mountains. Occasional sedimentary breccias with angular rock fragments represent the deposits of flash floods that carried scree down steep mountain slopes into and along desert valleys or wadis, a situation found today in the Middle East. Dating of unfossiliferous rocks is very difficult; in Scotland the Permian and Triassic are usually grouped together as the New Red Sandstone, since a definite Permian or Triassic age often cannot be deduced. Outcrops are found in Southwest Scotland in the Solway Firth, Dumfries, Stranraer, Lochmaben, Annandale, Thornhill, Mauchline and Arran basins (see Fig. 2.7), on Mull and near Stornoway, and as a narrow strip on the south side of the Moray Firth near Elgin (Figs 6.1, 6.2). The Hopeman Sandstone (as it is known in Moray) has been widely used as for building in northeast Scotland (e.g. the cloisters at Elphinstone Hall, University of Aberdeen), and for the Museum of Scotland in Edinburgh (Fig. 6.3). The Moray Firth contains very thick New Red Sandstone deposits. In the North Sea, the Hopeman Sandstone is an important oil reservoir rock.

On the Moray Firth coast, the Old Red Sandstone and New Red Sandstone rocks are in close proximity, although separated in time by 150 million years. In adjacent quarries, the rocks can look identical, and it was only the discovery of fossil reptiles at Hopeman in 1893 that finally proved that the Hopeman Sandstone was New Red Sandstone (Permian) and not Old Red Sandstone (Devonian). During the period when the Clashach quarries near Hopeman were reopened to provide stone for the museum (Fig. 6.3), new finds were made of reptile skeletal remains, including a skull, and trails of footprints across the fossil sand dunes. Large-scale bedding structures are an indication of these dunes (Fig. 6.4). Well rounded sand grains and wind-sculpted pebbles also indicate desert conditions on land. To the east of Scotland during this time, an arm of a shallow sea originating from the site of present-day Germany encroached on what is now the North Sea. During the Permian, the North Sea was a subsiding basin, partially filled with a shallow sea.

High temperatures meant that the water in this sea was constantly being evaporated, and therefore thick

Figure 6.1 Brodick, Isle of Arran: Permian dune-bedded red sandstone; Holy Isle (Tertiary) is in the distance.

Figure 6.2 Hopeman, Morayshire: Hopeman Sandstone, Triassic; showing disrupted bedding, picked out by weathering.

Figure 6.3 Edinburgh, Museum of Scotland, made of Hopeman Sandstone, showing natural colour variations.

salt, sodium chloride), gypsum and anhydrite (calcium sulphate), and potassium salts – are economically important in their own right, especially in northern England and Germany, and in the formation of salt-dome oil traps. High pressure attributable to the depth of burial of the salt, combined with earth movements, cause salt to move upwards, as it is very light compared to other rocks. Domes, mushrooms, plugs and sheets are buoyed up and move through overlying rocks, causing folding and faulting (Fig. 6.5), and behave in ways similar to igneous intrusions, although salt does not cause heating of any surrounding beds. Salt recrystallizes as it moves, and salt bodies can physically penetrate other rocks, but fractures in the salt are healed by crystal growth. Thick beds of salt can become detached from underlying rocks during earth movements by a process of gliding, which can cause enormous complications to the structure of oil and gas deposits. As salt is impervious, any oil and gas in surrounding reservoir rocks (usually sandstone) can be trapped beneath salt-dome structures, or when salt beds form a cap. Salt has particular seismic characteristics and is detected relatively easily during seismic exploration work; if the appropriate source and reservoir rocks are present, then salt domes can present promising targets as oil and gas traps.

The edge of the Zechstein Sea appears to have lapped onto the Moray Firth coast. Evidence for this is found in the Hopeman Sandstone, where contortions

salt deposits built up. Ocean water must have been able to seep into the sea from time to time, since over 1500 m of salt accumulated in what is referred to as the Zechstein Sea. These chemical sediments – halite (rock

Figure 6.4 Hopeman, Moray Firth: dune bedding in Hopeman Sandstone (late Permian to early Triassic); the broad open structure is the flank of a fossil sand dune.

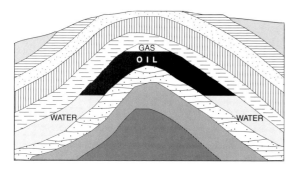

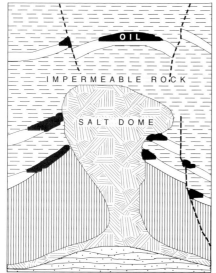

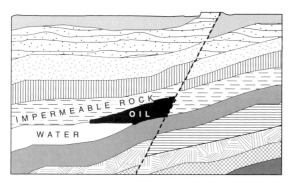

Figure 6.5 Various types of oil traps.

thickest (1200 m) sequences on shore is at Stornoway, where the red sandstones and pebbly conglomerates have at various times been referred to as Torridonian sandstone, Old Red Sandstone and New Red Sandstone. Although New Red Sandstone rocks (sometimes also referred to on maps as Permo-Triassic) are quite thin on land, offshore basins contain very thick deposits in fault-bounded troughs, including the Minch and the Moray Firth. These basins have over 2000 m of New Red Sandstone conglomerates and sandstones, derived from the Highlands and the Outer Hebrides. Rivers brought sediments into the subsiding basins from the upland areas and deposited sand and pebbles in floodplains and braided stream beds on low ground (Figs 6.6, 6.7).

Tropical seas of the Jurassic

The supercontinent of Pangaea existed for almost 200 million years from the time of the Variscan (Hercynian) orogeny, but it began to split apart not long after it formed. Rifts appeared, mainly in the south to begin with, and segments of Pangaea drifted away from one another. New oceans formed, including the Atlantic, and new mountain chains emerged at the sites of continental collision zones and ocean/continent trenches. These eventually formed the young mountain belts of

Figure 6.6 Burghead, Moray Firth: Triassic river channel (at top) cutting sandstone and conglomerate of the Burghead Beds.

and deformation structures caused dune bedding to be destroyed (see Fig. 6.2). Sea water rapidly moved in and expelled the air trapped between sand grains, causing disruption. In some places, the dunes themselves are perfectly preserved, suggesting that they were drowned almost instantaneously. One of the

Figure 6.7 **(a)** Triassic conglomerate (horizontal, at top) lying unconformably on Moine flagstones (on left, dipping to the right); **(b)** close-up of the conglomerate, showing rough horizontal bedding; Ardmeanach Peninsula, western Mull.

the modern Earth, including the Alps, Himalayas and Andes.

One consequence of this rifting had important economic implications for Scotland: the North Sea rifts formed and, as the rift valleys subsided, they were filled with young organic-rich sediments that were to be the source for the future oil and gas deposits. Stretching of the crust caused the North Sea floor to develop a Y-shape pattern of rift valleys or graben structures: the Viking Graben between Shetland and Norway, the Central Graben between Scotland and Denmark, and the Moray Firth Basin (Fig. 6.8). The graben structures have an orientation that was inherited from the time of collision between Laurentia, Baltica and Avalonia at the end of the Caledonian orogeny, some 400 million years ago. Sutures marking continental collision zones were later re-activated and formed fault-bounded troughs that were filled with thick sedimentary deposits, when the grabens sank

down to accommodate the sand and silt brought in from the Highlands and from Scandinavia.

During the middle of the Jurassic period, the gradual subsidence of the North Sea Basin was interrupted by a volcanic episode about 170 million years ago that lasted for about only 5 million years. This was caused by the intrusion of a plume of magma from the upper mantle, which domed up the crust and eventually ruptured through to form volcanoes and the Y-shape rift system. Basalt lavas poured from the volcanic highlands, located at the fork of the Y, at the northeast extension of the Caledonian Iapetus Suture – another indication of the role played by older weaknesses in the crust in controlling subsequent events. Rivers flowed northwards along the floors of the newly formed rift valleys, and a large delta built up at the north end of the Viking Graben. The resulting Brent Sandstone that was deposited in this, the Brent Delta, is an important reservoir in the Brent Field of the

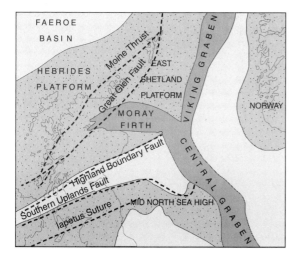

Figure 6.8 Sketch map showing graben structures in North Sea.

sank to the bottom, where they were partially decomposed by bacteria that could live in water that lacked oxygen and light. Occasionally, these bottom waters would suddenly overturn and mix with surface waters, thus introducing a rich organic soup, which in turn encouraged still more algal blooms to flourish in the surface waters, and the increased productivity led to yet more organic sediment piling up on the sea bed. These upwellings of bottom water may have been triggered by movements on faults cutting the sea floor, which also caused slumping of sands from the continental shelf into deeper water, down the continental slope. All in all, considerable thicknesses of organic clays (source rocks for oil and gas) built up quickly. These were buried by porous sands, which eventually became the sandstone reservoir rocks.

North Sea. During the late Jurassic, 150 million years ago, the Kimmeridge Clay was deposited in the graben structures. This organic-rich mudrock was subsequently cooked at depths of 3–4 km, to form oil. At this time, the North Sea was located at about 35° north of the Equator, and most of Scotland was land extending eastwards in the middle of the North Sea (the Forties High and Mid-North Sea High), with the deep Kimmeridge Sea beyond, to the north. Warm climatic conditions and high nutrient levels favoured abundant algal activity in the upper waters of the sea, and a constant "rain" of dead planktonic organisms

At the end of the Jurassic period and into the early Cretaceous (140–130 million years ago), stretching of the crust in the North Sea caused faults to develop, which moved in a way that created tilted blocks of crust. These fault blocks rotated like a hinge, so that the edge farthest from the fault was dropped down deeper than rocks close to the hinge. Such activity was responsible for the formation of structural oil traps. Fault movements meant that rocks of quite different ages were juxtaposed, and the Kimmeridge Clay source rock found itself adjacent to reservoir sandstones and limestones older and younger than itself.

The overall stretching of the crust during the

Figure 6.9 Eigg, Inner Hebrides: Laig beach, Jurassic sediments in foreground (beneath cliffs), looking to Tertiary basalt cliffs and Sgurr of Eigg Tertiary pitchstone lava flow above.

Cretaceous and Tertiary periods, in response to tension set up by the continued break-up of Pangaea, caused the Atlantic Ocean to open to the west of Scotland, which had been adjacent to North America and Greenland. This was marked by the start of seafloor spreading in the North Atlantic. Iceland appeared relatively recently, some 25 million years ago, in the middle of the Atlantic Ocean as a result of a hotspot rising from the top of the mantle, to produce abundant basaltic lava, which forms the Atlantic Ocean floor. This event is covered in detail in the Chapter 7, but suffice it to say here that the intrusion of the magma caused an overall tilt of the Scottish landmass to the east, and much sediment from Precambrian to Cretaceous age was stripped off from the mountains in western Scotland and transported to the North Sea by large east-flowing rivers.

While the North Sea was a marine basin during the Jurassic period, fringed by river deltas and coastal swamps that eventually formed coal seams (including the Brora coal in Sutherland), an arm of the shallow tropical sea encroached on the west coast of Scotland. The climate was warm and humid. Global sea level was relatively high, and Scotland was located at 40° north of the Equator. Jurassic limestone, sandstone, shale and ironstone are found in rhythmic alternations around the coasts of the Inner Hebridean islands, principally Skye, Raasay, Mull and the nearby Ardnamurchan Peninsula. Small outcrops are scattered around Rum, Eigg and Arran (Figs 6.9–6.12). The best outcrops are at Staffin on Skye, where the marine sediments are rich in fossil ammonites (see Ch. 1), which are important in dating the rocks and allow correlations to be made with the narrow strip of thin Jurassic rocks in the east, on the shores of the Moray Firth at Brora, Helmsdale, Easter Ross and Lossiemouth. Skye has an almost complete succession of Jurassic rocks, formed 210–150 million years ago (Fig. 6.10). Jurassic sedimentary rocks off the west coast of Scotland were deposited in two fault-bounded basins, which controlled deposition in the sense that the normal faults were active during the period in which sediment was being brought in from the Highlands. This resulted in thicknesses of 2500 m being formed, much greater than the 500 m of Jurassic sediments exposed on Skye. A northern basin was bounded by the northeast–southwest Minch Fault and to the south by a ridge of Precambrian basement

Figure 6.10 Kilt Rock, Staffin, Skye: Jurassic sandstones and shales, cut by a Tertiary sill at top of cliff, showing vertical cooling joints.

rocks (on Skye–Rum–Tiree–Coll). To the south lay the Inner Hebrides basin, bounded to the northwest by the Camasunary–Skerryvore Fault and to the southeast by the Great Glen Fault–Colonsay Fault. These faults were all initiated at the end of the Caledonian orogeny; during the later stretching of the crust they were re-activated and became normal faults, bounding the younger sedimentary basins and exerting active control on deposition.

On Skye and Raasay, the Jurassic rocks are alternating marine shales and sandstones, with occasional limestone beds. Ironstone on Raasay was worked for iron ore from 1914 to 1920. The ore is siderite (iron carbonate) and chamosite (a complex iron–magnesium mineral belonging to the chlorite group of silicates, with a sheet-like structure and related to clay minerals). In detail, the iron forms tiny rounded egg-shape particles called ooliths, which probably came out of solution when fresh water containing dissolved iron flowed into the warm shallow Jurassic sea, 190 million years ago. Sedimentation rates were very low, and only thin clay deposits formed extremely slowly. The iron content of the ironstone is about 25 per cent; this is considered to be a low-grade ore, and the 2.5 m-thick bed has probable reserves (estimated in 1940) of 10 million tonnes. During the lifetime of the mine, about 250 000 tonnes of ore were extracted from underground workings.

Throughout the Jurassic period and into the Cretaceous, there was an overall rise in world sea level, related to increased ocean-ridge activity as Pangaea broke up and new oceans formed – principally the Atlantic and Pacific. Locally, in the Hebrides and North Sea Basins, sea level fluctuated considerably, as a result of tectonic activity associated with fault movements on adjacent crustal blocks, which tilted and sank, thereby affecting sedimentation in a major way. In addition, the global climate became warmer and drier. Corals flourished in the shallow seas west of Scotland, when sea level was high. During times of lowered sea level, there were usually brackish shallow lagoons in which shales were deposited. The abundant mica in the shales was derived from the weathering of the Moine schist in the Highlands, which, together with the Hebrides, formed upland areas. By the end of the Jurassic period, 150 million years ago, there may have been a connection between the Hebridean region and the North Sea, with a shallow sea flooding the Midland Valley. Deepwater conditions prevailed in the North Sea rifts. Rifting in the North Sea stopped at the end of the Jurassic period, and the continental crust there did not rupture and split apart. The volcanic activity related to the intrusion of a plume of magma from the mantle ceased, and the bulge in the Earth's crust that resulted then subsided. The crust relaxed and the floor of the North Sea sagged, to create a sedimentary basin in the form of a broad downwarp. Sediment draped over the older structures to form a blanket. The unconformity caused by this event is marked in all the North Sea seismic profiles and geological cross sections. In structural terms, this event is known as thermal subsidence; that is, the crust cooled, shrank and formed a sag in place of the bulge that had been caused by expansion. Rifting, which had begun in the southern part of Pangaea, continued during the Jurassic, and eventually caused the North Atlantic Ocean to open, to the west of Scotland. New oceanic

crust was created and Scotland began to separate from Laurentia. Had the North Sea rifting continued, it is possible that an ocean ridge could have emerged there, and Scotland might then have remained attached to Laurentia.

The famous fossil sea stack at Portgower near Helmsdale in east Sutherland was a remarkable find. This is a strip of red sandstone embedded in Jurassic shales and mudstones. The fragment is a 50 m block of Caithness flagstone (Middle Old Red Sandstone) containing fossil fish, examples of which were collected by Hugh Miller. It was considered at first to have been a sea stack that stood along a former shoreline and which collapsed and fell into the Jurassic sea, so that now the bedding planes are arranged vertically. However, it is now interpreted as a segment of rock that became detached from the Helmsdale Fault as a result of earth movements. Many other blocks occur nearby, and the shore section consists mainly of the Helmsdale Boulder Bed, which formed as a rockfall deposit adjacent to the fault line. This section of the Moray Firth coast is highly instructive in terms of giving us some glimpses of the economically important sections immediately off shore in the North Sea oil and gas province. Although the coastal strip of Jurassic rocks is only 1.5 km wide at most, many of the formations that make up the North Sea floor are present on land, including the Kimmeridge Clay, source rock for most of the oil. Coal beds of Jurassic age give rise to gas. Coals formed when the area was a shallow lagoon that

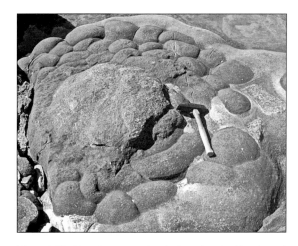

Figure 6.11 Limestone concretions in Jurassic rocks, Eigg, formed by lime-rich algal growths on the sea floor in warm tropical waters.

Figure 6.12 Eigg, Inner Hebrides: the view from the Sgurr to Jurassic sediments beneath Tertiary lava flows.

filled up with drifted tree trunks and plant debris from a nearby lush and dense forest in which *Equisetum* (horsetails), *Ginkgo* and the precursors of modern conifers grew. The lagoon eventually dried up and iron carbonate (siderite) formed cement in the mudstones that were then deposited. Associated claystones were once exploited at Brora for use in the local brickworks, which is where most of the coal was destined. Just off shore, 20 km to the east, is the Beatrice oil field (discovered in 1976), the platforms of which are usually visible from land. The oil is thought to have been derived in part from the Devonian fish beds and from oil shales associated with the Jurassic Brora coal. Deposition of Jurassic sediments in the Moray Firth was mainly controlled by the Helmsdale Fault, which was active then, as seen from the boulder beds and breccias along the fault line, the trace of which effectively controls the present outline of the coast. The present shape of both sides of the Moray Firth coast more or less coincides with the outline of the Jurassic basin. At Helmsdale itself it is possible to see folds in the Jurassic rocks that were the result of fault movements on the Helmsdale Fault after the deposition of the sedimentary rocks (see Fig. 2.12).

Chalk seas of the Cretaceous period

The Cretaceous period of Earth history is named after the Latin word for chalk, and in Britain the White Cliffs of Dover are the most famous example of this rock. On mainland Scotland, outcrops of Cretaceous rocks are rare, although there are abundant rocks of this age off shore in the Sea of the Hebrides and the North Sea, where the chalk forms an important oil reservoir rock. One extremely large block of Cretaceous sandstone that formerly existed at Leavad in Caithness was removed entirely by quarrying. It had been carried on shore from just off Dunbeath by glaciers during the most recent ice age. Sea level continued to rise during Cretaceous time, to such an extent that, by 90 million years ago, levels may have been up to 300 m higher than at present. Much volcanic activity caused very large volumes of carbon dioxide to be pumped into the atmosphere, resulting in greenhouse conditions on Earth. Temperatures were high everywhere and, even quite far north and close to the North Pole, dense forests flourished and coals were formed. Polar icecaps were absent during this period.

From the end of the Jurassic and during the early part of the Cretaceous (approximately 150–100 million years ago), Scotland was entirely above sea level. By about 95 million years ago, global sea-level rise meant that a shallow warm sea existed in the area of the Hebrides. Conditions on land were very arid; no terrestrial sediment was brought into the sea and the deposits that did accumulate were organic sediments derived from carbonate shells of small floating planktonic animals and calcareous algae. On death, their remains disintegrated as they fell to the sea floor, and fine white chalk gradually built up. This was further accentuated by the fact that the plankton was the preferred diet of small crustacean animals. Their faecal pellets contributed to the chalk deposits. Many other animals lived in the chalk seas, including ammonites, brachiopods, crinoids (sea lilies) and echinoids (sea urchins). These also have carbonate shells and their fragmented remains contributed to the build-up of chalk. Around the edge of the sea were some shallow sandy bays. By good luck, evidence for this is preserved at Loch Aline in Morvern, opposite the coast of Mull. Here, a remarkable mine has been operating in 12 m of pure white Cretaceous sandstone since 1941. The rock consists of 99.95 per cent pure silica in the form of round polished grains of quartz that were carried onto the beach by winds blowing off an interior desert. The sandstone has virtually no cement, and extraction is carried out by high-pressure water blasting (see Ch. 9, Fig. 9.14). This raw material is so pure that it can be used unrefined in the manufacture of high-precision optical instruments.

Tiny remnants of Cretaceous rocks are found in other parts of the Hebrides, including Mull, Eigg and Skye. Some of the chalk on Mull is cut by joints filled with desert sand; flints are also present, representing silica nodules formed on the sea bed, flint being a very common constituent of the English Chalk. In the Central granite on Arran (see p. 161), a small outcrop of chalk has been preserved at Pigeon's Cave, having fallen into a caldera around a volcanic intrusion. Chalk, being calcium carbonate, is soluble in rainwater, hence the formation of the cave.

Cretaceous sedimentary rocks are widespread in the North Sea. During early Cretaceous time (140–100 million years ago), shale and some sandstone were deposited, mostly in the deeper parts of the Viking Graben and the Central Graben, to a thickness of 600 m. The sandstones acted as hydrocarbon reservoirs in places. These rocks were laid down directly on top of the Jurassic Kimmeridge Clay source rock. Chalk was deposited to much greater thicknesses (up to 1500 m in the middle of the North Sea) in deep basins during late Cretaceous time (100–75 million years ago). Water depth in these basins may have been over 1000 m, caused by the worldwide rise in sea level. Surface waters are estimated to have been 25°C, and marine life flourished in abundance. Faults that bounded the edges of the deep basins were active, and this caused slumping of the chalk, which became fractured and moved down slope, to be re-deposited in deeper water in central parts of the basins. Movement of the chalk as a slurry of mud meant that the sediment had a more open structure than the carbonate mud that settled out and was not slumped. The re-deposited chalk is a valuable oil-reservoir rock, whereas normally chalk has had the pores sealed by the growth of carbonate minerals during diagenesis, resulting in greatly reduced porosity. Sealing of pore spaces occurs at relatively shallow depth of burial by sediment, about 600 m. Unless oil moves into the chalk soon after deposition, the formation of cement can effectively mean that the rock has little reservoir potential.

Towards the end of the Cretaceous period, 75–65 million years ago, the coastal regions of Scotland and the Midland Valley were probably submerged under a warm shallow sea, while the Highlands and Southern Uplands were islands. West of the Outer Hebrides, volcanic activity began, heralding the much more extensive igneous events of the following period, the Tertiary. Evidence for volcanic activity in the late Cretaceous is provided by the Anton Dohrn seamount (11° west, 57° north), which rises over 1000 m above the sea floor, in the middle of the Rockall Trough. This seamount is a drowned volcanic remnant, whose flat top is covered by Cretaceous chalk, 70 million years old. However, the main phase of volcanic activity to affect western Scotland began some 10 million years later, in the Tertiary, when the North Atlantic finally opened. This is the subject of the next chapter.

Fingal's Cave, Isle of Staffa: Tertiary basalt lava with columnar jointing (photograph by Pat & Angus Macdonald).

CHAPTER 7
Tertiary volcanic rocks

Introduction

Scotland's most recent volcanic rocks, and indeed youngest solid rocks, formed 60–55 million years ago on the west-coast Inner Hebridean islands, principally Skye, Mull, Rum, Canna, Eigg, Muck and Arran, as well as on the Ardnamurchan Peninsula and, west of Lewis, the St Kilda archipelago. These localities form part of the extensive North Atlantic Tertiary Igneous Province, which stretches for some 2000 km from the west coast of Greenland to the British Isles. Many volcanic centres have also been found during oil exploration in offshore areas to the west and north of Scotland (Fig. 7.1). Ailsa Craig is the southernmost outcrop of rocks of this age in Scotland, but the province also includes the Antrim lavas, the Mourne Mountains and Slieve Gullion in Ireland, and Lundy Island in the Bristol Channel.

Volcanic activity in Scotland at this time was related to the opening of the North Atlantic Ocean by seafloor spreading. Rifting of the Pangaea supercontinent moved into its final stages, and North America, Greenland, Britain and Scandinavia separated during this break-up. At the same time, Scotland drifted from a latitude of 45°N at the end of the Cretaceous period to 50°N then 55°N through the Tertiary (65–2 million years ago), eventually reaching its present position. During the early part of this period, the climate in Scotland was considerably warmer and wetter than it is today, and much rock was removed by erosion following intense periods of deep chemical and physical weathering. Some of the results of this episode still

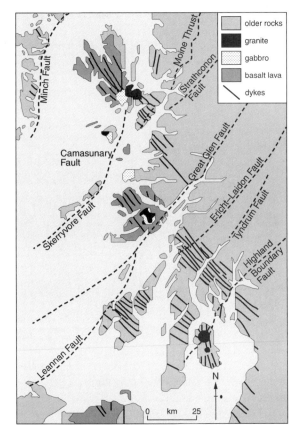

Figure 7.1 The Tertiary volcanic district of western Scotland.

remain, such as the tor formations on the granite summits of Bennachie (near Aberdeen), the Cairngorm Plateau (see Fig. 8.3) and the extensive planation surfaces of the Highlands. These rare landforms are now protected as Sites of Special Scientific Interest. The ice later removed most other tors and remnants of

Figure 7.2 Tertiary lava flows and trap topography: the Ardmeanach Peninsula on the west coast of Mull. The Ben More Central Volcanic Complex is visible in the background (right).

Tertiary landforms, especially those on the wetter west of the country.

The reason for the volcanic activity was the intrusion 62 million years ago of a large plume of magma from the upper mantle into the base of the crust beneath Greenland and the Faeroe Islands. This plume of high-temperature magma heated and expanded the crust, causing a large bulge to form in what was about to become the North Atlantic region. Molten magma then escaped to the surface as basalt lava flows, and eventually the bulge completely ruptured 55 million years ago. Seafloor spreading then caused Greenland to split away from Europe and, as part of this event, the Faeroe Islands, the Inner Hebrides and later Iceland (at 25–15 million years ago) were born. Today, Iceland sits astride the Mid-Atlantic Ridge, above a hotspot responsible for continuing volcanic activity. Iceland is rather like an upside-down saucer with a crack across the middle, along which lava escapes.

In Scotland, the volcanic activity began 62 million years ago with the eruption of the Eigg lavas, which were poured over a landscape of Jurassic and Cretaceous sedimentary rocks. It is thanks to the relative toughness of the lavas that the older sedimentary rocks beneath them are preserved at all. At 58 million years ago, there were major eruptions from volcanoes on Skye, followed at 55 million years ago by activity on Mull. Lavas are most extensive around Skye, Mull and Antrim; they filled the older sedimentary basins containing Jurassic rocks. Eruptions appear to have been shortlived events of possibly less than 200000 years, during which many flows were poured over the surface from vertical fissures, which are now preserved as dykes. Lavas tend to be about 5 m thick on average and the total pile is about 2 km thick. Basalt lava has a relatively low viscosity and can flow far across the surface, and this is then known as flood basalt or plateau basalt.

Weathering of the lava piles has created the highly distinctive trap landscape feature of the Inner Hebrides, conspicuous in northern Skye, western Mull and on Eigg (Figs 7.2–7.5). This trap topography has been referred to earlier in relation to the Clyde Plateau Basalt lavas in the west of the Midland Valley, of Carboniferous age (340 million years old). As with the Carboniferous lavas, the trap feature of the Tertiary basalts is accentuated by the presence of soft red fossil clay soil between individual lava flows. These soil horizons (also known as red bole) represent periods of subtropical weathering of the surface of the lava flows between eruptions. Rain led to deep weathering and leaching of various components that were carried downwards, and the red clay is a residue known as laterite, rich in iron and aluminium.

Forests were able to establish themselves during quiet periods between eruptions, and many of the red boles contain abundant plant remains, which have been used to date the lavas. The fossil plants at Ardtun on the west coast of Mull are famous for the leaves of temperate climate trees – hazel, oak, plane and ginkgo, the maiden-hair tree, together with magnolia, sequoia and the remains of some insects – exquisitely preserved in a temporary stagnant lake, surrounded by marshes. Thin coal seams are also present on Skye and Mull, and the poor-quality coal (lignite) was used for domestic purposes from the early eighteenth century. Mull has the most famous example of Tertiary plants, in the form of MacCulloch's Tree, which occurs at the

Figure 7.3 Horizontal basalt lava flows at Gribun, western Mull; deep clefts are weathered joints.

Figure 7.4 Basalt lavas, showing columnar joints beneath an agglomerate plug; Ardtun, west coast of Mull.

Burg on the Ardmeanach Peninsula at the Wilderness centre. In 1819 Dr John MacCulloch discovered it and described it in his famous book, *A description of the western islands of Scotland.* This tree, a member of the cypress family, is still in its upright position, indicating that it was rapidly overwhelmed by very liquid basalt. Igneous materials from the lava have replaced most of the upper part of the 12 m-high trunk, but the lower part did preserve fossil wood, now sadly removed by collectors and filled in with cement. Surrounding the tree is fine-grain basalt displaying characteristic columnar jointing. Columns form in lava at right angles to the cooling surfaces and, in the case of horizontal flows, the joints are therefore vertical. In three dimensions the joints form hexagonal columns, because of shrinkage as the uniform lava cools evenly (Fig. 7.6). Probably the most famous examples of basalt columns in the world are Fingal's Cave (on the island of Staffa; see pp. 140–141) and the Giant's Causeway (Antrim). The Kilt Rock, between Portree and Staffin on the Isle of Skye, is an equally impressive landform feature, the result of cooling joints in a horizontal dolerite sill that intrudes Jurassic sandstones (see Fig. 6.10, p. 136).

144

Figure 7.5   Tertiary basalt cliffs, showing landslip features now covered by bracken and heather; Cleadale, Isle of Eigg.

Figure 7.6   Sgurr of Eigg pitchstone lava flow, showing columnar jointing.

Eigg, Muck, Canna, Sanday and Rum

The group of islands lying west of Morar is referred to as the Small Isles (Fig. 7.7). Dominating the skyline along of the western seaboard are the magnificent mountains of Rum (sometimes called the Rum Cuillin; Fig. 7.8) and the extraordinary landmark of the Sgurr of Eigg (Fig. 7.9). Geologically, the islands have much in common and form part of the Tertiary volcanic province. Canna, Sanday, Muck and Eigg are made

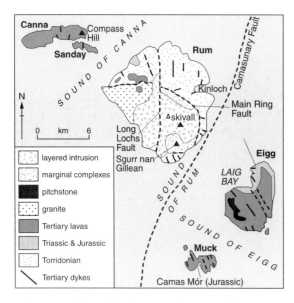

Figure 7.7 Rum and the Small Isles (based on BGS sources).

predominantly of basalt lava flows, whereas Rum is a central complex, intruded into Lewisian gneiss and Torridonian sandstone basement rocks.

Eigg and Muck

Eigg and Muck display the earliest evidence for volcanic activity in the entire Tertiary province. Here, the basalts have been dated at 63 million years old; they

would probably also have extended to Rum, but have since been eroded from there. Plateau basalts are about 500 m thick on the Small Isles, and cover an area of some 50 km², from a total of over 2000 km² for all the lavas seen on land in the Scottish part of the Tertiary province. Antrim has twice this area of lava, and there are also large tracts in the Sea of the Hebrides (1500 km²) that are covered in basalt. It is possible that the Eigg and Muck basalts were erupted originally from the Mull volcano, located to the south, although the main lavas on Mull are rather younger, at 59–60 million years old. As on Mull, the flat-lying basalt flows of Eigg and Muck were poured onto a surface of eroded Jurassic and Cretaceous sandstones, thus forming a shield that protected the sedimentary rocks from erosion. Laig Bay on the west side of Eigg has been carved into these rocks (see Figs 6.7, 7.5). On Muck, little of the older rocks remains, but, at Camas Mór on the south side of the island, lavas lie on top of Jurassic limestone, and both are cut by a very thick gabbro dyke, which caused high-temperature metamorphism of the limestone to produce an attractive multicoloured streaky rock at the contact zone where the magma became contaminated by the sedimentary rocks. This is a classic location on account of the 20 or more different silicate minerals, all rich in calcium, that were produced in the contact metamorphic zone. Muck is a low flat-top island where the land slopes in

Figure 7.8 Rum Cuillin mountains from the summit of the Sgurr of Eigg; basalt lavas in the foreground.

Figure 7.9 Eigg, An Sgurr: pitchstone lava above basalt lava (flows in lower ground).

accordance with the lava flows, which dip gently east-wards away from the escarpment of Beinn Airein.

On Eigg, the two main landscape features are the high basalt cliffs and the Sgurr. At Cleadale the basalts have collapsed because of landslip caused by movement along joints in the lava; blocks then slid over the soft underlying Jurassic shales (see Fig. 7.5). The Sgurr of Eigg is a prominent landscape feature in the south-west of the island. It represents a surface flow of highly acidic lava, now black glassy pitchstone with superbly displayed radiating cooling joints. This lava flow filled a valley excavated into the older basalts (Figs 7.6, 7.9). It is possible that the pitchstone may have originated as a flow of high-temperature ash, to form a rock known as a welded tuff, in which the volcanic material was bonded together during the eruption of the flow. Subsequent erosion removed the surrounding weaker basalt lavas, to produce a posi-tive feature in the form of the very hard and resistant pitchstone. Like Preshal Mhór on Skye, this is an example of inverted topography: what was a valley is now a ridge. The nose of the Sgurr clearly cuts across the older lavas, and at the base of the flow there is a conglomerate of Tertiary igneous fragments and peb-bles of Torridonian sandstone, laid down by a river. Both the Sgurr itself and the long flat ridge forming the tail running north-northwest from the nose create a very impressive feature like an upturned boat that is in sharp distinction to the rugged peaks of nearby Rum (see Fig. 7.8). Dating of the Sgurr pitchstone has shown that, at 53 million years old, it is the youngest of the Hebridean volcanic rocks and is younger than the northwest–southeast regional swarm of basic

dykes. These dykes continue to Rum, where they are cut by the igneous complex of layered gabbro and peridotite. The Sgurr of Eigg was once the subject of controversy. When Sir Archibald Geikie mapped the island in 1897, he concluded that the pitchstone was a lava flow in a former valley. In 1908, Alfred Harker described it as an intrusion. Then in 1914 Sir Edward Bailey came out in support of Geikie, but Harker remained adamant that the supposed valley was not like the flat terraced surfaces of the surrounding basalts and must have had a different origin. The mat-ter was finally resolved only in the 1980s, when the conglomerate at the base of the Sgurr was reinvesti-gated and found to have basalt pebbles present. A more recent question has now arisen with respect to the nature of the flow itself: was it viscous liquid lava or an ashflow? The presence of flattened tongue- or flame-like shards supports the conclusion that it is welded tuff. The only other place where this pitch-stone is found is on the small skerry of Oigh-sgeir (Heyskeir), lying 14 km west of Rum.

Hugh Miller wrote an eloquent description of the Sgurr of Eigg in his travel journal *The cruise of the Betsey* (1858). On this trip, he found a fossil tree at the base of the pitchstone, as well as bones of a fossil fish-eating plesiosaur in Jurassic shales in the north of Eigg. Plesiosaurs (Greek: primitive lizard) were marine reptiles up to 15 m long, with long necks, short bodies, large flippers and short tails. They had long jaws stud-ded with rows of conical teeth and they fed on fish. Abundant and widespread in the Jurassic and Creta-ceous, they became extinct along with the dinosaurs at the end of the Cretaceous, although some would have

us believe that one is still alive and well and living in Loch Ness.

Canna and Sanday

Canna and Sanday are outliers of the Skye lava field, which is magnificently exposed in the cliffs at Compass Hill on the east coast of Canna. An important feature of the lavas here is that they contain beds of river-lain sediments and windblown volcanic ash between the flows. Pebbles of igneous rocks found in the Rum Complex are contained in the conglomerates, which implies that the Rum rocks are older than the Cuillin Complex on Skye. Rum must have been uplifted and eroded (or unroofed) to provide material for the rivers flowing to Canna. It is known that the Skye and Canna lavas were erupted before the intrusion of the Cuillin Complex. Hence, Rum has the oldest basic and ultrabasic complex in the Hebridean province; it was eroded rapidly between 58 and 55 million years ago.

The conglomerates on Canna are the best developed of any in the British Tertiary volcanic province. It was Sir Archibald Geikie who in 1897, in his two-volume masterpiece *The ancient volcanoes of Great Britain*, first described these rocks as being laid down in river channels that had cut down into the lavas. Pebbles in the conglomerates, which are very thick in the sea stacks and cliffs of Compass Hill and nearby Sanday Island, include Lewisian gneiss, Torridonian sandstone, Tertiary lavas and examples of most of the igneous rocks on Rum that had been intruded deep in the crust. Equally impressive are the thick lava flows on Compass Hill, with their very clear columnar cooling joints. Compass Hill is famous for the pronounced magnetic fluctuations caused by the presence of large amounts of magnetite in the basalts; ships' compasses can be affected up to 5 km away.

Rum

Of all the Small Isles, Rum is certainly the most impressive, not only because of the grandeur of the southern mountains, but also because of its special geological interest as a site of world-class importance in the study of igneous intrusions. Igneous activity on Rum began 63 million years ago, with the eruption of basalts, part of the Eigg lava formation covering Eigg and Muck, but now preserved in Rum only in the southeast inside the main ring fault, which separates the igneous complex from the Torridonian sandstone

Figure 7.10 Rum Igneous Complex from Harris, west coast of Rum.

of the north and east of the island (Figs 7.7, 7.8, 7.10).

The Rum Complex is the oldest in the Hebrides and it contains the greatest range of ultrabasic igneous rock types in Britain. It is the eroded top of a magma chamber. It was intruded 60 million years ago into basement rocks of Lewisian gneiss and Torridonian sandstone, which were overlain by Triassic and Jurassic sedimentary rocks, few of which are preserved. Intrusion had the effect of doming the surrounding rocks and elevating them by as much as 2000 m, in the initial stage of the Central Complex. The country rocks then developed a circular fault (the main ring fault) that marked the formation of a caldera structure, which then collapsed into the crust. During this stage, acidic magmas produced rhyolite lava and volcanic breccia at the surface, and granite intrusions at depth, including the Western Granite and the Papadil Granite in the extreme south. The volcanic activity associated with the formation of the caldera was probably explosive, judging from the volcanic breccia (rock shattered by gas blasting and by the collapse of the steep caldera walls) and the banded ash-flows resembling rocks formed from burning clouds

(ignimbrites). The largest outcrops of these rocks are found in the southern mountains, in the Sgurr nan Gillean (764 m) to Ainshval (781 m) ridge, close to the main ring fault. A similar suite of acidic rocks, including volcanic breccias, occurs along the northern margin of the fault. Lewisian gneiss is found in the southern mountains, close to the Sgurr nan Gillean rhyolite lavas; melting of the gneiss may have been partly responsible for the formation of the acidic volcanic rocks. The granitic rocks on Rum are 15 km^2 in area, the largest body being the Western Granite. In relation to the volume of the basic and ultrabasic rocks making up most of southern Rum, the granites are minor. The granites have developed smooth rounded outlines, in contrast with the more rugged Rum Cuillin. Orval is the highest of the granite hills (571 m), and the Western Granite forms the westernmost of the four points of the diamond-shape outline of Rum. The first stage in the formation of the Rum Complex ended with uplift on the main ring fault, and intrusion of many small bodies or plugs of gabbro and peridotite, and a swarm of dykes radiating from the Central Complex. The gabbro dyke at Camas Mór on Muck referred to above (pp. 146–147) belongs to this swarm.

Figure 7.11 Allival from Haskival, Rum: layered ultrabasic igneous rocks.

During stage two of the evolution of the Rum Tertiary Complex, much layered peridotite and gabbro were intruded. These rocks form Hallival and Askival, the main peaks of the Rum Cuillin (Fig. 7.11); weathering and erosion of the layers has produced a distinctive landscape of steps and terraces. Rum is of exceptional geological importance, because it provides a section through the top part of a magma chamber and is relatively accessible. It is a truly world-class intrusion and was first put on the map, so to speak, by Alfred Harker in 1908. The total area of the basic and ultrabasic rocks is 30 km², the largest such body in Britain. Only the Skaergaard intrusion in southeast Greenland (of the same age) comes anywhere near Rum in terms of importance and, although larger, it is much less accessible. Layering in the Skaergaard intrusion was attributed to gravity settling of dense crystals (olivine, pyroxene, iron-ore minerals and chromium-rich minerals) from liquid magma, and this notion stimulated work on the Rum Complex, with a particularly fertile phase of research in the

1980s and 1990s, when several dozen important papers were published. The layered rocks are divided into three intrusions – east, west and central – the last of these cutting the two older intrusions as well as the main ring fault. Layering is present on various scales, from rhythmic units 1 cm thick formed from alternations of olivine, pyroxene and plagioclase feldspar crystals, to much thicker units, 10 m thick or more. Very thin layers (laminae) of the mineral chromite are common in the complex. Chromite (iron–chromium oxide) crystallizes first from the melt, at a temperature of 1200°C and, being very dense, it will often form the bottom of a layer. Erosion and weathering out of chromite laminae and olivine layers has led to the build-up of heavy-mineral deposits on the sea floor, off shore from Rum. There are substantial amounts in Harris Bay in the west and Dibidil in the southeast. These rich natural concentrations are potential ore deposits that could be dredged or vacuumed up.

One quite remarkable feature of the layered rocks is the presence of structures, similar to those found in sedimentary rocks, that indicate slumping, crumpling and sliding, as though crystal layers were being dislodged from the walls of the magma chamber tumbling down to deeper levels or being carried away by currents. Individual layers appear to have been broken up and lie contorted and twisted in a seemingly random jumble, surrounded by large crystals of pyroxene (Fig. 7.12). Many varieties and textures of ultrabasic rocks exist on Rum, and these were given local names, such as allivalite and harrisite. Allivalite consists of olivine and anorthite feldspar (calcium-rich), whereas harrisite is a highly distinctive variety of peridotite in which large lustrous black olivine forms branching crystals that appear to grow across the layering. The composition of layers is quite uniform throughout the Rum Complex, and this appears to suggest that 16 regular pulses of fresh magma of the same composition were injected from below and immediately began to cool and crystallize, with the heavier chromite and olivine forming first, and so making up the base of each layer. Altogether, there are about 1000 m of layered rocks. It has been estimated that all the events associated with the formation of the Rum Complex were completed in less than a million years (i.e. by 59 million years ago). Uplift and erosion

must have been relatively rapid, since the lavas on Skye, Canna and Rum itself (on Bloodstone Hill, Fionchra and Orval) have conglomerates at the base containing pebbles of rocks found in the Rum Central Complex, as well as pebbles of the basement Lewisian and Torridonian rocks. These are river deposits, representing valleys that were eroded before the lavas of Skye were erupted. This makes Rum unique in another sense: it is the only place in Scotland where lavas are younger than a central intrusion.

The Rum lavas are mainly basalts. In places, notably on Bloodstone Hill, they have cavities and fissures filled with banded agate and a type of pale greenish silica (chalcedony) speckled with red jasper flecks – the famous bloodstone, used in prehistoric times as an alternative to flint. Bloodstone arrowheads were fashioned by Rum's first settlers, 8000 years ago. Somewhat later, Queen Victoria was presented with a coffee table made from slabs of polished Rum bloodstone.

It is worth making one final comment on Rum. At Kinloch, the famous castle is made not of the local red Torridonian sandstone but of New Red Sandstone imported from Arran, on account of its more attractive hue, and soil was taken by ship from the Ayrshire coast to create the gardens around the castle. The

Figure 7.12 Rum: cross bedding in layered peridotite.

castle was built in 1897–1901 by Sir George Bullough, a textile magnate from Lancashire. As well as the imposing Edwardian castle, Bullough constructed an incongruous mausoleum in the form of a miniature Greek temple, overlooking Harris Bay on the west coast of Rum. In 1957, the family donated the island and the castle to the Nature Conservancy Council (now Scottish Natural Heritage) and since then it has been a wildlife reserve, where important research into red deer has been conducted, and where the white-tailed sea eagle was re-introduced in 1975 (Bullough had them all destroyed in 1909).

Skye

The Isle of Skye is renowned for its spectacular and varied scenery, a result of the great variety of rock types and ages (Fig. 7.13). At 3000 km², Skye is the largest of the Inner Hebrides. Its three distinctive landform regions are the stepped lava flows in the north and northwest and the sills intruding Jurassic sedimentary rocks along the Trotternish cliffs; the intrusive igneous rocks of central Skye – the Cuillin

Hills and the Red Hills; and the subdued topography of the Sleat (pronounced as "slate") Peninsula in the southeast. Rocks found on the Sleat Peninsula represent the basement into which the Tertiary igneous rocks were intruded, and the area is one of great complexity, because of Caledonian thrusting. In this regard, it closely resembles the situation in Assynt, part of northwest Sutherland (see pp. 44–46). Lewisian gneiss is the oldest rock, and is overlain by two groups of Torridonian rocks, an older Sleat Group (over 3000 m thickness of sandstones and shales laid down by rivers flowing from the west into shallow lakes), followed by 1500 m of red Torridon Group sandstones. Cambrian quartzite and Durness Limestone lie on top of Torridonian rocks, and all these basement units have been affected by three major thrust faults: the Kishorn, Tarskavaig and Moine thrusts. Between the first two thrusts is the Kishorn Thrust sheet; this contains the Sleat Group, which was driven westwards over the basement. Between the Tarskavaig and Moine thrusts is the Tarskavaig Thrust sheet, which represents a block of Moine schist that was carried westwards over the Kishorn sheet. Finally, the Moine Thrust was responsible for transporting Lewisian gneiss over the top of the Moine schist, to create an upside-down stratigraphy in which old rocks lie above younger rocks, and slices of basement rock have become detached and interleaved with younger material and thrust over the basement lying to the west. There is then a gap in the geological record, for no rocks now exist for the 250-million-year interval between the deposition of the Durness Limestone (Ordovician, 470 million years old) and the Triassic sandstones and conglomerates (220 million years old) found in southeastern Skye. Jurassic rocks lie above the Triassic, and these sedimentary rocks form important outcrops on Skye, around Broadford Bay, which is excavated into soft shales, and in Trotternish, where the sandstones and shales are riddled with igneous sills that were intruded along weak bedding planes. Skye has an almost complete succession of Jurassic rocks, covering the interval 205–145 million years ago (see p. 136).

The northern and western peninsulas (Duirnish, Waternish and Trotternish), and indeed most of central Skye, are made of basalt, the earliest of the Tertiary volcanic rocks, erupted 58 million years ago. The flat

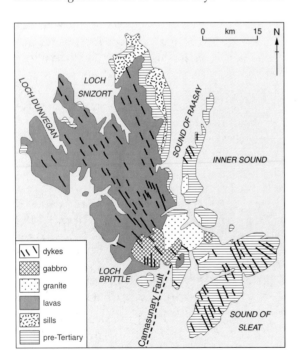

Figure 7.13 Skye (based on BGS sources).

stepped plateau features, typified by MacLeod's Tables in Duirnish, are part of the Skye basalts, tuffs (ashfalls) and thin sedimentary rock horizons. Differences in hardness have been picked out by erosion, to produce the characteristic stepped landscape. Volcanic activity began with an explosive phase, when ash and basalt lava bombs were erupted over a surface of eroded Jurassic sedimentary rocks. Ash fell into shallow lakes, as indicated by the presence of fossil tree leaves in shales among ash beds. Hot springs around the early volcanoes caused some of the ash to be converted to clay. Pillow lavas are a further indication of eruption into water. The main lava pile of northern Skye was over 1000 m thick. Individual lava flows are 10 m thick on average, usually with rubble at the base and gas cavities at the top. Often, the centres of lava flows, particularly the thicker ones, show vertical columnar cooling joints. Gas bubbles floated to the top of molten lava, and eventually the cavities were filled with minerals – usually carbonates and silicates, especially the zeolite family. These are soft, pale cream, yellow and green minerals. Because of the rather flattened shape of the cavities, lavas bearing these minerals are usually referred to as amygdaloidal (Greek, amygdale: almond shape). The lavas were probably erupted quite rapidly, from deep fissures that tapped magma from the upper mantle. After the initial explosive phase, the main lava pile was quiet.

Sedimentary rocks between lava flows indicate that there were periods when erosion and deposition took place during quiet intervals between eruptions, allowing time for soils to form and plants to colonize the area. Rivers from Rum, where the inner core of a volcanic intrusion was exposed to weathering and erosion, transported rounded fragments of igneous rocks in conglomerates. Valleys that had been eroded in the early-formed lavas were later filled when other eruptions took place. Sometimes the valleys themselves were eroded subsequently, leaving the lava fillings now standing proud. This is termed inversion of relief: valleys become ridges. One example is Preshal Mhór near Talisker on the Minginish Peninsula on the west coast. The Preshal Mhór lava, erupted 55 million years ago, has a composition similar to the ocean-floor basalts that had begun to build up the Atlantic during seafloor spreading, 500 km to the northwest of Skye. The Preshal Mhór basalt may represent a solidified

BILL BROOKER

Figure 7.14 The Needle, Quirang, northern Skye, part of an ancient landslip in Tertiary basalts.

lava lake. Shales bearing fossil plants in the thin sedimentary units between the lavas, and similar to the Ardtun leaf beds on Mull, indicate a moist subtropical climate. It has been estimated from the age of the plants that the entire lava pile was poured out over a relatively short timespan of less than 200 000 years.

Northern Skye boasts the best ancient landslides in Britain: the Storr rocks on the Trotternish Peninsula, north of Portree. Basalt lavas, now forming an escarpment, were erupted onto Jurassic shales and clay-rich rocks, which are much weaker than the basalt. Joints in the basalt allowed blocks to become detached from the main body of lava, and to slip and rotate on the softer sedimentary rocks beneath, and eventually slip some way down hill to form the Quirang, the Old Man of Storr, the Needle, and many other highly distinctive landscape features (Fig. 7.14).

Undoubtedly the most spectacular scenic feature on Skye is the Cuillin range of hills, a 12 km-long horseshoe-shape ridge of 30 or more peaks, about

Figure 7.15   The Black Cuillin mountains, Skye, from Applecross (Torridonian in the foreground).

1000 m high, with sharp ridges and deep rocky corries. The Cuillin is made of a series of arcuate igneous intrusions – gabbro (basic) and peridotite (ultrabasic) – that form the base of an eroded volcanic centre. Intrusion of the Central Complex took place after the eruption of basalt lavas at the surface. Because of deep and rapid erosion during Tertiary times, the volcanic root was uplifted and exposed at the surface. Many dykes oriented northwest–southeast cut the gabbro, and weathering of these has given the ridge its characteristic rugged appearance (Figs 7.15, 7.16, 7.18). Intrusions become younger towards the centre of the Cuillin Complex, and many of the rocks show distinctive layering that resulted from dense olivine crystals settling out from the molten magma, followed by layers of pyroxene and feldspar. These minerals have different resistance to weathering, with the result that, in particular, softer olivine weathers out, leaving other crystals standing proud; this provides a firm grip for climbers' boots, even on steep slopes. Layering in the 8 km-wide intrusion dips inwards and increases in steepness towards the centre.

Theories of gravity settling of crystals in layered igneous intrusions were first developed in the 1950s, on the basis of fieldwork in the Tertiary central complexes of Scotland, primarily Skye and Rum. The layered rocks on Skye form an outer and an inner series, with a total thickness in excess of 6000 m. Layered gabbro is clearly visible at Druim Hain, and peridotite at An Garbh-choire. The main Cuillin ridge is an arc joining Sgurr nan Gillean and Gars-bheinn, and an eastern ridge joins Blaven (Bla Bheinn) and Garbh-bheinn. The gabbro and peridotite central intrusions were then cut by many basalt cone sheets, circular at the surface and dipping downwards to a point beneath the main intrusions.

To the north and east of the basic and ultrabasic rocks of the Cuillin lie the Red Hills (Fig. 7.17). These granite intrusions are younger than the Cuillin and they cut sharply across the eastern edge of the Cuillin horseshoe. Scenically, the Red Hills stand out as being very different in shape from the Black Cuillin: they are lower and have smooth rounded shapes,[*] with pink scree forming an apron around the base. Altogether there are three separate granite intrusions, all more or less circular in plan. In terms of age, the oldest is the

---

[*] Feldspar and mica in granite break down readily to clay minerals, a fact that contributes to the landforms being round hills with smooth slopes.

Figure 7.16 Skye, summit of Black Cuillin ridge: frost-shattered gabbro peaks.

Srath na Creitheach centre, consisting of three dome-shape or sheet-like intrusions – Meall Dearg, Ruadh Stac and Blaven. The Western Red Hills centre was intruded later. It is formed of the Glamaig, Marsco and Loch Ainort granites, as well as a gabbro at Marsco and also unusual and distinctive hybrid rocks that resulted from the mixing of basic and acidic magmas (i.e. gabbro and granite) in the molten state. The granites were intruded as ring dykes, following the collapse of a caldera structure (through cauldron subsidence) at the end of an early explosive volcanic event, which produced agglomerate and volcanic breccia. Last to appear was the Eastern Red Hills centre, west of Broadford, which includes an outer

Figure 7.17 Skye, Red Hills: Tertiary granite as rounded summits clothed in scree.

155

Figure 7.18 Glen Brittle, Skye: gabbro hills of the Cuillin ridge.

granite from Ben na Cro to Beinn an Dubhaich and Creag Strollamus, an inner circular granite at Beinn na Caillich–Beinn Dearg, plus various gabbros and hybrid rocks in ring-dyke structures, such as at Kilchrist. The Beinn an Dubhaich granite, dated at 55 million years old, is intruded into Durness Limestone (Cambrian to Ordovician, 550–500 million years old) and contact-metamorphic effects include the formation of skarn, a type of marble resulting from the action of heat and fluids from the granite passing into the highly reactive limestone. Reactions between the igneous rock and dolomite (calcium–magnesium carbonate) produced silicate minerals such as talc, tremolite, diopside and forsterite. The result is a green, yellow and white streaky marble, which is an attractive decorative stone when polished; the Skye marble is quarried at Torrin. Iron ore (as magnetite crystals) is also a product of this contact metamorphism; magnetite is a magnetic variety of iron oxide that forms octahedral crystals. During the early 1940s, the Skye magnetite deposit was investigated as a possible source of iron ore. Part of the Eastern Red Hills Granite Complex includes sills that were intruded along bedding planes in Jurassic sedimentary rocks. These are composite intrusions, made of two distinct parts. Basalt was injected first, followed immediately by felsite, which makes up the middle of an igneous sandwich. The basalt (basic) and felsite (acidic) mixed together at their junction, to form an intermediate rock type. The Rubh' an Eireannaich (Irishman's Point) is the best-known example of a composite sill.

One feature on Skye that is common to all the Tertiary igneous provinces is the dyke swarm. Many northwest–southeast vertical dykes are associated with the Cuillin centre. Weathering of the dykes has contributed to the sawtooth shape of the Cuillin ridge (Fig. 7.18). The Inaccessible Pinnacle on Sgurr Dearg, a vertical wall of rock, is an example of one such dyke, and is well known to climbers. Erosion of the dykes by the sea in the north of Skye has given rise to many of the peninsulas, trending northwest–southeast, and narrow inlets, such as Loch Dunvegan. The eastern edge of the Trotternish Peninsula is characterized by a series of dolerite sills intruded into Jurassic sedimentary rocks, the best known being the Kilt Rock, with its exceptionally well formed cooling joints. Extensions of this set of sills occur to the north and northwest of Skye, as far as the Minch Fault. Occasionally, because of marine erosion, they appear above water as flat-top skerries such as the Ascrib Islands, Flodigarry, Trodday, Fladda and the Shiants. Dolerite sills on the Shiant Islands display very clear evidence that early-formed olivine crystals settled out by gravity, with lower-temperature crystals above, so that the composition of the top of each sill is markedly different from the base.

Geophysical studies of the Central Complex have shown that it can be modelled as a near-vertical cylinder, 20 km in diameter, of very dense gabbro and peridotite, rich in iron–magnesium silicates, and extending down into the crust for at least 15 km, and probably as far as the upper mantle. The granites on the other hand are relatively minor bodies, about 2 km thick and closer to the surface. Basic and ultrabasic rocks formed from very high-temperature magmas (1300°C), and it is possible that the granite bodies originated by partial melting of the Lewisian and Torridonian rocks that existed in the crust at the place of intrusion. In very broad terms, Lewisian gneiss and Torridonian sandstone have overall compositions similar to this granite.

During the volcanic activity, Skye and the other igneous centres in the Hebrides were affected by hot fluids circulating through the lavas, the central complexes, and the surrounding country rocks into which the complexes were intruded. The main source of the water-rich fluids was rainwater, which percolated underground and was heated up by the volcanic rocks. Extensive circulation patterns were set up through cracks, fissures, joints, bedding planes and pores, all part of an interconnected system. Chemical reactions, assisted by heat, produced alteration products. Close to the hot centres, epidote and chlorite formed, whereas farther out in cooler regions, the zeolite family of minerals form the commonest alteration product. Zeolites are also present as cavity fills in bubble holes (or vesicles) in lava flows.

Mull

In addition to Tertiary igneous rocks, Mull boasts an impressive series of older rocks, from the Moine schist, with a wide array of metamorphic minerals, and the Caledonian granite on the Ross of Mull, and in the east around Loch Don and Loch Spelve there are Dalradian slate and limestone, Old Red Sandstone conglomerate, Triassic conglomerates, highly fossiliferous Jurassic sandstones, shales and limestones, Cretaceous chalk with flints, and Tertiary sedimentary rocks (Fig. 7.19). The Ross of Mull granite has been used as a building stone world wide, and is particularly famous for its use in Scottish lighthouses, including the Skerryvore Light, built in 1838–44 by Alan Stevenson, uncle of the author Robert Louis Stevenson.

The igneous rocks of central Mull make up probably the most complex geology anywhere in Britain, with lavas, volcanic vents, cone sheets, ring dykes, plugs, radial dykes of various ages, compositions and

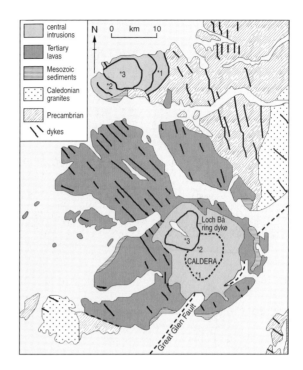

Figure 7.19 Mull and Ardnamurchan (based on BGS sources).

orientations intersecting and overlapping in a bewildering manner (Figs 7.19, 7.20). The original survey was carried out by Sir Edward Bailey, whose 1924 map and memoir have been described as "classic". In general terms, the sequence of events on Mull resembles that on Skye. Basalt lavas were erupted about 55 million years ago to form the plateau features of north and west Mull, which are particularly clear at Ardmeanach (see Figs 7.2–7.4). Staffa (see pp. 140–141) and Ulva formed at the same time as the Mull plateau lavas, which also spilled over onto the Morvern Peninsula on the mainland. Off shore there are connections to the basalts on Eigg and Muck, and flows extend west to Coll, where they are cut by the Caledonian Skerryvore–Camasunary Fault and thus are separated from the Lewisian gneiss of Coll. The main plateau lavas, over 1000 m thick, are covered by another 1000 m of slightly younger basalts in central Mull that were associated with the formation of a caldera. Some of these show pillow structures, indicating eruption into water, possibly a caldera lake. The caldera was about 4 km in diameter, and subsidence caused a collapse of about 1000 m. For the most

157

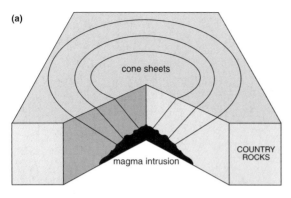

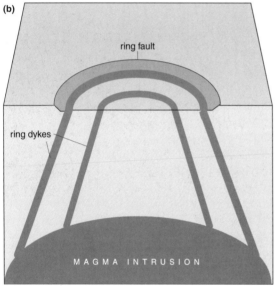

Figure 7.20 Ring dykes and cone sheets.

after the other (Figs 7.19, 7.22). The oldest of these is the Glen More centre, consisting of early granitic rocks and explosive acidic material, followed by basic cone sheets and layered gabbros at Coire Uisg and Ben Buie. An early caldera formed as a result of surface collapse. The Beinn Chaisgidle centre to the north intrudes the first centre. Here too, the first rocks formed from explosive vents around the edge of the caldera that formed earlier. Cone sheets, layered gabbros and ring dykes were then intruded in a complex sequence of events. The third and last centre, located to the northwest of the second centre, was the Loch Bà centre, which formed a late caldera. Compared to the other centres, rather larger volumes of acidic rocks (granite, granophyre and felsite) make up this centre. Cone sheets and ring dykes were intruded in abundance, both basic and acidic. Magma mixing led to the formation of intermediate or hybrid rocks. Activity ended with the intrusion of the Loch Bà ring dyke, a perfect example of such a structure; it is made of pale felsite, a brittle rock with a glassy texture (Fig. 7.22).

Granite in the first centre was derived by partial melting of Lewisian gneiss basement rocks, because of the intrusion of very hot basic magma, whereas in the other centres the granite bodies resulted from the crystallization of the last and lightest fraction of basaltic magma. When basalt begins to cool, denser and more iron–magnesium–calcium-rich silicates form first, at high temperature (over 1100°C). Towards the end of the cooling period, only 5 per cent of low-temperature melt may remain, which on crystallizing will form granite. As on Skye, geophysical measurements on Mull point to the existence of a large dense cylinder of basic and ultrabasic rock, 8 km wide and at least 15–25 km deep, beneath the Central Complex; granites, on the other hand, are at most 2 km thick. Alteration of the rocks in the Central Complex was extensive and resulted from the underground circulation of surface water. Temperature variations in bubble cavities filled with minerals indicate that the Mull lavas originally extended for some 25 km east of the present outcrop. Erosion later in the Tertiary removed most of the lavas, leaving only a remnant around Morvern.

One result of the intrusion of the Central Complex was the formation of concentric folds in the surrounding country rocks: Dalradian limestone and slate, Devonian lava, Triassic conglomerate, Jurassic shale

part, eruptions took place along fissures, without explosive activity. Some of the individual lava flows, which are 10–15 m thick on average, can be traced horizontally for great distances. Columnar cooling joints are common at the base of the pile, especially in southwest Mull (Fig. 7.4), and most famously on Staffa (Fingal's Cave).

As in Skye, pauses in volcanic activity allowed red soils (bole: a type of soil formed under subtropical conditions) to form, on which forests could grow. The Ardtun leaf beds referred to above (p. 143) are famous for providing an insight into the Tertiary climate, as well as giving the age of the basalts above and below. Eruption of lava at the surface was followed immediately by the intrusion of three volcanic centres, one

PAT & ANGUS MACDONALD

Figure 7.21 The Ardnamurchan Central Igneous Complex, showing its concentric ring structure.

and Cretaceous sandstone, which are particularly well exposed around Loch Don and Loch Spelve in southeast Mull. The Great Glen Fault, which reaches Mull from nearby Loch Linnhe, was also diverted to the southeast by the outward pressure caused by doming of the crust during intrusion (see Fig. 7.19). One of the latest events in the igneous activity was the intrusion of several sills, those around Loch Scridain being particularly noteworthy, for they contain sapphire crystals as a result of thermal metamorphism. Sapphire is a blue variety of corundum (aluminium oxide) formed when clay minerals in blocks of shale that were trapped in the magma chamber as xenoliths (Greek: foreign rock) were held at a high temperature (over 1200°C) and then injected as part of the sill complex. The sapphires are not of gem quality: they are full of inclusions of other minerals and, although up to 2 cm long, are very thin and of microscopic size.

Ardnamurchan

The long finger of the Ardnamurchan Peninsula, lying immediately north of Mull, consists of a well exposed and perfectly preserved Tertiary Igneous Complex, considered to be one of the finest in Britain. The igneous rocks were intruded into Moine schists of the Caledonian basement and thin Jurassic sedimentary rocks (Fig. 7.19, 7.21). As in the other Hebridean igneous complexes, activity began with the eruption of basalt lava flows onto a surface of Moine, Triassic and Jurassic rocks. The volcanic episode began about 60 million years ago and was relatively short lived. Most of the activity was centred around Ben Hiant, where there are volcanic vents filled with agglomerate, ash and lava. This sequence is clearly exposed at MacLean's Nose, a headland formed of volcanic breccia produced during an explosive phase of volcanic activity that created large craters. The Ben Hiant vent was then intruded by dolerite and gabbro. Most of the

Figure 7.22 Outcrop of Tertiary ring dyke at Loch Bà, Mull, forming an arcuate outcrop to the right.

plateau lavas on Ardnamurchan were removed by Tertiary erosion, so that only 100 m of lava remains.

Ardnamurchan is most famous for its three intrusive centres, which cut one another in succession, as the focus of activity shifted with time (see Fig. 7.19). Centre 1 includes the Ben Hiant vents and intrusions, and also had its focus near Meall nan Con. Activity then shifted 5 km to the southwest of centre 1, to a focus at Loch Aodann. Many ring dykes and cone sheets were intruded concentrically around centre 2, mostly gabbro and dolerite in composition. The third and youngest centre is located 3 km northeast of centre 2, with a focus near Achnaha. Here, basic and ultrabasic ring dykes surround various more acidic

Figure 7.23 Ben More Central Igneous Complex, Mull, showing basalt lava flows at the top.

rock types. Centre 3 forms the largest and most complete set of ring-shape igneous intrusions in the country. The most prominent feature in Ardnamurchan is the ring dyke known as the Great Eucrite, which forms a high circular ridge in the centre of the peninsula, surrounding the inner part of centre 3. Eucrite is a coarse gabbro, with large crystals of plagioclase feldspar, augite and olivine. The form of the crags around the eucrite makes the influence of geology on topography more obvious on Ardnamurchan than anywhere else in the Tertiary volcanic province. Centre 3 intrusions have very sharp boundaries cutting centre 2 bodies and are therefore clearly younger. Dating of the igneous rocks on Ardnamurchan has shown that the activity, including basalt eruptions and the intrusion of the three centres, lasted less than a million years, and was completed by about 59 million years ago. At centre 3 there are possibly as many as ten or more circular intrusions nested one inside the other, with the youngest bodies at the centre being intermediate rocks, in contrast to the outer bodies, which are all made of gabbro. Despite the amount of attention devoted to Ardnamurchan, an unresolved controversy exists as to the exact form of the centre 3 complex. Rather than consisting of nested ring dykes, the intrusion have a funnel shape. Geophysical surveys indicate that the main body in Ardnamurchan is associated with centre 2, where a cylinder of basic and ultrabasic rocks 12 km in diameter extends vertically downwards for 5 km, which is much less than the Skye centre. Scenically, the Ardnamurchan Complex is the most impressive in terms of the relationship between geology and landscape.

Arran and Ailsa Craig

People are often heard to remark that Arran is "Scotland in miniature". This epithet is in no small measure because Arran's highly contrasting landscape features reflect the varied geology. The northern and Central granites have a Highland appearance; the lower cultivated ground is reminiscent of much of the Midland Valley on the nearby mainland. Arran has a considerable variety of rocks: Dalradian schists north of the Highland Boundary Fault, Highland Border Complex rocks in the fault zone, Devonian (Old Red

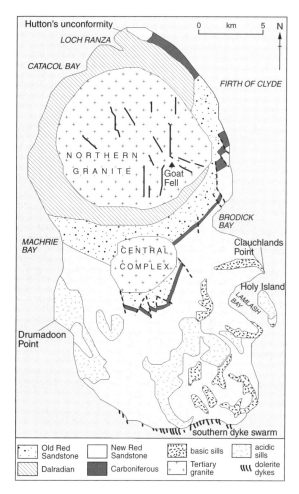

Figure 7.24 Arran (based on BGS sources).

Sandstone), Carboniferous, Permian–Triassic (New Red Sandstone) sedimentary rocks and Tertiary intrusions in abundance (Fig. 7.24). Within the Central granite, small fragments of Jurassic and Cretaceous sedimentary rocks have been preserved, as well as remnants of Tertiary basalt lava. Arran was made famous by James Hutton, who in 1795 first described an unconformity between the top of Old Red Sandstone rocks and the older Dalradian schists in the north of the island, similar in nature and significance to his other finds on the east coast of the mainland at Siccar Point (p. 95).

Tertiary igneous rocks abound on Arran. The most prominent is the Northern Granite, with its serrated peaks, the sharp ridges of Cìr Mhór ("the big comb")

and rock corries (Figs 2.6, 7.25). Goat Fell (874 m) is the highest mountain on the island and it commands impressive views not only of the low ground in New Red Sandstone rocks around Brodick but of much of the Firth of Clyde mainland as well. The Northern Granite is an almost circular body, 10 km in diameter, and is composed of an outer coarse-grain silver-grey granite and an inner and slightly younger fine-grain granite. Intrusion of the outer granite 60 million years ago caused the Dalradian slates, schists, grits and quartzites to be pushed aside and folded as the granite dome made its way up into the crust. The Highland Boundary Fault also appears to have been bent around the intrusion, and the boundary between the granite and surrounding country rocks is vertical and very sharp (Fig. 7.25), so that the granite body resembles a cylinder. The outer granite forms the rugged topography in north Arran, and the inner intrusion gives rise to lower and more rounded hills. Evidence is hard to find for heating effects (i.e. thermal metamorphism) in the country rocks around the granite, suggesting that the granite could have been intruded when it was nearly solid and had almost finished crystallizing. Cavities in the granite are often filled with quartz, feldspar, topaz and beryl.

In the very heart of Arran lies the Central Igneous Complex, a circular intrusion 5 km in diameter containing many rock types, arranged roughly in ring shapes. The Central Complex is younger than the Northern Granite: the dome formed by that granite is clearly cut by the margin of the Central Complex. A dome formed at first, which then collapsed and subsided 1000 m into the central caldera. Blocks of Jurassic shale and sandstone and Cretaceous chalk occur, often in jumbled masses. At least four volcanic cones grew on the floor of the collapsed caldera. These volcanoes produced agglomerate, ash and lavas, ranging in composition from basalt to andesite. Several different intrusions also forced their way in, frequently with arcuate shapes, and including early gabbro (basic) and later granite (acidic) to the east of the ring of eroded volcanic cones, around the margin of the complex. Hybrid igneous rocks are common, an indication that molten acidic and basic magmas mixed together. Arran is almost unique among the Tertiary igneous complexes, because erosion has not removed the entire volcanic edifice, and evidence for

caldera collapse is well preserved in the central ring complex. The other example is at Ben Hiant, Ardnamurchan (see pp. 159–160). On Rum, in contrast, the erosion level is very deep, allowing us to see into a magma chamber that would have fed lavas at the surface, but these have been removed long ago. Geophysical work on Arran – gravity surveys that determine the density of rocks below the surface – indicates that the Arran Complex, like the other Inner Hebridean complexes, in underlain by dense basic rocks, probably gabbro.

The southern half of Arran, south of the Central Complex, consists mainly of many igneous sills intruded into Permian sandstones, and a swarm of northwest–southeast dolerite dykes, particularly well exposed on the south coast, where they form fingers, ribs and reefs running out to sea across the wavecut platform. This is the best example of a dyke swarm anywhere in Britain, and some would claim it is of world importance. In sharp contrast to the dykes on the south coast, those that cut the Northern Granite have weathered out preferentially because of differences in hardness between the granite and the dolerite. The resulting notches and clefts give the ridge of A'Chir its highly distinctive character (Fig. 7.25). Ceum a'Chaillich ("the witch's step") is the most prominent of these notches and is well known to climbers. A stream in the magnificent U-shape glacial valley of Glen Sannox has eroded just such a dyke, to form a narrow gorge. The near-horizontal sills in the red sandstone are mostly dolerite (whinstone), and many display fine vertical cooling joints, including those at Dippin Head and Drumadoon. A sill around Lamlash Bay and Holy Island forms a broken circle, with the edges dipping radially inwards towards Lamlash, suggesting that it is a cone sheet. At Corrygills, the rock making up the sill is a black volcanic glass known as pitchstone, which was used in prehistoric times for arrowheads and stone knives that were exported to all over the west coast of Scotland. At Drumadoon, on the west coast of Arran, are Judd's Dykes, named after John Wesley Judd (1840–1916; Professor of Geology at Imperial College London), who studied them in great detail in 1893. This is a set of composite dykes, that is, they have dolerite (basic rock) at the edges, and pitchstone or felsite (acidic rock) in the interior. The acidic magma was injected

Figure 7.25 Goat Fell, Arran, Northern Granite, from summit; the low ground in the distance is the Carboniferous rocks of Ayrshire.

along the same fissure immediately following the intrusion of dolerite. Intrusion of so many dykes – well over 500 in a few kilometres – caused the crust to be extended laterally by at least 10 per cent.

Ailsa Craig, a landmark in the Firth of Clyde, some 20 km away from Arran, is one of the smallest Tertiary igneous intrusions. It formed about 60 million years ago, close to the start of Tertiary igneous activity. It is described as a boss – the deeply eroded root of a volcanic plug – and is made of fine-grain granite, containing a range of unusual minerals, including riebeckite (a sodium–iron member of the amphibole family of silicates), which give the rock its characteristic bluish colour. Columnar jointing is very prominent around the circular island, as are the many quarries from which the rock has been extracted to manufacture the famous polished curling stones (known as ailsas). Its distinctive composition, almost unique in Scotland, apart from the Rockall granite of the same age, has been useful in tracing the path of the most recent ice sheet across Britain. Boulders of the rock have been found as glacial erratics in parts of England and Wales, and around Cork harbour in Ireland.

St Kilda and Rockall

Lying 65 km west of North Uist is the small archipelago of St Kilda, the deeply eroded fragment of a Tertiary central intrusion. The complex consists of layered gabbro (the Western Gabbro) about 60 million years old, cut by a later granite to the east (the Conachair granite), which has been dated at 55 million years old. Between these two bodies in the centre of Hirta, the main island, is the Mullach Sgar Complex, which shows evidence that basic and acidic magmas existed simultaneously and became intermingled. Gravity surveys around St Kilda indicate the existence of a cone of basic rocks extending down for at least 20 km into the crust. No lavas are found on the islands: erosion has been very intense and any rocks that may once have formed on the surface, such as lava flows, have been completely removed. The St Kilda Complex was probably intruded into Lewisian gneiss of the continental crust. Marine erosion has created the spectacular cliff scenery of the stacks and pinnacles (Fig. 7.26). The seacliffs of Conachair form the highest sheer cliff face in Britain (430 m). St Kilda is a world heritage site on account of its geology, landscape, wildlife and cultural heritage.

Rockall, which rises to a mere 20 m high above sea level, is the last fragment of land west of Scotland,

DONALD PATERSON

Figure 7.26 St Kilda Complex, Boreray: strongly jointed igneous rocks under massive attack by the sea.

before the continental shelf ends. It is made of an unusual variety of granite, containing rare minerals, and has been dated at 52 million years old. The granite appears to have been intruded into older basalt lava and gabbro (65–60 million years old).

Several other large igneous centres lie beneath the North Atlantic, including seamounts. They are mostly circular; they represent the remains of volcanic structures that formed at the end of the Cretaceous period (65 million years ago) and into the early Tertiary (to 52 million years ago), in response to the opening of the Atlantic Ocean. Between the Outer Hebrides and the Faeroe Islands lie the Anton Dohrn, Rosemary Bank, Geikie, Darwin and a few other volcanic centres. North of Shetland are the Erlend and Brendan centres. Closer to the mainland is the Blackstones Bank centre, southwest of Mull (see Fig. 7.19). Scuba divers have explored the Blackstones Bank rocks, which are in shallow water. This complex is made of gabbro, intruded by dykes, dated at 58 million years old.

Tertiary dykes

One of the most prominent and widespread features of the Tertiary volcanic province are the swarms of dykes trending northwest–southeast (Fig. 7.1). Most of these are dolerite and many are likely to have fed fissure eruptions of basalt lava at the surface. Dykes are associated with all the central complexes and they appear to have formed over a long period, before, during and after the intrusion of the complexes. Some swarms radiate out from the centres, but the existence of so many closely spaced centres had an effect on the stress pattern within the crust, so that dykes from one centre may be deflected around another centre. The dykes can be very extensive; for example, the Cleveland dyke on the border with North Yorkshire originated from the Mull magma chamber (dated at 58 million years old) and may have travelled the 370 km along a rapidly propagating fracture in a matter of a day or two. This dyke, 25 m wide, has been extensively quarried for roadstone. Dykes from the Mull centre also extend to the Outer Hebrides and the mainland (Figs 7.1, 7.27). As they cut through all the intrusions and lava flows, it is likely that lavas were being poured out at the surface almost continually somewhere or other in the province during the 12 million years or so of igneous activity. However, each centre appears to have been active for no more than a million years, and the early lavas were erupted in a few thousand years before the central intrusions were forced into place. Dating the sequence of events has been done using fossil plants found in soil horizons between lava flows, as well as known magnetic reversals, together with radiometric dating – primarily argon/argon and potassium/argon methods. An accurate timetable has been established for normal and reverse episodes of the Earth's magnetic field over the past 150 million years. Dyke swarms are orientated perpendicular to a regional northeast–southwest stress field related to crustal thinning caused by the intrusion of the large mantle plume in the North Atlantic region. The location of the main central complexes appears to be related to the presence of northeast–southwest Caledonian faults. Skye and Rum are close to the Camasunary–Skerryvore Fault, and on Rum there is also the north–south Long Loch Fault, which may also have played a role; Ardnamurchan is near the Strathconon Fault; the Great Glen Fault crosses Mull, and the Highland Boundary Fault crosses Arran near the Northern Granite. Farther south, in Antrim, the Slieve Gullion and Mourne

Mountains centres may relate to the southwest extension of the Caledonian Iapetus Suture. Once again, there is evidence of the importance of early-formed structures in controlling later geological events. Magma moved from the upper mantle along lines of weakness that allowed easy access. The Caledonian faults are major structures that penetrate the entire thickness of the crust and would have been intersected by the mantle plume that caused the crust to be domed up, thinned and stretched. Dyke swarms would have found an easy passage, followed by the intrusions. The general pattern, seen on maps of the Tertiary complexes, is of circular ring dykes, cone sheets, plugs and radial dykes, like the spokes of a wheel, formed by pulses of magma meeting cold brittle crust (e.g. Lewisian gneiss). Similar features, reflecting the stress pattern, can be seen in a thick plate-glass window that has been struck by a stone.

One of the most important features associated with the Tertiary volcanic episode was the considerable amount of uplift along the west coast of Scotland. An escarpment of Torridonian sandstone was pushed up to face the Atlantic. Deep erosion in the humid climate led to the Torridonian being worn down rapidly and the products were transported to the Minches and beyond on the west, leaving isolated mountains standing above the Lewisian gneiss basement (e.g. Figs 3.9, 3.10). The entire land surface was uplifted by

Figure 7.27   Tertiary dolerite dyke (dark) cutting Dalradian quartzite, Jura.

at least 500 m and tilted to the east. It was at this time that the river system became established; most of Scotland's main rivers flow east into the North Sea and Moray Firth. Significant thicknesses of Tertiary sediments were deposited in the Central Graben. For the remaining 50 million years of the Tertiary period, relaxation of the crust meant that the North Sea subsided to accommodate the debris carried in by the precursors of the Spey, Tay, Forth and Tweed. Most of the Tertiary sediments are the sand, shale and mud that filled the North Sea Basins. Some was also transported south along the coast as far as Southeast England, where heavy minerals derived from Highland rocks are found in the Tertiary sediments. Impure coals (lignite) are also present in the Moray Firth, the Dornoch and Beauly coals (55 million years old) forming in a stagnant delta that extended eastwards. Ash from the Hebridean volcanoes was carried east by the wind and deposited in the North Sea to form clay. Some of these beds form key marker horizons, important in oil-exploration work. Oil deposits are found in Tertiary sediments at many places in the North Sea and on the Atlantic margin, such as in the Faeroe–Shetland Basin and the Rockall Trough. In 1970 the Forties Field, the first giant oilfield, was discovered in Tertiary sandstones, located east of Peterhead. It contains 500 billion barrels of oil from Kimmeridge Clay source rock in the buried rift valley where the Moray Firth Basin and the Central Graben meet (see p. 134). Other giant fields of this age are the Foinaven and Schiehallion fields on the Atlantic margin, west of Shetland.

Volcanic eruptions in western Scotland coincided 55 million years ago with a period of global warming, but the temperature began to fall steadily thereafter, exacerbated by the northward drift of Scotland by almost 10°, towards its latitude today. Huge amounts of lava were erupted as the North Atlantic opened rapidly to create new ocean floor and, eventually, Iceland. This would have produced sulphate aerosol particles from the sulphurous volcanic gases. Once in the stratosphere, aerosols and fine volcanic dust have an adverse effect on climate (i.e. they contribute to global cooling). Continental drift led to the clustering of large landmasses around the North Pole. The Arctic Ocean became almost an enclosed sea, and warm waters from equatorial latitudes were unable to reach very far north into it. By about 15 million years ago, close to the end of the Tertiary period, icecaps were established in Antarctica and, by 3 million years ago, in Greenland. The appearance of the Northern Hemisphere ice in Greenland, North America, Scandinavia and northern Europe marked the start of the Quaternary period and the beginning of the most recent ice age, 1.8 million years ago.

While Scotland during the Tertiary was an upland area with high volcanic mountains in the northwest, England in complete contrast was low lying with a shallow sea in the southeast, surrounded by dense temperate forests. Alpine movements, centred in the Mediterranean region (creating the Alps and the Himalayas), began to cause crustal warping in southern England and the North Sea from about 25 million years ago. Eventually the North Sea Basins filled up and ceased being places of extremely thick sediment deposition. Mainland Scotland was scarcely affected by the Alpine orogeny, save for movements on some older faults and general overall uplift. Erosion during the Tertiary removed 1–2 km of material from the surface, before the onset of the Ice Age, the main effect of which was to modify the Tertiary landscape by further erosion.

Landscape evolution

The early Tertiary volcanic episode described above began with the eruption of basalt lava across an eroded surface of basement rocks, in response to the opening of the North Atlantic. Scotland lay at the edge of a mantle plume, centred on what is now the east

Figure 7.28 Barns of Bynack, Ben Avon, Cairngorms: remnants of tor formation in granite, formed by weathering along two joint sets.

coast of Greenland. Intrusion of central igneous complexes took place into old, cold, brittle crust and led to the elevation of the land surface. Movement along faults – some of them of ancient origin (the northeast–southwest Caledonian faults and the older northwest–southeast-trending faults), others related to ring fractures around the central intrusions – caused differential movements. Thus, in the area of the Minches, there are several relatively small isolated sedimentary basins filled with Mesozoic rocks (Triassic, Jurassic and Cretaceous), separated by upfaulted basement highs. In some places the intrusions were forced up; elsewhere the lava plateaux were downfaulted. The overall effect was to create an elevated surface, rather higher and steeper in the west, and gently warped and tilted to the east.

As Greenland separated from Europe, Scotland came under the influence of more humid climatic conditions, with important consequences for the evolution of the landscape. The newly elevated surface was immediately acted on by the forces of erosion: mass movement of material down slope, downcutting by rivers, and removal of the products of chemical weathering and erosion, which were spread out on valley floors and carried in the direction of the sea. Valleys became wider rather than deeper, as slopes retreated because of the intense tropical weathering.

Huge volumes of sediment were deposited in the Minches and the North Sea during the Tertiary. On land, remnants of Tertiary conglomerates are found in the Hebrides and in the Buchan district of Aberdeenshire. In the west, these contain lava pebbles and, in the east, reworked flints from the eroded Chalk. Jurassic and Cretaceous rocks occur beneath the protective shield of lava flows on the west coast, and as thin strips around the Moray Firth, but there is no evidence that these sedimentary rocks were more extensive on land, or that they once covered the entire surface and were all removed in the Tertiary, as used to be thought. At most, there may have been arms of the sea along the length of the Great Glen and through low-lying parts of the Midland Valley. Most of the country, though, has been above sea level since the end of the Caledonian orogeny, apart from a brief incursion of the tropical shelf seas of early Carboniferous times into the Midland Valley and the Borders region around Duns (the Merse).

Also, Scotland since then has been in latitudes where the climate was warmer than today, and when weathering and erosion would have been much more intense, in particular during the Tertiary when the climate was more humid. This situation prevailed for almost 50 million years. Evidence for the climate is provided by the lignite deposits (low-grade brown coal) found in the Hebrides between lava flows, and in the Minches and North Sea basins and around Lough Neagh in the north of Ireland, where there are important reserves of lignite 25 million years old. Fossil leaves at Ardtun near Bunessan on Mull are of magnolia and other species that require much warmer conditions than today's climate affords.

Eruption of lava had virtually stopped by 55 million years ago, and then followed an extensive period of deep chemical weathering of the rocks exposed at the surface to produce a deep rotted layer, almost regardless of the underlying rock types. A series of pulsed uplift movements on faults, particularly in the early Tertiary, provoked further downcutting by streams. These pulses were followed by long periods of stability, during which weathering and erosion continued unabated. Valleys were widened by slope retreat, when weathered rock collapsed and was transported away. The main result of this pattern of events was the creation of isolated steep-sided mountains in the north and west, broad open basins in the centre, and flat plateau structures in the east. Most of the evidence of Tertiary landforms was removed by ice in the west, where glacial scouring was very intense. However, in the east, particularly in the Buchan district and on parts of the Cairngorm plateau, which were not affected by deep glacial erosion, a few remnants remain in the form of Tertiary sediments, deep weathering horizons and mountain tors on granite summits (see Fig. 7.28). These exceptional areas are Sites of Special Scientific Interest, as they shed important light on the age and evolution of the landscape. Analysis of Tertiary sediments recovered from the North Sea during oil exploration has also provided an insight into the rate of erosion of material from the mainland and the rate of uplift of the surface during the Tertiary. Glaciation may have created some of Scotland's most spectacular landforms, especially in the west, but the effects were relatively minor compared to the events of the Tertiary period.

Glencoe: a U-shape valley, excavated by glacial erosion, in the Caledonian Glencoe Igneous Complex; Aonach Eagach ridge on left, Loch Achtriochtan in centre, the Three Sisters on the right, showing andesite lava flows and sills; Loch Leven in distance at top left (photograph by Pat & Angus Macdonald).

Introduction

Two million years ago, Scotland had arrived at its present northerly latitude. Simultaneously, there was a marked deterioration in the world's climate, especially in the North Atlantic region: Scotland had entered the Ice Age. The most recent period of Earth history is the Quaternary (Latin: fourth; a reference to a previous subdivision of geological time into four units). The boundary between the Quaternary and the preceding Tertiary period is put at 1.8 million years ago; the Quaternary is further subdivided into the Pleistocene, which includes the glaciations, and the Holocene, from 10000 years ago, when the ice finally melted from Scotland, up to and including the present. Investigations of past climates in the Quaternary clearly indicate that the present warm stage is merely the most recent of relatively shortlived interglacials between glacial stages (or full ice ages) and that, if the pattern continues, we are heading for another ice age within the next few thousand years. The present global warming is related to burning of fossil fuels (wood, peat, coal, oil and gas) and intensive agricultural practices: much carbon dioxide and methane are produced, which enhance the natural greenhouse effect, and this may delay the onset of the forecast ice age. On the other hand, we may be in for a surprise, since it is not yet fully understood how all the components of the climatic system interact, and disturbing the system may well produce quite unforeseen effects that could not be predicted.

For human beings, the Quaternary is the most significant period in Earth history, since that is when modern humans first emerged. The most recent ice sheets put the finishing touches to the landscape of Scotland, and the soils developed only after the ice melted. It is also the best-understood period, since nearly all the animals and plants found as fossils in the Quaternary exist today, and geological events in the Quaternary can be dated with remarkable accuracy.

In the annals of geology, Scotland is well known as being the first place where the existence of former glaciers was deduced, following on the observations of the Swiss geographer Louis Agassiz (1807–1873), who visited Scotland in 1840 and studied various localities, notably Edinburgh, where he observed scratch marks on rocks in the Braid Burn, and Glen Roy, where he interpreted the Parallel Roads as the signs of water levels in a former ice-dammed lake. (Incidentally, Agassiz was a palaeontologist who specialized in studying fossil fish, and he was influenced by Hugh Miller's collection from the Black Isle.) Many of the landforms produced during and after the most recent ice age are important internationally, including in particular the suite of landforms in the Cairngorms. Here we find remnants of the pre-Quaternary landscape still preserved, as well as superb examples of glacial-erosion features, such as corries and U-shape valleys, glacial deposition related to deglaciation (moraines, meltwater features), and more recent changes associated with permafrost activities and the establishment of plant communities and environmental changes over the past 10000 years.

The worsening climate

We have already seen in previous chapters that the basic outlines of Scotland were established many millions of years ago, long before the ice sheets blanketed the land. Already at the end of the Caledonian orogeny 400 million years ago, the Highlands, Central Lowlands and Southern Uplands had formed as basic features, and so too had the main features of the coastline, as dictated by the major faults. Apart from a relatively brief incursion of the sea in the early Carboniferous (340 million years ago), when coral limestone was deposited, most of the period since the uplift of the Caledonian Mountains has been characterized by erosion from the land and deposition in sedimentary basins around Scotland, particularly in the North Sea graben structures and in the Minches. Additionally, continental deposits were laid down in river valleys and at the foot of the mountains during the Devonian (Old Red Sandstone) and Permo-Triassic (New Red Sandstone) periods. Subsequently, the next important event (and the last time solid rocks were formed) was the volcanic and intrusive activity of the early Tertiary period (60–55 million years ago), when the central complexes formed off the west coast. As a consequence of the intrusion of a mantle plume that gave rise to this volcanism, the land was elevated in the west, and the eastern part was tilted towards the North Sea. The river system that became established then has remained with us until the present, albeit in modified form.

In the popular view, the present-day landscape of Scotland resulted from the work of ice: deep erosion on the mountains, and thick deposits of glacial till ("drift") across the lowlands. However, this would be to ignore the evidence that the landscape is in fact much older. For most of the past 400 million years, Scotland has effectively been above sea level and has also been in much warmer and at times wetter climatic belts than at present. This has meant that chemical weathering and physical erosion have dominated the evolution of the landscape. The once-high Caledonian Mountains were rapidly worn down in the Devonian to produce the Old Red Sandstone sedimentary rocks in the lowlands, and many deep granite intrusions were exposed at the surface then (or "unroofed"), which is inferred because Old Red Sandstone conglomerates contain Caledonian granite pebbles, and the sediments themselves rest directly on top of these granites. In the Northwest Highlands, the landscape is even older. In Chapter 3 we saw how the characteristic knock and lochan topography of Lewisian gneiss has been systematically unearthed as the overlying Torridonian sandstone has been removed. The landscape we see there today is 1000 million years old and,

although planed off by Tertiary erosion, it must have resembled the landscape when the Torridonian sandstone was being deposited. During the period of the Tertiary intrusions along the west coast, uplift was also marked, and the subsequent erosion was intense, rapid and deep. It was then that the distinctive form of the landscape in the north was shaped. This is particularly noticeable in the isolated mountains that rise abruptly from a low plain near sea level, to give the characteristic inselberg (German: island mountain) or monadnock shapes. The word "monadnock" is derived from the type locality, Mount Monadnock in New Hampshire, USA. Particularly impressive are the mountains of the Inverpolly Reserve, Stac Pollaidh (Stack Polly, Fig. 8.1), Cùl Mór and Cùl Beag, and, slightly farther north, Suilven, to name only a few (see Figs 3.9, 3.10). Farther north still, Ben Hope and Ben Loyal are similar, and there are also such hills in Caithness, including the Maiden Pap and Scaraben (see Fig. 5.8). The isolated mountains in northwest Scotland are remnants of a once extensive and continuous pile of sedimentary rocks that was progressively denuded during the Tertiary period, when west-facing slopes were eroded eastwards.

When the ice finally did engulf Scotland, it encountered a landmass much the same as we see today, with

Figure 8.1 Stac Pollaidh, Assynt: an isolated hill (monadnock) of Torridonian sandstone resting on a basement of Lewisian gneiss.

dissected mountains and valleys in the northwest, upland plateaux in the northeast and in the Southern Uplands, and, in the Midland Valley, low ground that contained several important areas of high ground, the Campsie Fells, Renfrewshire Heights, Pentland Hills, Hagshaw Hills, Ochil Hills and Sidlaws being the most significant. The coastal outline and the pattern of river valleys had already been established long since.

Causes of ice ages

The Earth has been affected by glacial or icehouse conditions several times during its long history, most notably in the Precambrian at 1250, 900 and 650 million years ago, at the Ordovician/Silurian boundary 440 million years ago, in the Carboniferous–Permian period 350–250 million years ago, and finally the most recent period, the "last" ice age, which began late in the Tertiary, when a global period of cooling began in earnest about 10 million years ago. Ice ages are characterized by thick continental ice sheets over much of the globe, extending to much lower latitudes than during ice-free conditions.

In the glaciations of the past 2.5 million years, there have been about 20 major shifts in climate between cold glacial periods and warm periods or interglacials. The general pattern is one of gradual cooling to create a long-lived ice age (40000–100000 years long), which is brought to an abrupt end by a sudden and rapid warming that culminates in a shortlived interglacial (about 10000 years long). Large and frequent temperature fluctuations are a main characteristic of the most recent ice age. At its maximum extent 18000 years ago, ice sheets many kilometres thick covered a third of the Earth's land surface, principally in the Northern Hemisphere and along the length of the Andes mountain chain in South America and Antarctica. A wide variety of natural causes combined to produce the recent glaciations. This combination demonstrates in an elegant and satisfying way how planet Earth acts as an incredibly complex and intricate system in which all the components are inextricably linked together. Ice ages result from climatic changes, and the climate in turn is ultimately affected by the amount of the Sun's energy that reaches the Earth. Climate owes its origin to the interactions of the atmosphere, the oceans and any existing continental ice sheets, which today are found only in Antarctica and Greenland. When the ice sheets were at their maximum, so much water was held as ice that global sea level was 120 m lower than it is today.

If you look down at the North Pole on a world globe, you will notice the clustering of large continents around the pole: Canada, Greenland, Scandinavia and Siberia. The Arctic Ocean is more of an inland sea than a true open ocean, there being only two narrow connections with the North Atlantic and North Pacific. Moreover, the cluster of large landmasses contains high mountain ranges and elevated plateau table lands. This configuration is the key to the origin of the most recent ice age, and it resulted from the workings of plate tectonics. About 160 million years ago, the supercontinent of Pangaea began to break up and the circulation patterns in the world's oceans were radically altered. Australia separated from Antarctica, and a southern ocean opened up. India drifted northwards very rapidly, to collide with Asia at the same time as Africa collided with Europe, thereby closing the east–west Tethys Sea, which had existed around equatorial latitudes. The Atlantic opened progressively, from south to north; South America separated from Antarctica, leaving it isolated, and collided with North America, again sealing off east–west currents in the area of the present-day Caribbean. Continental collision created the Pyrenees–Alps–Himalaya chain and elevated the great inland plateau of Tibet, which now stands at over 5000 m high. Ocean/continent collision led to the formation of the Rockies and the Andes in the Americas, which are extensive cordilleras of high mountains, as well as broad high lands in the southwestern USA. One immediate consequence of this mountain building was a sudden and marked increase in erosion as the mountains began to be worn down. During rapid weathering and erosion, much carbon dioxide becomes locked in sediments and oceans, having been removed from the atmosphere because of the effects of gravity: high young mountains experience huge landslips and avalanches, usually caused by earthquakes and the lack of plants and soils that stabilize lower and gentler slopes. Carbon dioxide is a natural greenhouse gas and high levels of it are associated with global warming, lower levels leading to cooler climatic conditions. Extensive areas of high land had a pronounced effect on wind patterns, on the location of high-pressure and low-pressure belts, and on the amount of precipitation falling on land.

However, the tectonic effects that led to continents being clustered around the North Pole do not in themselves explain the Ice Age. We must still look towards changes in the amount of solar radiation reaching the Earth. An explanation for variations in this factor was proposed in 1867 by James Croll (1821–90), from Cargill in Perthshire, who worked for the Geological Survey. Croll argued that climate is influenced by ocean currents and by the distribution of oceans and continents. These ideas were further elaborated by the Serbian mathematician, astronomer and geophysicist Milutin Milanković (1879–1958), who studied astronomical cycles, which he believed were responsible for variations in climate. Milanković realized that the amount of solar radiation received by the Earth depends on latitude and the Earth's distance from the Sun. The tilt of the Earth's axis varies over a 40 000-year period and its orbit around the Sun varies from circular to elliptical every 95 000 years and 400 000 years. In addition, the gravitational effects of the Sun and Moon cause the Earth's axis of rotation to wobble around in a complete circle every 27 000 years. These three cycles act simultaneously, so that at certain times the Earth is much closer to the Sun than at others. Milanković calculated the precise mathematical effects of these interactions to produce solar radiation curves for the past 650 000 years, which he compared with observed climatic cycles. There is a remarkable agreement between these cycles, and the pattern of ocean temperature curves obtained from studying deep-sea sediments and the composition of foraminifera, minute planktonic animals with carbonate shells. It is the Milanković cycles that are of greatest importance in determining when glacial conditions will be initiated. When the cycles combine to give reduced levels of solar radiation during the summer in the Northern Hemisphere at 60–65°N, where there is most continental land mass, then snow from the previous winter can remain in a cool summer. The difference in radiation can be as much as 30 per cent between summer and winter, when the Earth's orbit is at its most elliptical. As few as 20 cold summers can lead to such a build-up of unmelted snow that more of the Sun's heat is radiated back from the white surface into space. This is known as the albedo effect (Latin: white) and it produces positive feedback to the climate system, acting in such a way that the situation becomes exaggerated and the climate rapidly deteriorates. Up to 90 per cent of the Sun's radiation may be reflected back and so cannot heat the Earth's surface. This fact was also first recognized by James Croll. A complication arises because of the possibility that the Sun's heat engine does not operate at a smooth rate with time. The amount of heat generated is known as the solar constant, and this fluctuates in approximately hundred-year cycles. Evidence for this is available from studying tree rings, which can give some information on solar energy, and from radiocarbon dating, but only for the past 10 000 years.

The mechanism that finally triggered the most recent ice age appears to have been the change in ocean-current circulation resulting from the collision between South and North America, which was completed by 2 million years ago. Today, there is a worldwide conveyor belt of surface and deep currents moving between the Atlantic and Pacific oceans, and constrained by the distribution and shape of the continents, as well as the position of mid-ocean ridges and deep abyssal plains. Warm water flows in shallow currents from southern regions westwards, then northwards into the Atlantic. The well known Gulf Stream (really the North Atlantic Drift) from the Gulf of Mexico flows up into the North Atlantic, thereby warming Scotland to a level much higher than other places at the same latitude. High evaporation and the many rivers discharging into the Atlantic cause the ocean to have high salinity. Cold salty water is very dense and sinks to the bottom, forming the North Atlantic deep water, the downward movement of which adds a vertical link to complete the conveyor belt of currents. Deep water travels south, around Africa, then close to Antarctica, before moving up around New Zealand and Australia, where it is warmed near the Equator and eventually surfaces in the North Pacific to complete the worldwide pattern. As the oceans and atmosphere are closely interlinked, a change in one will have an effect on global climate. In the North Atlantic, the currents are liable to rapid fluctuations in speed and latitude, and it has been shown that the climate can become glacial when the currents are slow moving. The controlling factor is the position of the Polar Front, which is the boundary between cold and warm North Atlantic water. At present, the front lies roughly along a line between Newfoundland and the west coast of Iceland. During glacial periods, the Polar Front is pushed much farther south, to about half way along the west coast of Portugal, and at such times sea ice is able to float into the Bay of Biscay. Shifts in the position of the Polar Front can occur extremely rapidly and, when it moves south, the warm waters of the North Atlantic Drift are blocked from entering northern waters. Small changes in any of the intricately interconnected and inter-dependent components of the climate system will result in knock-on effects, sometimes sufficient to plunge the world into a glacial period.

Landforms created by glacial erosion

For the past 1.5 million years, Britain has been in the grip of ice ages. During the ice ages, the climate was generally drier than in interglacial stages, and then, as now, the greatest precipitation was on the west, over the highest mountains, regardless of their geological composition. Ice built up gradually over the high ground, filled valleys between mountains, and even-tually covered the tops of the highest mountains. For most of the period, valley glaciers and icecaps on the western Highlands were the norm, with extensive ice sheets covering the entire country on only two occa-sions. Because of the erosive power of moving ice, evidence of earlier activity has mostly been obliter-ated, and we now see mainly the products of the most recent glaciation of 22000–13000 years ago (the Late Devensian). In eastern Scotland, over the Cairngorm and Monadhliath mountain plateau, the ice was much thinner than in the west, because snowfall was less. Also, the east was much colder than the west, the North Sea was smaller, and the east of Scotland had a climate more continental than that of today (Fig. 8.2). These west–east differences had another major conse-quence: as ice is a relatively good insulator, the base of a thick ice sheet can actually warm up from the input of heat from the Earth's surface (geothermal heat). The heat is not dissipated: it is trapped, as it were, so warms up the ice at the rock/ice contact. Also, the ice

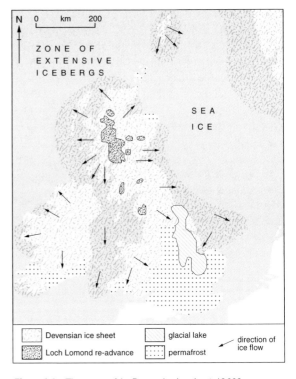

Figure 8.2 The extent of the Devensian ice sheet, 18 000 years ago.

sheet in the west was close to the sea, so that the ice had a very thin film of water at the base (similar to the effect of pressure on ice skates allowing a skater to glide on water). In the west the air temperature was not as low as in the east, and the ice was thicker because of higher snow accumulation; the pressure at the base of a thick ice sheet is greater than for a thin sheet. The combined effect was to produce a film of water at the rock/ice interface. Such ice sheets are known as warm based, and they move across the sur-face, scouring the rocks and removing loose material. This can happen at rates up to ten times faster than in ice sheets that remain frozen to bedrock (the cold-based glaciers). Friction generated by ice movement releases more heat and tends to accelerate the process. For this reason, the west side of Scotland has suffered much more deep erosion and glacial scouring than the east, and the effects are clearly seen in the landscape contrasts. Knock and lochan topography is very characteristic of the Northwest Highlands and Outer Hebrides (see Fig. 3.4), where warm-based glaciers slid across the surface and scoured extensive areas,

Figure 8.3 The summit of Cairngorm: flat slabby boulders of weathered granite forming the pavement (recent frost action), leading towards remnant of a tor.

creating a rough rocky pavement of bare rounded rocks. In contrast, the thin cold-based glaciers of the east did far less damage and, in the eastern Grampian Mountains and the Buchan area (northeast Aberdeenshire), remnants of the Tertiary landforms (including tors on granite summits such as Bennachie and Cairngorm; Fig. 8.3) and sediments have been preserved, seen now as mountain plateaux and gently eastward-dipping hills or low flat coastal plains. A few corries have been bitten out of the edge of the Cairngorms (Fig. 8.4), and U-shape glacial troughs do exist, such as Lairig Ghru and Glen Avon (Fig. 8.5), but these hardly compare in scale and intensity with similar features on the much more dissected western Highlands and islands (see Fig. 7.18), where closely spaced valleys and summits rising direct from sea level create a very different picture. Another feature of glacial erosion that imparts to the west coast the highly indented nature of the shoreline is the effect of selective linear

erosion. Here, the ice flowed along deep narrow valleys picked out along northwest–southeast trending lines of weakness in the basement rocks. This direction is followed by shear zones, basic dykes (Scourie Dykes), pegmatite sheets and granite dykes (Laxford granites), and later faults and fractures in the Lewisian gneiss and overlying Torridonian sandstone bedrock in the Northwest Highlands, so that now we see many deep and narrow fjord-like inlets of the sea, such as Loch Laxford, Loch Inchard and Loch Maree (an inland loch). In the Southwest Highlands of Argyllshire, the linear pattern is more northeast to southwest, a reflection of the influence of Caledonian trends in folds and faults seen in the Dalradian bedrock. The coastline of Argyll, Jura and many of the deep sea lochs follow this trend; for example, Loch Fyne sits along the line of the Tyndrum Fault, and Loch Awe at the end of the Ericht–Laidon fault line (Fig. 8.6). Ice moving off the land also scoured out very deep

BILL BROOKER

Figure 8.4 The Cairngorm plateau, from Aviemore on the A9 road. Caledonian granite was intruded into early Dalradian rocks; small glacially eroded corries are visible, and Lairig Ghru U-shape valley.

Figure 8.5 Loch Avon, Cairngorms: glacially excavated rock basin in Caledonian granite.

erosional rock basins in the Firth of Clyde, the Minches, the Great Glen, Firth of Lorne, Sound of Jura and Loch Lomond. The island of Raasay has deep troughs on either side, in the Sound of Raasay, created by the Skye icecap moving off the Cuillin, and along the Inner Sound, which was gouged out by a branch of the Skye icecap, and the mainland ice moving off the Applecross mountains. Similar deep troughs surround Arran, as ice moving down Loch Long and Loch Fyne was swept south around the island. The vast majority of rock basins are in the west; one notable exception in the east is the deep channel in the

Figure 8.6 Loch Awe, Argyllshire: glaciated landscape in the Cruachan granite; the Cruachan pumped-storage hydroelectric scheme uses the corrie lochan for water storage; Lorne plateau lavas (Devonian) are in left foreground.

Firth of Forth, which is not in the middle of the firth, so that the Forth railway bridge has an asymmetric design (see Fig. 9.11).

The greatest extent of the Scottish ice sheet was 22 000–18 000 years ago, when almost the entire land surface was under a 2 km-thick sheet that extended south to the English Midlands. Only Orkney and the northeast corners of Caithness and Buchan appear not to have been glaciated then. The Scottish ice sheet extended into the North Sea, where it may have met the much larger Scandinavian ice sheet, which was pushing south from Norway. Off the west coast, the ice rapidly thinned at the edge of the continental shelf, around the position of the Flannan Isles (part of the Lewisian platform), and did not reach as far as St Kilda, which was surrounded by icebergs. It is likely that all the mountain tops were covered by ice, with only a few peaks on An Teallach and the Skye Cuillin remaining above the ice as nunataks (Greenlandic: isolated rocky pinnacles protruding from ice). The ice sheet itself had a complex shape, dictated by the underlying topography. Continental ice sheets are of dome shape, like upside-down saucers, thicker at the centre and thin at the edges. In Scotland, the main ice dome was very elongate, running nearly north–south from northwest Sutherland down to Loch Lomond, with more local, rounded domes over Harris, Skye, Mull, Arran and the Cairngorms, and an almost

east–west trending dome over the Southern Uplands (see Fig. 8.2). Shetland appears to have had its own local icecap, shaped in the same north–south orientation as the island chain, flowing out in all directions along the very pronounced Caledonian fold and fault trends. The maximum thickness of the mainland ice dome was centred on Rannoch Moor, from where the ice spread out radially, creating a pattern of deep narrow lochs and valleys, now used by roads. Rannoch Moor is surrounded by high mountains, and the area had been subjected to thick ice masses during all the glacial episodes, so that an extensive rock basin has been carved out, with a flat valley floor and steep hill faces around the edges. While moving, the ice tended to flow in streams along pre-existing valleys, which became wider and gained smooth profiles (Figs 8.6, 8.7 and see pp. 168–169). The main ice streams were along the Great Glen into the Moray Firth, the Spey and Firth of Forth into the North Sea, down Loch Lomond, and those off the west coast referred to above, moving principally down the Firth of Clyde and from the mainland into the Minch. In all these cases, deep rock basins were cut into the sea bed (between Arran and Bute the sea floor is 320 m below sea level) by the ice streams, which spread out into lobes or fan shapes once the narrow valley constrictions were breached. Loch Lomond is probably the best example of this, where the broad, shallow (10 m

Figure 8.7 Moffat, Borders region: glacially excavated U-shape valley following the line of the Moffat Water Fault; smooth, rounded hills of Silurian greywacke; photographed from Grey Mare's Tail.

deep) southern end of the loch, south of the Highland Boundary Fault where soft Old Red Sandstone occurs, contrasts sharply with the narrow, steep rocky northern part, where Dalradian schist and tough quartzite predominate; here the rocky base of the loch is down to 170 m below the surface.

Rapid warming between 14 000 and 13 000 years ago resulted in the disappearance of the ice over most of the country, save perhaps for some mountain glaciers remaining in the north and west. Glaciers had begun to retreat after 18 000 years ago, with a reduction in snowfall. Once starved of new snow, glaciers shrink, because melting exceeds supply even under cold conditions. A sudden climatic deterioration at 11 000 years ago led to the return of the ice on high ground, first as mountain-top, corrie and valley glaciers, which then merged to form a mountain icecap from Torridon to Loch Lomond. Small outliers of ice also occupied other isolated areas of high ground, principally the Tertiary volcanic complexes of Skye, Mull, Rum and Arran, and the Caledonian granite mountains of the Cairngorms and Galloway. This episode is usually referred to as the Loch Lomond re-advance (or stadial, a shortlived glacial interval). It lasted a mere thousand years, but was responsible for giving the mountain landscape in the west its rugged outlines, including the Arrochar Alps, when many peaks would have remained above the surface of the icecap, to be shattered by intense frost action in the severe climatic conditions. Many of the erosional features are the result of several successive episodes of glaciation, but dating these is not straightforward. In the Cairngorms, it is possible to find corries within corries. The extremely deep Loch Morar in the west (300 m below sea level) could have been gouged out only by multiple movements of ice sheets over a long time interval. The main reason to explain the Loch Lomond re-advance was the sudden influx of vast quantities of cold fresh water from the huge glacial meltwater lakes at the front of the North American continental ice sheet. This formed a lid above the normally salty water and so upset the currents in the North Atlantic conveyor-belt system that the Polar Front was forced southwards once more, and Scotland was plunged temporarily into glacial conditions.

In addition to scoured rock pavements, corries, rock basins, fjords and U-shape valleys, the ice was responsible for creating many erosional features on a smaller scale, including smoothed and shaped rock outcrops. Among these are the roches moutonnées,[*] formed when ice moved up and over a gentle rock slope, then plucked boulders from the steep face in the lee of the ice, as a response to pressure differences on either side of the hill. Good examples can be seen at Grantown-on-Spey, a Site of Special Scientific Interest. Large obstructions can cause ice to be deflected around them, creating features known as crag and tail. Possibly Europe's most famous example of this landform is the Royal Mile in Edinburgh, sloping away from the crag formed by the volcanic plug on which the castle sits (see Fig. 9.16). Ice moving eastwards to the Firth of Forth encountered the bluff of igneous rock and moved around the sides, scouring out deep canyons in the surrounding sandstone, and plastering the lee slope with glacial debris to form the tail on the east side that leads down to Holyrood Palace. The Arthur's Seat volcano in Edinburgh is another example, larger but in some ways less spectacular (see Fig. 5.10). The Necropolis adjacent to Glasgow Cathedral is located on another well known crag and tail, where a dolerite sill forms the crag. Features such as crag and tail and roches moutonnées give a clear indication of the direction in which the ice moved. Additional evidence is obtained from scratches on rockfaces, known as glacial striae or striations. Mention was made above of the striations along the banks of the Braid Burn in Edinburgh, which led Louis Agassiz to conclude in 1840 that they were the work of ice, and for the first time to propose that Scotland had been glaciated. Although this was met with scepticism, James Croll championed the work of Agassiz and went on to conduct important research on glaciation in Scotland.

As a scenic area, the Cuillin ridge knows no equal. This is partly because of its rocks and partly because it was frost shattered but not covered and smoothed by an icecap. Some of the best roches moutonnées are to be found around Loch Coruisk and on the floors of many of the corries in the Cuillin area. Corries meet back to back, forming jagged backbones or arêtes (French: fish bones); corrie lochans abound; and the

---

* French: for a hill shaped like a wig, worn in the eighteenth century; "mouton" refers to the fact that the wig was held on the head with mutton fat.

rugged peaks were sharpened by frost action in the cold zone above the melting icecaps (see Figs 7.16, 7.18). Frost shattering is an example of periglacial activity (i.e. in the cold conditions in the vicinity of a glacier). Other superb examples include the summit zones of Stac Pollaidh and An Teallach in the Northwest Highlands (see Fig. 3.10).

The brief Loch Lomond re-advance was cut short by a sudden and rapid rise in temperature of at least 7–8°C in no more than 50 years. Ice then began to melt very quickly and disappeared completely. During the melting episode (or deglaciation), some of the ice continued to move from high ground to low, melting at the edges. Active retreat can thus lead to a continuation of erosion. In other instances, large sections of ice became separated from the main icecap and melted in place (or "wasted") as stagnant ice.

Landforms created by glacial deposition

Apart from modern processes in rivers and along the coast, the most recent major event to influence the landscape of Scotland was the melting of the ice sheet. As happened throughout the Quaternary period, the deglaciation phase was affected by many rapid climatic oscillations between warm and cold extremes. Deglaciation, although rapid, was not smooth and uniform throughout the country. We see evidence instead for a series of pulses of warmth and ice melt, followed by standstills in colder periods. Clearly, melting of ice sheets that covered one third of the Northern Hemisphere landmass resulted in vast amounts of cold fresh water being returned to the world's oceans, sea level having fallen globally by about 120 m during the maximum extent of the ice. For Scotland, as elsewhere, this had two consequences. First, much of the North Sea was drained of water, and barren land existed instead, with bridges to Europe that allowed the southward migration of mammals no longer found here, such as reindeer, arctic fox, bear and wolf. Secondly, the large mass of ice weighed down on the crust, which sank to accommodate the extra weight. Crustal downwarping took place slowly, the mechanism for this being flow in the plastic mantle beneath the crust.

Local sea level then began to fall progressively, and

the net effect on the coastal landscape was to create a series of raised beaches marking different standpoints, when wavecut platforms and cliffs with sea caves were formed. In many parts of Scotland these former shorelines are now some way inland and higher than the modern shoreline. Good examples are seen in Islay, Jura and Arran (see Figs 4.12, 8.8), and in many places along the west coast, particularly the Firth of Clyde and Ayrshire. On the cliffs just outside St Andrews in Fife, the different postglacial sea levels are very clear from the caravans pitched on each successive former cliff top. And the town of Kirkcaldy in Fife is known locally as the Lang Toun for having been built as an elongate settlement on a strip of raised beach. Names such as Wemyss Bay and Pittenweem testify to the presence of raised beach caves (weem, Pictish: cave). The caves at East Wemyss in Fife contain Pictish carved symbols, first described in 1865 by Sir James Young Simpson (who is more famous for introducing the use of chloroform in surgery).

As the ice sheet moved down valleys from mountainous areas to the lowlands, the material picked up by the ice and used to grind, smooth and polish the surface was transported and deposited when the ice melted. The veneer of glacially derived debris that covers the bedrock is known as till, boulder clay or

Figure 8.8 Gruinard Bay, Wester Ross: a post-glacial raised beach in the foreground; a rounded plateau of Lewisian gneiss in the middle ground.

drift, most of which was deposited directly at the base of warm-based glaciers, and which can be about 50 m thick. Much of the till is locally derived, but there are boulders that have been carried over long distances.

One of the most useful pieces of evidence for ice-movement directions is from boulder trails, where glacier ice has plucked or picked up loose material from a source area and transported it within the ice sheet to some distant point. When the ice melts, the contained boulders are dropped onto the land surface. Because they are usually different from the under-lying bedrock, the boulders are referred to as glacial erratics. Often, their rock type is highly distinctive, and it is not difficult to trace where the erratics have come from, and to plot maps showing ice-movement directions. Some erratics are very large and form prominent landmarks in their own right, such as the Clochodrick stone in Lochwinnoch, the perched boul-der on Coll, the huge granite boulder near Brodick (Fig. 8.9), and the rounded basalt stone of St Anthony's Well in Edinburgh's Holyrood Park, which came from Craiglockhart, a few kilometres to the west. Some of the most highly distinctive erratics, which form trails from known outcrops in the source area, are the Inchbae augen gneiss east of Ullapool, which was transported along much of the Moray Firth coast, and the Lennoxtown essexite in the Campsie Fells, which was taken east along both shores of the Firth of Forth. Erratic blocks of Ailsa Craig granite found their way deep into the south of Ireland from the Firth of Clyde as the Scottish ice sheet merged with

an icecap over Ireland. The granites of the Galloway Hills have their own trails radiating out in all direc-tions, indicating the former presence of an ice dome in Southwest Scotland. There are many other examples of easily recognizable erratics of coarse-grain igneous rocks that were plucked off rock outcrops on high ground and carried within or at the base of the ice.

Around the low-lying coastal areas of Orkney, Caithness, the Moray Firth, Ayrshire and the Rhinns of Galloway, much of the till contains marine shells, indicating that ice moved from the sea onto the land in those areas. At Clava near Inverness, one large block of clay with arctic shells was carried up to 150 m above sea level as a raft, broken off from the Moray Firth floor; the clay was formerly worked for brick making. The Clava shelly clay provoked a fierce debate at the end of the nineteenth century, with some proposing a drowning of the land by over 150 m, and others insist-ing that the beds had been transported by ice, then left isolated when the glacier melted *in situ*. Radiocarbon dating and analyses of the shell structure indicate an age of 44 000 years for the clay beds. Further research in 1990 finally resolved the 100-year-old controversy in favour of the ice-raft theory.

In the final stages of deglaciation, some of the till was moulded by the remaining ice into streamlined low mounds, known as drumlins (Irish: smooth hill). Drumlins contain a wide variety of glacially trans-ported boulders, cobbles, pebbles, gravel, sand and clay, and the unconsolidated material was shaped and moulded beneath the ice. Clusters of drumlins give rise to the well known "basket of eggs" topography, which is a common feature in the lowlands. Particu-larly famous are the swarms in and around Glasgow, on both sides of the River Clyde; these are orientated northwest–southeast and are mostly about 1 km long and 30 m high, tapering to a tail at the southeast end. Streamlining is parallel to the Clyde, indicating that the ice stream moved up the valley. Some drumlins have rock outcrop at the core, with till moulded around the sides and over the top. The River Kelvin in Glasgow winds its way around the sides of drumlins, and Sauchiehall Street in the city follows a low shoul-der between two drumlins. Other sites include the impressive drumlin field near New Galloway in the southwest, and those around the Endrick Water south of Drymen. Since the meltwaters generally ran to the

Figure 8.9 Arran: glacial erratic boulder of granite resting on the beach at Corriegills, south of Brodick.

Figure 8.10 Ryvoan Pass, near Lairig Ghru, Aviemore, Cairngorms: granite boulder field in upper valley, marking fault excavated in Cairngorm granite, with valley continuing into distance.

east, most of the glacially deposited sediment is found in low ground around the North Sea. The material transported by meltwater streams is glacial outwash or fluvioglacial sediment, much of which is sand and gravel that has been sorted during transport. Quartzite pebbles and quartz sand form a high percentage of this outwash material, and it has been a valuable resource for building, to the extent that many of the landforms created by glacial meltwater have been totally destroyed by the extraction industry.

Landforms owing their origin to streams from glacial melting are particularly abundant in the east of the country, especially the Moray–Buchan area, Fife, East Lothian and the Borders, reflecting the general flow direction of meltwater rivers. In addition to outwash deposits, there are many examples of the erosive action of meltwater streams flowing under pressure beneath the ice. Such streams would have carried heavy loads of pebbles and cobbles that were picked up by the ice, and the combination of high-pressure water charged with rock debris produced extremely powerful localized erosion agents, acting like rock saws. Many such channels are deep, narrow and straight, such as Glen Valtos on Lewis, or the Corrieshalloch Gorge near Ullapool – the best example of a rock gorge in Scotland. The Corrieshalloch Gorge is cut into a joint in the tough quartz-rich Moine schist bedrock, which controlled the location of the gorge. It is over 1 km long and 10 m wide, with a drop of 60 m at the impressive Falls of Measach, one of several stepped waterfalls, which are progressively cutting back and extending the feature up stream. In many

cases, the meltwater channels are now dry and seem to be unrelated to the present topography. Such channels were driven by high pressure beneath the ice cover. Those in the east are often both thickly wooded and called dens, being quite abundant in Fife and Aberdeenshire. Dean Village in Edinburgh is another splendid example of a deep glacial gorge, cut by the Water of Leith. Others include the Ryvoan Pass near Aviemore in the Cairngorms (Fig. 8.10), the Black Rock Gorge at Evanton in Easter Ross, the deep gorge by the main A9 road at the Slochd summit near Inverness, and the Ailnack Gorge (Fig. 8.11).

Some of the largest meltwater streams spread huge amounts of sand and gravel across wide, flat valley floors. The Spey is one of the best examples, where the floodplain deposits formed sheets known as sandur (Icelandic: the flat sandy coastal plain in the south of Iceland). Such deposits build up rapidly by interweaving braided streams. Today, the River Spey itself is relatively narrow; the very wide strath was formed by a much larger Tertiary river, and the original valley

Figure 8.11 Ailnack Gorge, Cairngorms: a narrow and deep glacial channel cut into Old Red Sandstone; in the background, a granite plateau.

was then extended by ice streams moving off the Cairngorms, and finally by the meltwater river.

Rivers flowing in tunnels within the melting ice carried sediment that was deposited in long narrow ridges (eskers) that have a rather sinuous shape. In Scotland the finest examples of these are in the north-east, around Inverness (the Kildrummie eskers, incorrectly known as the Kildrummie kames), where the highest esker in Britain is found at Torvean (70m high), and at Carstairs near Lanark. Both these areas contain the best, most famous and most important sites of glacial features in Britain, including eskers, kames, kettle holes and outwash fans, which formed on top of or within large masses of stagnant ice that melted and left the sediment behind as a series of interconnected ridges and mounds, the hollows between being filled with stranded blocks of ice that then melted to produce kettle lakes. Previously, many of the esker systems were called "kames" (e.g. Carstairs kames and Bedshiel kames; Fig. 8.12). The difference is that kames are irregular mounds or ridges of stratified sand and gravel that formed at the margins of stagnant ice (i.e. ice that melted where it stood); the word "kame" is from the Scots for "comb". Eskers on the other hand are higher and more continuous ridges that formed in tunnels of ice more or less at right angles to the front edge of the melting glacier; the word is from the Irish "eisceir", meaning a ridge.

Melting glaciers left behind deposits at their edges along valley walls, and at the front (the snout). These piles of transported material are moraines, a word used in the French Alps to describe heaps of stony earth (Figs 8.13, 8.14). Internally, moraine is usually a jumbled mixture of boulders, cobbles, pebbles, gravel and sand, randomly assembled together. Adjectives are used to describe where the moraine formed: terminal moraine marks the front of a glacier, lateral moraine was deposited at the sides, and medial moraine can form where two glacier streams meet, so that two lateral moraines form a line along the join of the two ice streams. Medial moraines are rare in Scotland, but one fine example can be seen on the west coast of the island of Jura, where a trail of angular boulders of white Dalradian Jura quartzite originating from the Paps forms a straight line running 3.5km from Beinn an Oir down to a raised beach. The boulders probably fell onto the ice surface from quartzite

PAT & ANGUS MACDONALD

Figure 8.12 The esker at Bedshiel, Borders: a ridge of sand and gravel formed in a river tunnel within a melting ice sheet.

nunatak hills. In Gaelic this particular medial moraine is known evocatively as the Witch's Slide (Sgriob na Caillich). Some moraine fields are arranged in chaotic clusters, forming hummocky moraine, the best example of which is in Glen Torridon, in the Valley of a Hundred Hills (Fig. 8.14). Another and more accessible site is at the Drumochter Pass on the A9 road. Terminal moraines mark the limit of a glacier. These are common around Callander and other places in the vicinity of the ice stream that flowed out of Loch Lomond and spread outwards south of the Highland Boundary Fault. A site known since the nineteenth century, and the first to be recognized as such in Britain, is the terminal moraine on Skye, at Loch Coir' a Ghrunnda in Glen Brittle, near Loch Coruisk, where other spectacular examples of glacial action include heavy ice scouring over bare rock surfaces, now

Figure 8.13 The Cairngorms, near Glen Einich: a hummocky glacial moraine in the foreground, now being reworked in a stream flood plain.

moulded, polished and scratched. Scratch marks are known as glacial striae or striations. It was the discovery of these at Blackford Hill in Edinburgh that led Louis Agassiz in 1840 to conclude for the first time that glaciers were once present in Scotland, although his conclusions were greeted with considerable scepticism at the time.

The Cuillin area is one of the most important sites in Britain for interpreting glacial landforms of erosion and deposition by both mountain icecaps and corrie glaciers. There are examples of hummocky, lateral, medial and terminal moraines, and the entire complex of features exists in a relatively small area, hence its exceptional importance.

Periglacial processes

Periglacial features in the areas beyond the ice sheets include forms created during permafrost conditions. Permafrost is frozen ground, often extending to considerable depths beneath the surface, and is overlain by a thinner layer that melts in summer and freezes in winter. There is permafrost today in northernmost Siberia and the Canadian Arctic, and in Scotland such conditions were present around the decaying ice sheets, when winter temperatures were as low as –25°C. Initially, the ice decayed because it was deprived of snow, as opposed to melting because of rising temperatures. Constant freeze/thaw action has produced patterned ground in the form of large polygons (usually with six sides), separated by narrow wedges that penetrate the ground. The ice wedges

Figure 8.14 Glen Torridon, Valley of a Hundred Hills (Coir' a Cheud Chnoic): glacial moraine hummocks.

represent cracks, which then extended and inter-sected to form polygonal patterns. The polygons that were produced at the end of the Ice Age have mostly been destroyed by farming and forestry, although remnants do occur, for example on the top of the Ronas Hill granite in Shetland and in Buchan, north-east Aberdeenshire, where relics have been seen from the air, as distinct marks beneath crops in fields. Early inhabitants used some of the sites in Buchan for settle-ment purposes, and the natural patterns are overlain by the results of primitive farming. Evidence for ice wedges is more common, particularly in eastern parts of the Midland Valley (e.g. Edzell) and in Buchan, around the edges of the former ice sheet. These wedges appear as narrow V-shapes, cutting across sand and gravel bedding planes. On sloping ground, soil is able to slide down hill on top of the permafrost layer, to produce tongue-shape loops known as soli-fluction lobes, which are commonly seen on many rounded mountain tops. There are excellent examples at the top of Ben Wyvis, which consists of Moine schist; the schist weathers readily into small flat flakes that are easily moulded and moved by solifluction processes. Tongues of boulders also slipped down hill, and in cases where large quantities of boulders were trapped in ice, the collapsed boulder fields at the foot of steep slopes are known as rock glaciers. One of the best examples is at Strath Nethy in the Cairn-gorms. Examination of periglacial features suggests that permafrost formed during the Loch Lomond re-advance. The features are common in the east, where snow cover was thinner and the climate was much colder than near the west coast. Today on the highest mountain tops, periglacial activity is still taking place as freeze/thaw action in the main. Present-day forms relate more to the modern wet and windy climate, as opposed to the cold dry conditions of the end of the Ice Age. Permafrost is no longer present in Scotland, but high winds and frost do create patterned ground, including stone stripes and solifluction lobes. Rock-falls and avalanches are common on steep unstable slopes, and material slumps rapidly down hill in heavy rain and spreads out to form fans and cones on valley floors.

As the ice was melting, large blocks of ice became detached from the main sheet and continued to melt in place as stagnant masses. When they melted, they left hollows, not usually connected to the drainage sys-tem; hence they often contain ponds or small lochs (lochans). Much of the kame and kettle topography resulted from such conditions.

Many valley mouths were blocked by ice, and meltwater built up behind these dams, to form tempo-rary glacial lakes. Two very important sites are the Parallel Roads of Glen Roy and the terrace deposits of Achnasheen, both of which have been studied exten-sively for two centuries and more (Fig. 8.15). The Parallel Roads and the surrounding area of Glen Spean, from Loch Laggan to Loch Lochy in the Great Glen, are considered to be of outstanding inter-national importance in geomorphology (the study of landforms); they are among the most famous and best preserved features of late glacial landforms in Britain. Planning permission to bulldoze a hill track across them was refused in the 1980s. Thomas Pennant first drew attention to the Glen Roy locality in 1771, and described the features as roads used for hunting. In 1841, Louis Agassiz correctly interpreted the Parallel Roads as former shorelines of an ice-dammed melt-water lake. The shorelines were formed when notches were cut into Dalradian bedrock by waves and frost (freeze/thaw) action during periods when the lake level was constant. These lines are now at 350 m, 325 m and 260 m above sea level. Nearby, Glen Gloy and Glen Spean also have fossil shorelines, and all three adjacent valleys were filled by meltwater trapped behind glaciers from the Loch Lomond re-advance. The glaciers themselves also left behind impressive moraines. Different lake levels formed successively lower down the valleys as the ice retreated.

Eventually, the dams were breached and melt-water poured out in a series of catastrophic floods, forming the deep and narrow Spean Gorge, and into the Great Glen, eventually disgorging into the Moray Firth at Inverness. Gravel deposits were spread out in lower ground as a result of these floods, at Fort Augustus and Inverness, and older sand and gravel terraces were cut through, such as at Lochardil, Inver-ness. In Iceland, catastrophic meltwater floods are common, including those formed as a result of vol-canic eruptions beneath icecaps, and the term jökul-hlaup (Icelandic: glacier burst) is used to describe these examples wherever they occur. Much of the sandur of the southern Iceland coast referred to above

Figure 8.15 The Parallel Roads of Glen Roy. The three lines on the left, above the river bed, represent successively lower water levels of an ice-dammed lake that occupied the glen.

(p. 181) has been produced by jökulhlaup events. The possibility has been raised that the Glen Roy and Great Glen floods may in part be attributable to localized earthquake activity, triggered by the rapid loading and unloading of the crust by ice and water. Fossil landslides provide some evidence for this notion.

Landforms at Achnasheen in Strath Bran (between Kinlochewe in Wester Ross and Dingwall in Easter Ross) were also formed by ice-dammed lakes. They appear as an impressive series of flat-top terraces 30 m above the river floodplain and are clearly visible from the A832 road. The lake was dammed by a lobe of ice that extended out from the Fannich Mountains, which lie to the north. The stepped landscape formed as outwash terraces, which were cut across by outflow from the lake. Deltas are also preserved, forming as meltwater streams flowed into the lake and caused sediment to accumulate. Sediments on the floor of the lake show evidence of dropstones that fell out of floating icebergs. Like Glen Roy, the assemblage of features at Achnasheen is very important in geomorphology.

About 10 000 years ago, the Polar Front moved north to its present position, and warm waters of the North Atlantic Drift penetrated as far as Scotland once more. The Loch Lomond re-advance, which had been caused by changes in the North Atlantic currents forced by catastrophic flooding of meltwater lakes at the southern edge of the Laurentian (American) ice sheet, had come to an end, and rapid melting of the Scottish ice sheet produced copious amounts of water that initially caused a rapid rise in sea level and flooding of the coast. The global (eustatic) sea-level rise was soon compensated by isostatic uplift or rebound of the land surface following unloading when the ice melted, and this therefore led to relative sea level falling continuously. The glacial rebound theory was first proposed in 1865 by Thomas F. Jamieson, from Ellon in Aberdeenshire, who was a vigorous supporter of Louis Agassiz, and who made a major contribution to the study of glacial landforms in Buchan and Caithness. Uplift was most rapid at 13 500 years ago, but the Loch Lomond re-advance temporarily halted this, at least in western Scotland, where the ice was thickest; the rebound effect restarted from 10 000 years ago, but the effect was intermittent, and it tailed off exponentially and now is scarcely noticeable. Sea level around

185

Figure 8.16 Grey Mare's Tail, Moffat, Borders region: a hanging valley and waterfall descend into the lower main valley as a result of post-glacial uplift. The stream emerges from Loch Skene, a corrie lochan.

Scotland is rising at about 1.5 mm per year and the uplift rate in the western Highlands, where the ice was thickest, is 2 mm per year, giving a net rise in the elevation of the land surface of at most 0.5 mm per year.

In addition to leaving evidence of uplift around the coast in the form of raised beaches, rebound effects inland resulted in tributary streams being taken up to higher levels above the main valley floor, to create hanging valleys above the former confluence (where the tributary meets the main stream), and waterfalls are now a feature of such valleys. The Grey Mare's Tail at Moffat in the Borders is a spectacular example (Fig. 8.16) and the Eas a' Chùal Aluinn, near Loch Glencoul in Assynt, is also a hanging valley, from which Britain's highest waterfall descends 201 m into the glaciated valley below (Fig. 8.17). Renewed downcutting will also cause rivers to erode their own older terraces, creating stepped levels on valley floors.

Drowning of the landscape happened in areas farthest from the maximum uplift, which coincided with the location of the thickest ice. Thus, in the Outer Hebrides, Shetland and Orkney we see many embayments, flooded inlets and sinuous coastlines. Before the drowning of the coast, Orkney would have consisted of just one low, flat island, the high point being volcanic rocks on what is now Hoy. Channels between

the islands are very shallow and the Orkney coast is dominated by long gentle curves, characteristic of a drowned landscape. On North Uist in the Hebrides, Lochmaddy (Loch nam Madaidh) is the best example in Europe of a complex sea loch caused by postglacial drowning. It is an extraordinary lacework of islands, bays, lagoons and narrow twisting inlets, sheltering an internationally important ecological system. Turbulent tidal rapids result from the steps and sills (rock ledges) created by ice scouring across the Lewisian gneiss bedrock. If rugged areas are inundated, then the coastal outlines are very intricate, such as in Argyll. The Kintyre Peninsula is almost a series of islands, cut by the long narrow fingers of sea lochs.

Postglacial landscapes

As in warm periods (interstadials) in the Pleistocene (the earlier part of the Quaternary), the past 10 000 years (the Holocene) have been characterized by climatic instability. Rapid cooling occurred at 5500 and 2500 years ago, and had a marked effect on natural vegetation. In the first 500 years after the ice melted, the landscape would have been bare and rocky in upland areas, and waterlogged over low ground. Unstable rockfaces and melting of permafrost would have led to avalanches, rockfalls and mass wasting by gravity, probably exacerbated by earthquakes related to rapid uplift. Mosses and lichens were first to colonize the open landscape, to produce tundra (Finnish: moss-covered hill) that is typical of rather flat, featureless, boggy and treeless areas in northern Scandinavia, Siberia, Alaska and Arctic Canada. The mosses and lichens were quickly followed by juniper 9000 years ago, and by Scots pine some 500 years later on well drained sandy soils in the Northwest Highlands. Birch, hazel, rowan and willow followed, then oak, ash and elm south of the Highland Boundary Fault, to create woodlands and forests as the climate rapidly ameliorated. All the island groups in the north and west of Scotland were wooded with birch, hazel, willow and occasional pine, as can be seen from pollen records recovered by sampling mud on loch floors. By 4000 years ago, the woods on the exposed island chains of the Hebrides, Orkney and

Shetland had been obliterated by the early farmers, and the present-day landscape is one that has changed little since then.

A combination of cold wet conditions and forest clearance led to a decline in the native woodlands from 5500 years ago onwards, a decline that was to accelerate at the next cool event 2500 years ago, when blanket bog rapidly spread out across much of Scotland. Abundant evidence exists for deliberate felling and burning of forests over several millennia. Deforestation was complete by the eighteenth century, as trees were felled to create farmland, as well as timber for housing, shipbuilding, fuel, and to make charcoal for use in iron smelting and lime burning. The remaining woods were destroyed after the 1745 Jacobite uprising, to prevent Highland clans from seeking refuge. During the infamous Highland clearances, people were moved off the land to make room for sheep and deer. This had a devastating effect, for the small communities had been living in balance with nature,

ROGER JONES

Figure 8.17 Eas a' Chùal Aluinn waterfall, a hanging valley at Glencoul, Assynt, northwest Sutherland. It is the highest waterfall in Britain at 201 m. Cambrian quartzite scree in foreground, Lewisian gneiss on right of U-shape glacial main valley; Loch Beag in middle distance.

managing the land and the woods (by coppicing). Overgrazing by sheep and deer has caused widespread degradation of the forests and moors, made all the worse by the erosive effects of the animals' hooves: most hillsides are criss-crossed by animal tracks, and soil creep is evident in every part of the Highlands. Now all that remains of the legendary Caledonian pine forest, which once clothed much of the land, are a few isolated remnants in highland valleys inaccessible to deer.

Peat bogs are now the norm in many wilderness areas of northern Scotland. Stumps of Scots pine can often be seen poking out of bogs, such as on Rannoch Moor; the bogs are now too wet to allow regeneration of the forest. Peat bogs owe their origin partly to human interference with native woodlands: the bogs arose from soil erosion, increased runoff, waterlogging and acidification of soil, which prevented trees from re-establishing. Most of Shetland is now heather moorland, and the Flow Country in Caithness has nearly half a million hectares of blanket bog. The present treeless landscape is therefore not entirely natural. Somewhat ironically, proposals in the 1980s to reforest Caithness were met with vociferous protests.

In addition to human modification of the landscape, such as removal of peat lying on top of the post-glacial marine muds in the flat carselands* of the Clyde and Forth estuaries, which were previously flooded, catastrophic events strongly influenced settlement patterns and agricultural practices. For example, there were many eruptions from the Hekla volcano in Iceland, which caused local climatic change that affected vegetation in Scotland: pine and heather were reduced and blanket peat bogs expanded. Volcanic ash (tephra) from Iceland is easily recognized as fine black layers among peat, and these can be dated quite accurately. Settlements were abandoned in northern Scotland 3200 years ago, possibly as a result of volcanically induced deterioration of the climate. While the postglacial sea level was rising, the east coast was affected by a giant tsunami wave 7200 years ago. This event produced a clearly identifiable

sand layer around the Moray Firth shores, in Orkney and Shetland, and as far south as the Firth of Forth. The tsunami was caused by slumping of the Storegga slide off western Norway. This is the largest area of unstable sediment in the world, and the tsunami is reckoned to have built a 20 m-high wall of water that inundated the Scottish coastline in an instant. Any human habitations would have been carried away without leaving a trace. The cause of the slumping was the sudden conversion of frozen methane into gas. Methane emanating from the underlying bedrock froze in the permafrost layer; warming of the sea water converted the methane into gas, and the loose sediment collapsed catastrophically, creating a devastating tidal wave in its wake.

Coastlines are active zones that continue to be modified by the effects of waves and the wind (see Figs 4.16, 4.20, 4.22). One feature of the west coast, unique to northwest Scotland and Ireland, is the machair sand formation, created by onshore sea currents and winds, which transport shell debris onto the beach from the Minches (see Fig. 3.33). Cooling of coastal waters in glacial periods resulted in the mass mortality of planktonic micro-organisms. Their carbonate shells accumulated to form offshore sandbanks, added to by more quartz-rich mineral sand, carried by glacial meltwater streams. The most extensive machair lands form the narrow coastal strips around the Uists, Barra and Tiree, where the shell content is about 90 per cent. Beaches are fringed by dunes, stabilized by marram grass, and the blown sand, being carbonate rich, supports a great variety of plant life, creating Europe's rarest natural habitat. Fertile machair soils attracted some of Scotland's earliest settlers, and many archaeological sites are known from the Outer Hebrides, Orkney and Shetland, the most famous being Skara Brae on Orkney, which was overwhelmed by a sandstorm 4500 years ago and revealed in 1850 when the dunes were removed in another violent storm. Skara Brae is the best-preserved prehistoric village in northern Europe (see Fig. 9.3). On the east coast of Scotland, the dunes have a much higher quartz content, reflecting their origin from the erosion of granite, quartzite and schist, transported into the Moray Firth and other coastal outlets by large glacial meltwater rivers. The dune-backed beach areas in Easter Ross, Fife and East

---

* Carselands (kjaer, Old Norse: rough, scrub-covered marshy ground along a river estuary) have proved to be fertile farmland once the overlying peat is removed. In the east, they are famed for the soft fruit orchards alongside the River Tay. However, the clay soil tends to be sticky and is liable to waterlogging in wet weather.

PAT & ANGUS MACDONALD

Figure 8.18 Culbin Sands near Lossiemouth, Moray Firth: spits of shingle and sand, extending the coast in a westerly direction.

Lothian are known as "links".* There are very large dune systems west of Lossiemouth on the Moray Firth – these are the Culbin Sands between the mouths of the rivers Findhorn and Nairn (Fig. 8.18). Fluvio-glacial sand is constantly on the move by longshore drift because of current activity. Forest has now stabilized the dunes, some of which are 30 m high. Severe storms in 1694 totally buried Culbin village, the culmination of decades of dune drifting caused by human actions, in what was once known as the Scottish Sahara. Human settlements on this fertile coastal plain are known since 2000 BC (30 000 flint tools and other artefacts from Culbin are now housed in the National Museum in Edinburgh). Stripping of the turf layer (for use in house building) lying above the post-glacial sands exposed them to wind erosion, and to this day the sandy soils in the fertile Laich of Moray near Elgin are frequently moved by sandstorms, which create an ever-changing coastal landscape.

---

* A Scots word for gently undulating sandy ground near the shore, covered in turf and associated with golf links.

Oil rigs at entrance to Cromarty Firth, guarded by the North Sutor (Moine schist); Old Red Sandstone in foreground, Cromarty village, Easter Ross.

CHAPTER 9
Riches from the Earth –
Scotland's natural resources

Thanks to the variety of rock types formed in different plate-tectonic settings over 3 billion years of Earth history, Scotland has a wide range of natural materials with economic potential. The most significant, which are currently being commercially exploited, are North Sea oil and gas, coal and limestone from the Midland Valley, and baryte from Aberfeldy in Perthshire. Bulk materials in the form of crushed rock, and sand and gravel for use in the construction industry, are also of importance and are rather widely distributed in Scotland. Their bulk and low price make transport costs the most significant factor in their exploitation, so that most of these materials are used locally, close to where they are extracted. Other minerals exist, including metal ores, but they are generally in small deposits, and economic considerations have resulted in the closure of most such mines.

Mineral resources have been dealt with in the context of the regional geology. The details will not be repeated here, but instead this chapter presents an outline of how useful minerals form and why they occur in particular geological settings. It is such knowledge that allows new discoveries to be made, driving exploration forwards.

Introduction – what is a natural resource?

At its simplest, a natural resource is any material that could in principle be removed from the Earth and put to some practical use for the benefit of people. However, in practical terms it is more usual to consider reserves, which is the amount of a particular resource that could be extracted at a profit. For example, the total amount of coal in the United Kingdom has been estimated to be 200 billion tonnes, but only 4 billion tonnes of this could be mined at a profit. Hence the stock of coal, the resource, is 200 billion tonnes, and the lower figure constitutes the reserve. Economic factors could change the situation. If more efficient and cheaper ways could be found to mine coal, then some more of the 200 billion tonnes could become part of the reserves. One other term worth defining clearly is "mineral". In the economics of commodities, Earth resources are frequently referred to as "minerals" or "mineral resources", where the term can mean oil and gas, coal, metal ores, stone, crushed rock and other materials used ultimately in construction, agriculture and the chemical industry. However, the term "mineral" is used here for a naturally occurring solid with a crystalline structure, such as quartz, feldspar or calcite. "Natural resources" is therefore used in preference to "mineral resources"; "fuel minerals" are fossil fuels (peat, coal, oil and gas); and "bulk minerals" or "industrial minerals" are referred to here as aggregates or industrial materials.

Natural resources of geological origin (i.e. not including water, soil, wood or other organic materials) are grouped according to the type of material, under the following headings:

- Metal ores: these are minerals in the geological sense, and include oxides, carbonates, sulphides, and so on, of metallic elements, usually with other unwanted minerals (termed "gangue", e.g. quartz or calcite).
- Fossil fuels: peat, coal, oil, oil shale, bitumen, natural gas, which are extracted and burned; they are therefore non-renewable and cannot be replaced once used up.
- Industrial materials: a wide range including building stones, slate, polished stone, crushed rock for roads, and concrete aggregate, cement, brick clay, glass sand, agricultural and chemical raw materials (fertilizers, salt, sulphur, etc.).

Examples of all these categories occur in Scotland, and here we examine how the workings of the rock cycle and plate-tectonic movements have created conditions necessary for their formation in specific geological environments at distinct periods in time.

Metal ore deposits

Scotland has a wide variety of mineral deposits in rocks ranging in age from the Precambrian Lewisian gneiss to the Tertiary igneous intrusions. However, the distribution of the main ore mineralization is related to events surrounding the formation of the Caledonian Mountains. Ore deposits are found in igneous, metamorphic and sedimentary rocks, and result from processes in the rock cycle that concentrate small amounts of widely dispersed metals into potentially useful deposits.

The oldest deposits are found in the Loch Maree

Group, part of the Lewisian Gneiss Complex of the Northwest Highlands. These rocks represent a series of sedimentary rocks and basic lavas laid down 2200–2000 million years ago and metamorphosed to gneiss and hornblende schist. Closely associated with the metamorphosed igneous rocks are sulphide horizons, up to 4 m thick, containing iron, copper and zinc sulphides, together with native gold, and the body has been interpreted as a volcanic massive sulphide deposit laid down on thinned continental crust. The ultimate source for the igneous rocks – basic lavas, tuff, ash, sills and sheets – was the upper mantle. Once mantle material is introduced into the crust, it effectively remains there and useful minerals may become redistributed from this crustal reservoir by the workings of the rock cycle. During the eighteenth and nineteenth centuries, the Loch Maree ironstones were mined and smelted locally, using large quantities of native pine wood, which provided charcoal. Place names such as Furnace, and remains of smelters at Letterewe and small overgrown quarries, are the only reminders in the landscape of one of the very few heavy industries in the Highlands. Ironstone is also found in the Lewisian on Tiree and Iona. Other rocks of the Lewisian Complex – mainly gneiss and quartz–feldspar–mica pegmatite – are devoid of ore deposits. Several of the pegmatites have been exploited for use as abrasives, ceramic raw materials and insulators.

Some 750 million years of erosion followed, resulting in the Lewisian basement being uplifted from near the base of the continental crust to be exposed at the surface. Thick deposits of continental redbeds (the Torridonian sandstone) were laid down on this basement. Torridonian sedimentary rocks lack useful mineral deposits (in this context such rocks are referred to as "barren"), although blocks are used extensively as a local building stone. Pure sandstone (now white quartzite) and the Durness Limestone were laid down 550–500 million years ago, unconformably on top of both the Torridonian and the Lewisian rocks. The carbonate-rich sedimentary rocks have made useful fertilizers and, where they have been intruded by granitic rocks, they have acted as hosts to iron ores in skarn deposits. Magnetite on Skye belongs to this class, and the Broadford marble (Skye) and Ledmore marble (Assynt) are thermally metamorphosed Durness Limestone outcrops in the aureoles or contact zones of younger intrusions. These marbles have been quarried, then cut and polished for use as decorative facing stones.

East of the Moine Thrust and northwest of the Great Glen Fault lie the Northern Highlands, composed of the Moine schists, a thick sequence of shallow-water sedimentary rocks deposited on continental crust of normal thickness, then folded and metamorphosed 1000 million years ago. The dominant rock types are quartzite and mica schist, virtually barren of mineral deposits. Similar in composition and equally barren are the rocks of the Grampian Group at the base of the Dalradian, immediately south of the Great Glen Fault. However, the middle part of the Dalradian was laid down in a quite different environment, dominated by thinning and stretching of the lithosphere. This eventually culminated some 750 million years ago in rifting and continental rupture, which led to the creation of the Iapetus Ocean and an influx of heat and magma from the upper mantle. Interconnected faults and fractures on the scale of the whole crust allowed hydrothermal convection systems to become established; the giant barium–zinc deposit at Aberfeldy formed in this environment (see p. 83). As the ocean continued to develop, rifting extended and igneous activity increased, during which copper–zinc sulphide mineralization occurred in the Middle Dalradian sediments that were forming. Gold deposits in Argyllshire (e.g. Cononish) are associated with this period, and the lead–zinc–silver deposits at Tyndrum represent metals from the crustal reservoir that were remobilized (by being picked up in solution) and concentrated in veins, faults and fractures in late-stage events within the Dalradian terrain (Fig. 9.1). Farther to the northeast in Aberdeenshire, very large layered basic–ultrabasic igneous bodies were intruded into the Dalradian 490 million years ago during folding and metamorphism. Locally, for instance at Knock Hill, Huntly, there are layers of iron–copper–nickel minerals in these intrusions, which settled out as dense segregations during crystallization. By 450 million years ago, the Iapetus Ocean was beginning to close and the Moinian and Dalradian rocks in Shetland collided with a volcanic arc in the ocean, resulting in thin slices of ocean floor and upper-mantle rocks being thrust over the metamorphic basement rocks of the continental crust. This

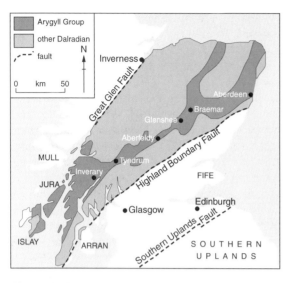

Figure 9.1 The location of ore deposits in Middle Dalradian rocks.

An event with major consequences for ore-deposit formation was the intrusion of many granite bodies in the end stages of the continent/continent collision that gave rise to the Caledonian Mountains. Most of the granites are concentrated in the Grampian Highlands, with rather fewer in the Northern Highlands and the Southern Uplands. The surface volcanic rocks, of andesite composition, accompanied the intrusive events in the Grampian Highlands and Midland Valley. The commonest type of mineralization associated with this end-Caledonian igneous activity was the formation of copper, molybdenum and gold in veins and shear zones in strike-slip faults adjacent to the granites. Late-stage uranium mineralization is found in several Caledonian granites, including Helmsdale in eastern Sutherland and Criffel–Dalbeattie in Dumfries and Galloway.

Ordovician greywackes in the Southern Uplands acted as the host rocks for the extensive lead–zinc–copper and gold–arsenic–antimony mineral veins (e.g. at Glenhead Burn on the Loch Doon granite), whereas the overlying, younger Silurian rocks are barren. An exception to this general rule is the Glendinning deposit, from which 200 tonnes of stibnite (antimony sulphide) was mined; stibnite is the main ore of the metal antimony, used in anti-friction alloys, batteries and the manufacture of fireworks. This deposit in Silurian greywackes has many metal ores, including iron, arsenic, gold, antimony, lead, zinc, copper and mercury.

The type of mineralization in the Caledonian mountain belt associated with the 400-million-year-old granites and andesite lava flows is referred to generally as being of porphyry style, wherein the main metal is usually copper, sometimes also molybdenum. In this style, the forceful intrusion of pipes into intrusive and volcanic rocks causes brecciation (fracturing), allowing extensive vein systems to be set up in which sulphides are concentrated in the veins, as well as in zones of alteration around the fractured igneous rocks. None of the many deposits has any economic significance, as they are on too small a scale, although they are of academic interest. Examples include the Ballachulish granite, and Glen Gairn in the Cairngorms, with tin, tungsten, molybdenum, copper, lead, zinc and rare topaz.

The important Lagalochan gold–silver deposit in

produced the Shetland ophiolite, now seen as fragments in Unst and Fetlar. The mineral chromite occurs as many pods and lenses in layered peridotite. Associated with the chromite, which was once mined on Shetland, are signs of platinum mineralization, which is not economic at present. Industrial talc, which formed within serpentinized mantle material, was also mined from this rock complex and exported from Shetland.

By about 435 million years ago, the Iapetus Ocean was being rapidly closed by the collision of three continents – Laurentia, Baltica and Avalonia. In the north of Scotland, sinistral (left-directed) major northeast–southwest strike-slip faults moved the Northern Highlands, Grampian Highlands, Midland Valley and Southern Uplands together along the Great Glen Fault, the Highland Boundary Fault and the Southern Uplands Fault. Subduction continued and thick wedges of sedimentary greywackes in the Southern Uplands were stacked together as Laurentia was pushed over Avalonia. Folding and low-grade metamorphism of the greywackes produced the Southern Uplands fold belt and eventually, by 395 million years ago, north and south Britain were joined along the Iapetus Suture, marked by the line of the Solway Fault in Scotland, which connects to the Navan–Silvermines linear structure in Ireland, which is a very important zone of lead–zinc mineralization.

Argyllshire is associated with the lavas and small volcanic pipes of the Lorne Plateau and has disseminations (widely scattered fine-grain mineralization) of lead, zinc, silver, gold, arsenic, antimony, molybdenum and copper. Reference has already been made to the Devonian hot-spring deposit at Rhynie, in Aberdeenshire (see box on p. 117), where the world-famous early plant fossils are preserved in chert that contains much gold, arsenic, antimony, tungsten and molybdenum. In eastern Sutherland, the Kildonan gold deposit, scene of the famous gold rush of 1868–70 at Baile an Òir ("gold town") is found in modern river sediments. This is known as a placer deposit, and the gold is alluvial, being a very dense metal, it is concentrated among the heavy sediments in the stream bed. The gold is thought to have been derived from a mother lode[*] in granite veins cutting Moine rocks in the vicinity of the mountain Scaraben, but the source itself has never been found and may actually have been completely eroded away.

The final event in the Caledonian evolution of Scotland was uplift of the mountain chain, followed by rapid erosion and the formation of thick continental redbeds deposited by rivers on a desert floor – the Old Red Sandstone of Devonian age. Although these sedimentary rocks are stained bright red by iron oxide (hæmatite), they are almost devoid of metal-ore deposits, exceptions being the high gold and silver content of the Rhynie chert and also uranium in Old Red Sandstone shales and conglomerates adjacent to both the Helmsdale granite (at Ousdale in Caithness) and to basement gneiss and granite in Orkney (Stromness and Graemsay).

The original source of the uranium mineralization is believed to be the granites of the Northern Highands. Many of the fossil fish beds in the Caithness Flags are highly radioactive, probably as a result of uranium salts entering the skeletons of the dead fish stagnating in huge numbers on the floors of shallow lakes. Uranium is also found in peat beds overlying the flagstones, and has been leached out into groundwater over thousands of years. Caithness and Orkney are considered to be a minor uranium province and, in case of strategic need, there could be some potential for strip mining. However, uranium concentrations

are low and large amounts of peat and bedrock would need to be removed, with untold consequences for the environment. Political considerations would override economic factors. Currently, there is a glut of uranium ore and the price is very low. Fine-grain sandstones of this age were used locally as building stones. Generally, it is only the igneous rocks of Devonian age that contain useful minerals. Caledonian granite has been used in bulk as crushed rock aggregate, most notably the Strontian granite at the Glensanda coastal superquarry in Loch Linnhe, near Oban (Fig. 9.2).

By the start of the Carboniferous period 354 million years ago, Scotland was approaching equatorial latitudes and the Midland Valley became flooded by a warm shallow shelf sea in which organic limestones formed. These limestones played host to extensive lead–zinc–silver–copper sulphide mineralization in Ireland, but in Scotland this event was minor in com

Figure 9.2 Strontian: a fault cutting the Strontian granite, and waste heaps from former lead–zinc mines along the fault line.

---

[*] The original ore body from which a placer deposit has been derived.

195

parison, and is recorded in only a few lead–zinc veins in Argyllshire and the Southern Uplands. The coral reefs were killed off later in the Carboniferous, as rivers brought down silt and sand onto coastal plains in response to the elevation of upland regions.

The next geological event was in many ways the most significant in terms of Scotland's economic development: the formation of thick coal seams and associated sedimentary ironstones. In mid-nineteenth century Scotland, 25 per cent of Britain's iron came from this source. The two main types of sedimentary iron ore in the Carboniferous rocks are blackband ironstone, made of finely interbedded iron carbonate (siderite) and black organic mudstone, and clayband ironstone, in which the siderite filled the pore spaces of unconsolidated clay sediments. Concretions and nodules without any bedding or banding make up the clayband (or claystone) ironstone, in contrast to the blackband ore type. Both types formed by the precipitation of iron compounds in fresh or stagnant water, or in swamps where fine sediment was deposited very slowly. Bog iron ore is a related type, but with a lower iron content; this was one of the first forms of iron ore to be worked in Scotland, and the extensive native forests were used in the seventeenth and eighteenth centuries for charcoal in the smelting process, at small local plants called bloomeries.

The first coal-fired furnace, the Carron works, opened near Falkirk in 1760 and the industry developed rapidly in the next hundred years, particularly near Airdrie and Coatbridge. It was the fortuitous occurrence of coal and iron ore together that provided the engine for the industrial revolution centred in the Midland Valley. By 1950, technological advances in smelting, coupled with the demand for richer and purer ores, resulted in the Scottish ironstones being abandoned. Ironstones extend through much more of the Carboniferous in Scotland than elsewhere in Britain. Blackband ironstone occurs immediately above coal seams and could be exploited at the same time as the coal was being mined. The black bands are thin layers of organic-rich sediment, and the high carbon content of this type of iron ore allowed it to be smelted with relatively ease. Like the coal, blackband ironstone seams repeatedly occur in a cyclical fashion as a result of periodic flooding of the coal swamps by fresh water. The siderite was probably precipitated from swamp waters, in the presence of organic matter.

Clay ironstone formed mainly in lake sediments, near sea level, and beds are usually in the middle part of a coal cycle (i.e. some distance above a coal seam). Iron carbonate formed a cement as water was being expelled from the fine silt deposits during burial. Bacterial action in the still, stagnant lake waters caused the iron-rich nodules to grow just below the sediment surface. Fine-grain lake mudstones and oil shales usually follow on top of the ironstone bands. Sedimentary iron ores with about 30 per cent iron were classed in Britain as ore reserves and were therefore worth exploiting; this figure is low compared with the 60 per cent cut-off value of Australian sedimentary iron ores, below which production is considered unprofitable.

Mineralization associated with Carboniferous and Permian igneous activity 300 million years ago is seen in the form of silver–cobalt–barium veins in the Midland Valley, for example the Muirshiel baryte mine in Renfrewshire. In the Ochil Hills there are silver veins in brecciated lava (see p. 113), and the Strontian lead–zinc mines also date from this period. The Strontian granite is a Caledonian intrusion 400 million years old, and the later vein mineralization was a response to crustal stretching 300 million years ago, related to an early phase of rifting in the area that was to become the North Atlantic Ocean. At Strontian, baryte was previously a waste (gangue) mineral during the period when lead and zinc were being mined, but later it attained value in its own right when North Sea operations required baryte for use as a heavy drilling mud (baryte, barium sulphate, is also known as heavy spar because of its high density) to prevent pressure blowouts in oil wells. Occasional native copper veins are found in the Clyde Plateau basalt lavas of Lower Carboniferous age, but are of no economic significance.

By the Permian period, Scotland was part of the Pangaea supercontinent, created by the collision of Laurentia and Gondwanaland (the southern continents). Being just north of the Equator in the middle of a large landmass, the climate was that of a hot dry desert. All of Scotland was a land area, and sedimentation was dominated by dune-bedded sands. In the Permian–Triassic interval, the New Red Sandstone beds were formed; these were, and they remain, an important source of building stone. Metal ores are almost absent. However, in the North Sea area, shallow salt seas were present and high evaporation led to salt deposits (gypsum, anhydrite and halite) forming, which are extremely important in their role as oil and gas traps. Rifting of the area, following the collapse of a large thermal dome, allowed sedimentary basins to form, which filled with fine organic-rich muddy sediments in the Jurassic period. The thick sediments were buried and became the source material for North Sea oil. Around the margins of the Inner Hebridean islands, there are outcrops of Jurassic sedimentary rocks such as limestones, sandstones, mudstones and shales, including several ironstone horizons. One of these beds was worked in Raasay; here the deposit is of layers of iron carbonate and iron oxide, often in oolitic rocks (i.e. with small round egg-shape grains).

These rocks formed in shallow warm basins surrounded by low hills, from which rivers occasionally transported iron-rich material from the land. Sedimentation rates were very low and the iron was gradually concentrated in the sediments, as small sand grains and shells were gently winnowed back and forth on the sea bed and iron-rich layers built up around tiny particles, which may initially have been made of iron silicate that was later replaced by siderite (iron carbonate). Siderite then grew to fill pore spaces between the ooids during burial, compaction and diagenesis. This particular ironstone is of less value as a resource because of the presence of calcite and phosphate with the iron minerals.

By 60 million years ago, the North Atlantic Ocean had finally begun to open by rifting, in part because of the formation of a large thermal dome beneath eastern Greenland, Scotland and Norway. This was the Tertiary volcanic episode, which gave rise to the basalt lavas and central igneous complexes of the Inner Hebrides. During this event, layered ultrabasic bodies were intruded in Skye and Rum, some containing layers of chromite and olivine. Erosion removed these dense minerals, which were later sorted by marine currents to form placer deposits off the Rum coast (see p. 150); these are potential reserves for future exploitation. Tertiary granites in the Red Hills of Skye have intruded Durness Limestone in places, causing thermal metamorphism and the formation of skarn deposits of iron ore (magnetite, an iron oxide) within marble.

Parallel to the northwest–southeast trend of dolerite dykes emerging from the Tertiary centres are various metalliferous mineral veins in the Midland Valley, including in particular baryte, which was mined at Muirshiel in Renfrewshire and Gasswater near Muirkirk in Ayrshire. These mines yielded about 800 000 tonnes before being closed in the 1960s. Baryte occurs in veins cutting Clyde Plateau basalt lavas at Muirshiel and in Carboniferous sandstones at Gasswater. It is possible that the fractures are older than the dykes, and the dykes followed earlier lines of weakness. The veins are probably related to igneous and hydrothermal activity, which was a phase in the early stages of the opening of the North Atlantic Ocean, 240 million years ago, when tensional stresses developed that eventually led to continental rifting. Baryte is also commonly found with the metallic-ore mineralization in the Ochil Hills, near Alva, but it is considered to be a waste product or gangue mineral, much less valuable than the silver, nickel, cobalt and copper.

Almost all of Scotland's ore deposits are small and insignificant; many were surveyed during the two world wars of the twentieth century as potential strategic reserves, but there was little exploitation at the time. Since the 1980s there has been a slow worldwide decline in the price of metals, and many materials (e.g. lead in car batteries) are recycled or alternatives have been found (e.g. copper and plastic piping to replace lead pipes in domestic water supplies). Lead-free petrol is now standard for use in cars, so the demand for new sources of lead ore (galena) is greatly reduced.

Industrial raw materials

Although the popular idea of treasures from the Earth doubtless includes gold, silver and precious stones, there are more mundane and utilitarian resources to be obtained, and in Scotland these raw materials have greater economic importance than metal-ore deposits. Such materials include building stones, aggregate (crushed rock used in bulk for concrete and road construction), fertilizers, brick clays, glass sand, and many others used in chemical and manufacturing industries. Scotland is fortunate in having a wide variety of these resources, some of which, particularly crushed stone and limestone for cement, are available in significant quantities.

Figure 9.3 Skara Brae, Orkney: 5000-year-old houses and furniture made of Devonian flagstones.

Stones for building

In prehistoric times, stone was used for tools, particularly the glassy volcanic rocks (pitchstone, obsidian and bloodstone) found in the Tertiary Volcanic Province on Arran, Rum and Eigg. Flint from Cretaceous (Chalk) deposits at Boddam in Aberdeenshire was also used for making tools. People have used local stone to build their houses in Scotland for at least the

past 5000 years, and some of these constructions were so sound that they have withstood the ravages of time and the elements, such as the settlements at Skara Brae (Fig. 9.3) and Maes Howe in Orkney and Mousa Broch in Shetland (see Ch. 4), which are made of durable flagstones of Devonian age. Ancient stone monuments and ritual sites include the standing stones of Callanish, Lewis, made of slabs of Lewisian gneiss (Fig. 9.4), and the equally awe-inspiring Ring of Brodgar on Orkney was constructed from thin sheets of flagstone (Fig. 9.5). Dùn Carloway on Lewis is built of Lewisian gneiss cubes, directly onto a gneiss outcrop on a knoll that commands a magnificent view (Fig. 9.6). The original quarries for these constructions are relatively nearby and in all cases they are no more than 10 km away, but to have moved the rock to its site of construction must have been a considerable feat 5000 years ago. St Magnus Cathedral in Kirkwall (twelfth century), Orkney, is a superb example of the use of local stone (red and buff sandstone) in Scotland's only cathedral fro the early medieval period that remains intact.

Quarried stone used in construction is referred to as dimension stone. In Scotland, the commonest dimension stone is sandstone, of various ages, from

Figure 9.4a Callanish, Isle of Lewis, Outer Hebrides: standing stones (second only to Stonehenge in importance) of Lewisian gneiss slabs; an early example (5000 years old) of the use of local materials in construction; the vertical faces are foliation planes.

Figure 9.5 Ring of Brodgar, Orkney: prehistoric stone circle made of Devonian flagstone slabs (27 remain of the original 60; Neolithic, 5000 years old); the vertical faces are bedding planes.

Figure 9.4b Standing stone: hornblende pod forming the "eyes" of the mythical people that the stones were supposed to represent.

the Midland Valley and the northeast coastal fringes. Apart from its use in prestigious buildings, most stone has been used locally, mainly because of the high cost of transport. Irregularly shaped random blocks have been used as rubble walls, and only the outer façades of buildings tend to have dressed stone. Since local stone predominates, the character of architecture in different parts of the country is strongly influenced by the geology of the region. In Orkney, Caithness and Shetland, the thinly bedded Caithness Flagstones are

Figure 9.6 Dùn Carloway, Isle of Lewis, Outer Hebrides: a 2000-year-old broch or Atlantic roundhouse, made of local Lewisian gneiss, on which the broch is built; occupying a prominent place in the landscape.

used extensively for house walls and roofs, and as field boundaries, giving a highly distinctive feel to the built landscape (Fig. 9.7). From 1825, the first flagstones were shipped from Castletown in Caithness to the Midland Valley for use as paving slabs, and many more went by rail. The flagstones are well bedded fine-grain sandstones and siltstones that were deposited in a lake that occupied the area around the Moray Firth, Caithness, Orkney and Shetland during Devonian times (380 million years ago). The rock is easily worked – use is made of bedding planes and natural

Figure 9.7 Caithness Flagstones used as a field boundary; between Halkirk and Thurso, Caithness.

vertical joints – and is available in large quantities of uniform shape, size and colour. Being very well cemented, it is a tough and durable stone.

Where the bedrock is of schist and gneiss, as in much of the Highlands and islands, the building styles are very different from those of the northeast. These crystalline rocks are much more difficult to work, and the tendency was to use irregular rounded blocks and boulders from beaches and from glacial deposits, which were then packed with smaller stones and rendered with mortar on the outside.

In southern Scotland, Old Red Sandstone, Carboniferous sandstone and New Red Sandstone have been used to great effect in all the large towns and cities, as well as for prestigious buildings and viaducts (Figs 9.8, 9.9). Most of Edinburgh is constructed of cream Craigleith sandstone, a particularly resistant sedimentary rock that hardens on exposure to the atmosphere and becomes much tougher. This is as a result of the grains becoming progressively more firmly cemented as the freshly quarried sandstone dries out. So good was the Craigleith sandstone that it was exported to Europe and to London, where it was used in the construction of Buckingham Palace. The main properties of this building stone are the uniform bedding, even grain size, strong silica cement, consistent colour, and

Figure 9.8   Leaderfoot rail and road bridges over the Tweed; made of local Old Red Sandstone; former river terraces are visible on the left.

Figure 9.9   Glamis Castle, Angus: built of local Old Red Sandstone, mainly in the seventeenth century, it was the home of the late Queen Elizabeth the Queen Mother, and birthplace of the late Princess Margaret.

the presence of joints and bedding planes that facilitated the extraction of large blocks, with minimal use of explosives. Such sandstones are referred to as free-stone (liver stone), meaning that it could be worked equally in all directions. Much of the red sandstone used in Glasgow and Edinburgh was quarried from the Permian rocks of Dumfries, especially the Mauchline basin. These sandstones were laid down on a desert floor and they contain traces of dune bedding. Sand grains in the rock are even and very well rounded, having been shaped by the wind. Some of the quarries operated in deposits that are up to 1000 m thick. Permian–Triassic (New Red) sandstone has also been quarried extensively around the south shores of the Moray Firth, especially at Hopeman near Elgin (see Fig. 6.2). The variety in Morayshire is much paler than the deep purplish-red New Red Sandstone from Dumfries, Sanquhar and Lochmaben, and it resembles the underlying Old Red Sandstone, with which it was confused until fossil reptiles and their footprints were found, proving a younger age. Hopeman Sandstone was used in the construction of the Museum of Scotland in Edinburgh (Fig. 6.3), built in the 1990s, as well as the cloisters of Elphinstone Hall, Aberdeen University (built in 1931). Off shore in the Moray Firth, this sandstone is an important reservoir rock for North Sea oil.

The use of granite as a building stone in the Grampian Highlands and in Dumfries and Galloway is greatly influenced by its occurrence locally. Aberdeen in particular is famous for its extensive use of the local silver granite, so much so that it is known as the Granite City (Fig. 9.11). Most of the material was derived from a single quarry at Rubislaw, opened in 1741 and once known as the deepest hole in Europe. As well as granite's use as a building stone for houses and public buildings, it was also in demand as a load-bearing stone in bridges and docks. The Forth Railway Bridge piers (1890) are made of Rubislaw granite (Fig. 9.10), and the Liverpool docks were constructed of the Criffel–Dalbeattie granite (sandstone from Brodick on Arran was also used). Much was also used for kerbstones and setts (cobble stones), and the strength and durability gave such added value that granite was exported profitably beyond the confines of local quarries. It has also been used as an ornamental polished stone, for plinths, facing stone (i.e. thin slabs used to clad the outer façade of a concrete or other construction) and gravestones. Granite from the Ross of Mull was exported by ship to many countries around the world, including China, New Zealand and America. It was used in London (Westminster Abbey, Albert Memorial, Blackfriars Bridge), Liverpool Docks, and Scottish lighthouses at Skerryvore, Ardnamurchan, and several others.

The crystalline nature of granite (and other igneous rocks) allows it to take on a high polish, which adds to its durability and enhances its appearance.

Figure 9.10 The Forth railway bridge: built in 1890, the northern piers are located on outcrops of Midland Valley Sill (a Carboniferous-age dolerite intrusion); the southern piers in the foreground are built of Aberdeen granite from Rubislaw quarry; the bridge has an asymmetric design in order to avoid sinking piers in the 170 m-deep glacially scoured channel in the river bed.

Figure 9.10 Marischal College, Aberdeen University: an imposing granite façade, made of local Rubislaw silver-grey granite (Caledonian); the second largest granite building in Europe after the Escorial Palace in Spain.

Natural joints in granite intrusions, often nearly vertical and horizontal, were utilized to great effect in quarrying. Other igneous rocks used as stone blocks include dolerite and basalt (whinstone), mainly as kerbstones and setts. Dolerite and basalt tend to have closely spaced cooling joints and are difficult to work into large dressed building blocks.

The main metamorphic rock used in construction has been slate from the Caledonian mountain belt of the Southern Highlands, principally the quarries at Aberfoyle, Easdale and Ballachulish (see pp. 80, 82, 83). Slate was originally a fine-grain sedimentary rock (mudstone, shale or claystone), which was recrystallized during folding and metamorphism, when slaty cleavage was developed, to produce evenly spaced planes of weakness allowing thin sheets to be split. Slate from the Highlands was distributed extensively across Scotland. More locally, slate of poorer quality was used in the Southern Uplands, to complement the architectural style of the local Ordovician and Silurian greywacke blocks, and Devonian red sandstone window and door lintels, also locally derived (Fig. 9.12).

Figure 9.12 Contrasting local building stones in a nineteenth-century cottage: dark irregular greywacke in the walls; light soft, workable Old Red Sandstone around the window (Lauder, Borders region).

The "slates" in the Southern Uplands are actually thinly bedded fine siltstones, not true metamorphic slates, and the cleavage is along bedding planes, as opposed to the slaty cleavage of true slates; hence, the individual slabs are thicker and heavier. In Caithness the situation is similar, where very thin flagstones, larger than slates, are used as roofing material.

Other metamorphic rocks used in ornamental work include the greenstones from Argyllshire. These are metamorphosed basalt lavas from the Dalradian rocks. The intricately carved Kildalton Cross (eighth century) on Islay is a magnificent example (see Fig. 1.5). Other well known materials are marble (metamorphosed limestone) from Iona, Tiree (both are Lewisian), or the metamorphosed Durness Limestone from Broadford (Skye) and Ledmore (Assynt). The Portsoy marble from Banffshire is not a true marble but a green and red serpentinite, which was an ultrabasic igneous rock, metamorphosed in a shear zone. These rocks are relatively soft and take on a high polish; they are mostly used as slabs internally within buildings, for example, the altar in Iona Abbey, and that in Westminster Abbey, which is of Iona marble.

The number of quarries in Scotland still producing dimension stone diminished sharply from the mid-twentieth century onwards, but there are still some small-scale operations where Old Red Sandstone, Carboniferous sandstone and New Red Sandstone are being extracted, including the Caithness Flagstones. Defunct quarries are sometimes temporarily reopened (e.g. Ross of Mull, Hopeman, Kemnay) when restoration work is being carried out on historic buildings. Building-stone operations are labour intensive and the work has to be carried out more or less by traditional methods to select, split, cut, lift and dress the stone; hence, it tends to be relatively expensive.

Stone aggregate

Although the demand for cut and shaped load-bearing building stone may have declined because of high cost, the opposite is the case for crushed rock or aggregate, which is used to make concrete, roadstone and railway ballast. In Scotland, almost all the crushed rock comes from quarries in igneous and metamorphic rock: dolerite, granite, quartzite and greywacke. The dolerite is mainly derived from Carboniferous sills in the Midland Valley, and the greywacke comes from the Southern Uplands. Dolerite, which is very abundant, is ideal for roadbuilding purposes. The interlocking three-dimensional crystal lattice structure is particularly strong; the random-shape chips pack well and can bear a heavy load. Also, the minerals in dolerite are impervious to rain and chemical spills, and the rock bonds well with asphalt; the surface has excellent skid resistance and does not weather smooth.

Other rocks can be used for the layers beneath the top surface of a road (known as the wearing course), except for metamorphic rocks such as slate or schist, which have a very strong preferred orientation and would therefore tend to form planes of weakness when compacted. Basalt has a composition similar to dolerite, but is of finer grain and often has gas cavities filled with soft minerals such as calcite, carbonates or zeolites. Not infrequently, lavas are more weathered than dolerite and hence are weaker. Because dolerite is common and inexpensive, production has always tended to be local, and so there are countless small roadstone quarries scattered all across Central Scotland. The red felsite chips in Lanarkshire (from the Tinto felsite) give the roads there a characteristic appearance. However, felsite has a lower skid resistance than dolerite, and is now used much less.

Transportation costs are too high to make it worthwhile to move crushed rock great distances by road. However, this situation is changing, as the greatest need for aggregate is in Southeast England and the largest supplies are in Scotland, so that prices have risen to accommodate the significant costs of transport. One form of transport that is cost-effective is the ore-carrying ship (which can carry aggregate for 1 per cent of the cost of road transport), and this has stimulated the development of coastal superquarries. The only one currently operating in Scotland is the Glensanda superquarry on Loch Linnhe, which is working the Strontian granite. Rock is blasted and crushed on site and loaded onto tankers for export to North America, Europe and Southeast England; much was used to line the Channel Tunnel. A superquarry is defined as one that can provide at least 5 million tonnes of rock per annum, and has reserves in excess of 150 million tonnes, equivalent to 30 years of activity. This means that large igneous bodies near the coast may be targets for exploitation. The Scottish Executive envisages that, by the year 2009, there could be a maximum of four superquarries in Scotland, including Glensanda provided they are in widely separated rural areas. Well over 20 sites around the coast have been identified, where rock is available in adequate quantities and there are suitable deepwater facilities (mostly ice-scoured sea lochs).

Since 1995, a controversy has been raging about planning permission to establish a superquarry at Rodel in South Harris, where an anorthosite body in the South Harris Igneous Complex of Lewisian age has been identified as a suitable location. The Glensanda superquarry was predicted to be producing 10 million tonnes per annum by 2000, but in fact has never yielded more than half that amount, and questions are being raised about the need for more such quarries in Scotland. Glensanda is an interesting example of the multiple use of a resource at different periods in time. The Strontian granite contains lead–zinc veins in which baryte is a waste product. When the metal-ore mines became unprofitable, North Sea oil was in full production and the demand for baryte increased, so the mines were reopened, this time for baryte. But eventually these mines became unprofitable and were closed in the 1980s. Shortly after that, the superquarry opened. Strontian is also well known as the type locality of the mineral strontianite, which gave the name to the element strontium.

Superquarries are major installations, which have enormous impact on the landscape: the quarry itself, approach roads, storage facilities, crushing plants, berthing, as well as visual intrusion, noise, dust, waste, and pollution of sea water, although they do provide a livelihood for some.

The other major source of crushed-rock aggregate is sand and gravel from glacial deposits in the northeast lowlands, Fife and the Midland Valley, where glacial meltwater streams brought large quantities of igneous and metamorphic rock fragments from the Highlands. Most of the production is used in concrete manufacture, and the sand fraction is particularly valuable, since it has been washed clean by fresh water. Sand dredged from offshore deposits has to be washed before it can be used in construction. From the point of view of geological conservation, the downside of sand and gravel extraction is that fluvioglacial landforms (eskers and kames) are lost in the process

and the depleted sites often do not enhance the rural landscape. Like crushed rock, the price of sand and gravel is low and it is worked and used only locally.

Clay

Clay is required for the manufacture of bricks, and in Scotland most of the supply has been derived from mudstone and shale waste produced in coal mining areas, hence the frequent association between brickworks and mines, and miners' cottages built from bricks. Small amounts of clay were also previously obtained from claypits in glacial deposits, especially in the east of Scotland. A different type of clay, used for fire-resistant bricks in iron smelters, is fireclay, which is mainly the aluminium-rich clay mineral kaolinite. Fireclay occurs in Carboniferous Coal Measures rocks and it originated as soils on which coal swamps became established. The muds that originally formed the soil came from the weathering of igneous rocks, in which the feldspar broke down in humid climatic conditions, to kaolinite. Bentonite is a type of clay that is present in sedimentary rock successions of several different ages in Scotland, being a soft white material derived from the breakdown of volcanic ash; it has uses in filtration and purification processes, but there are no deposits currently being worked.

Limestone

Limestone has been worked in Scotland since the agricultural revolution of the mid-eighteenth century, for use as a soil improver, and lime kilns dot the landscape of Central Scotland. The limestone is mostly of Carboniferous age, and it was burned with coal or charcoal to make slaked lime, which was then spread on fields from which acid peat had been stripped. Some limestone has been used as an aggregate for concrete or roadstone. None was ever used as a dimension stone, although some varieties of fossiliferous limestone have been cut and polished and used as a simulated marble in ornamental fireplaces, for example. Thermal metamorphism of the Durness Limestone in Assynt and on Skye has produced marbles used as ornamental polished stone (see p. 61). Currently, the main use for limestone in Scotland is in the manufacture of cement, and the very large quarry at Dunbar in East Lothian is fortuitously located on an outcrop of interbedded Carboniferous limestone and shale in exactly the right proportions for cement. The beds are almost horizontal and a conveyor-belt operation removes material in a continuous process; then the excavated ground is filled in with waste, which is subsequently landscaped and restored to agricultural use. The raw material is transported by conveyor to a kiln, where the rocks are ground, mixed and heated to make cement. Estimates made in 2002 suggest that the remaining resource in the fields around Dunbar is sufficient to last until 2030.

Another type of soil improver and slow-release fertilizer is the potash-rich shale from the Northwest Highlands, the Fucoid Beds of Cambrian age (525 million years old), found in a strip of land from Loch Eriboll (on the north coast) to Loch Broom; the rock forms part of the shallow-water succession that includes the Durness Limestone. The bright green grassy swards in the Northwest Highlands are often located on outcrops of the Durness Limestone, as at Elphin in Assynt, and they form distinctive landscape features in contrast to the heather and bracken on basement gneiss rocks.

Limestones are also present in the Dalradian rocks, and these too have been worked locally in Shetland and the Grampian and Argyll Highlands, for both roadstone and agricultural use, as at Dufftown, Tomintoul, Portsoy, Banff, Fort William and Blair Atholl. Limestone in the Highland Border Complex (see p. 89) was worked at Aberfoyle and converted to agricultural lime at nearby Dounans. Remnants of the industry are still preserved in the area of the Highland Boundary Fault nature trail at the David Marshall Lodge visitor centre, Aberfoyle. As in the case of the Durness Limestone, some of the Dalradian limestone outcrops have given rise to more fertile pasture land, an example being Kilmartin in Argyllshire, which is well known for its many prehistoric sites, and which provided an important area of human settlement for several millennia. On Orkney, nodules in some of the Devonian fish beds in the flagstones have a high lime content and were formerly quarried for use as building lime. Soil fertility on Orkney owes much to the lime content of the bedrock.

Sand

The two main uses for sand are in concrete manufacture and glass making. Glass sand (or silica sand) is

Figure 9.13 Loch Aline, Argyllshire, near Ardnamurchan; a Cretaceous glass sand mine.

obtained from the pure white Cretaceous sandstone at Lochaline on the coast of Morvern, opposite Mull (Fig. 9.13). This is 99.9 per cent pure silica and was formed when desert sand was blown into a warm shallow tropical sea some 95 million years ago. It is so pure that it was once used to make glass lenses. Silica sand is also obtained from clean Carboniferous sandstones and recent windblown sand. In addition to its use in glass manufacturing, sand is also used for making mouldings in the iron and steel industry, although demand for this purpose is now sharply reduced.

Other materials

Although Scotland is known to have over 500 different minerals, the country is not particularly well endowed with precious stones. Minerals such as cairngorm (smoky grey or black quartz) are very rare, and other semi-precious stones are few and far between. However, there are many sites known to collectors, where individual specimens have been obtained of an extremely wide range of minerals. Examples include sapphire at Loch Scridain on Mull, formed by the thermal metamorphism of sedimentary rocks trapped in igneous sills. In the Midland Valley, silica-rich Devonian lavas contain agates. Many localities are to be found along the east coast, especially at Montrose and in north Fife, and several on the Ayrshire coast. Carboniferous and Tertiary basic lavas sometimes contain zeolites in gas cavities, as well as occasional agates, chalcedony, jasper and bloodstone.

The first compilation of minerals known in Scotland was published in 1901 as *The mineralogy of Scotland* (two volumes) by Matthew Forster Heddle, from Hoy on Orkney. Heddle was Professor of Chemistry at the University of St Andrews, and his collection, housed in the Museum of Scotland in Edinburgh, is judged to be one of the finest of any single country. Heddle's glossary included 162 minerals, half of them silicates. Some famous minerals first described in Scotland and named after Scottish places include strontianite, caledonite, lanarkite, leadhillite, greenockite, tobermorite, mullite and kilchoanite. A complete listing of Heddle's minerals and more recent discoveries, with localities, is given in the *Scottish Journal of Geology* (1982 and 1993).

Talc is found in economic quantities only in Shetland, mainly on Unst in the ophiolite complex of serpentinized ultrabasic igneous rocks that represents a fragment of the ocean floor thrust up among continental rocks deformed during the Caledonian orogeny (see p. 92). The Unst talc was formerly quarried and exported for industrial use; it is not pure enough to be used in cosmetics. Other deposits are present at Cunningsburgh in southern Shetland. Soapstone is a form of talc that was once extracted from outcrops of altered ultrabasic igneous rocks in the Lewisian Complex near Scourie (at Badcall Bay), where a prehistoric quarry has been found; the rock is extremely soft and was probably used for stone carvings.

Diatomite was once extracted from deposits in the Inner Hebridean islands of Skye, Mull and Eigg at the end of the nineteenth century, and at the Muir of Dinnet in Aberdeenshire, between Aboyne and Ballater, which was also exploited at the same time. Diatomite is a soft fine-grain porous material formed from the remains of diatoms, a type of single-cell alga with a silica structure. The material was used in filtration processes, and as an absorbent of nitroglycerine in the manufacture of dynamite.[*]

Feldspars and mica, abundant in pegmatites in the Lewisian gneiss and some Moine schists, have potential uses as ceramic raw materials and insulators respectively. These have been worked intermittently in the Highlands and islands, especially in South Harris and near Durness (feldspar) and at Knoydart in Wester Ross (mica).

---

[*] In this context it is known as kieselguhr, from the German for loose crumbly rock.

Fossil fuels

Fossil fuels are natural resources extracted from the Earth and burned to produce energy. These are coal and oil shale, formed in the Carboniferous period 300 million years ago, and oil and gas from Jurassic rocks 140 million years old in the North Sea Basins. Once exhausted, these resources can never be replaced – they are non-renewable. Burning them produces carbon dioxide, water and other greenhouse gases, which are building up in the atmosphere and are implicated in global warming. Scotland is fortunate in having enormous reserves of fossil fuels, exploitation of which created the engine for the eighteenth-century industrial revolution, based on coal; oil and gas continue to contribute huge amounts of revenue to the national economy.

Peat
It has been estimated that 10 per cent of the land area of Scotland is covered in peat. Many areas of the Highlands, islands and Southern Uplands have peat bogs, formed since a deterioration in Scotland's climate 5500 years ago. Most of the bogs are small and have been worked by hand for use as local domestic fuel and in whisky distilling and horticulture. Large tracts of land in lowland Scotland, especially Fife, were cleared of peat during the agricultural improvements of the eighteenth century, and the acid soils were improved by drainage and by the addition of lime from nearby Carboniferous limestone workings. Peat has a low heat value and high ash content. Attempts in the middle of the century at mechanical excavation of flatter, more extensive and accessible bogs (e.g. around Altnabreac, above Bonar Bridge) proved costly and were abandoned. The total resource is substantial, particularly in places such as Caithness, Yell and mainland Shetland, but conservation pressures have put a halt to large-scale exploitation. In a human lifespan, peat is a non-renewable resource.

Coal
Coal has been worked in Scotland for at least a thousand years, initially as a fuel in salt making, then later in the iron-smelting industry. Peak production was 43 million tonnes in 1913, when 150,000 people (10 per cent of the working population) worked in coal

mining, but modern production is a mere tenth of that figure. The last deep mine was closed in early 2002, and all coal is now extracted from opencast pits, mostly small workings with a lifespan limited to three to five years. Large quantities remain, but the cost of production, for use as a fuel at any rate, does not justify extraction from deep mines, where drainage and ventilation are costly. This illustrates the point that a resource need not necessarily be classed as a reserve: economic, political and environmental factors have to be taken into account. Coal is now carried by bulk in ships from overseas very cheaply compared to road haulage in Scotland. However, coal is an important raw material in the chemical industry, and it will always be in demand for uses other than electricity generation.

The oldest coals in Scotland are of Carboniferous age (325 million years old) and they formed in dense forests on swampy ground at sea level when Central Scotland was in equatorial latitudes (p. 120). Before the Carboniferous, land plants were not abundant enough to produce the peaty soils and plant debris that are required to make coal. Good-quality coal is low in volatiles and ash, and high in carbon, so that it has high heat value. This generally means that older, more deeply buried coals are better. Deep burial has enabled compression to drive out water and gases (volatiles).

Before the establishment of the Coal Measures at the top of the Carboniferous 305–315 million years ago, the sea became shallower and retreated, and limestones were replaced by river deltas on which forests were established from time to time, most notably in the Westfield basin in Fife. Here, 60 m of the 150 m of rock consists of thick economic coal seams that were worked for over 20 years, until 1987. During that period, more than 20 million tonnes of coal was extracted from a 200 m-deep opencast mine, one of the largest and deepest in Europe. The Westfield Basin is a relatively small coal basin, extending to 3 km², and it has been entirely removed by mining, thanks to the abnormally thick development of coals. Some of the coal seams, the Bogside coals, were almost 10 m thick. Bed thickness was greatest in the central part of the basin, which was folded into a syncline, and the coals thin out very rapidly away from the fold axis. The Westfield coal was a low-grade fuel and much was converted to gas, and at one time it contributed 20 per cent of Scotland's demand, until the arrival of North Sea gas in the early 1970s.

Carboniferous coals are present mainly in the Midland Valley, in three synclinal basins: Ayrshire, Lanarkshire (central) and Fife–Midlothian in the east (Fig. 9.14). Other fields of this age outside the Central Belt are those in the south, in the Douglas Valley, Sanquhar and Canonbie, and there are smaller ones near the Highland Boundary Fault at Machrihanish on the Kintyre Peninsula of Argyllshire (which closed in 1967 after a disastrous fire) and in the north of Arran, which was worked in the eighteenth century. The Arran coal was used for heating saltpans to produce salt from sea water, and also to fire limestone kilns. Remnants of this industry are still present in the landscape at the Cock of Arran, where mineshafts and the panhouse (where the iron saltpans were kept) remain, with ruins of the miners' cottages.

Currently, coal is being mined in opencast mines across the Midland Valley; the last remaining deep mine near Clackmannan was closed in early 2002. Deep coal was formerly mined beneath the Firth of Forth in the basin connecting the Fife and Lothian coalfields. Considerable reserves of coal remain, much of it deep at 50–120 m beneath the surface, which are not economic to mine at present. Shallow coals with less than 50 m of overburden (other rocks lying on top of coal seams) are however amenable to opencast working.

Folding of the sedimentary rocks in the Midland Valley at the end of the Carboniferous period produced synclines or basin-shape folds in which the coal-bearing rocks are preserved. Around the edges (i.e. on the fold limbs) the shallowest coals have generally been worked out and the remaining reserves tend to be concentrated at deeper levels in the cores of the synclines, hence non-productive strata separate the individual coalfields.

The Romans first used coal in Scotland for fuel, 2000 years ago, although the earliest written records date from the twelfth and thirteenth centuries in charters granting mining rights to the monks in the abbeys at Newbattle, Holyrood, Dunfermline and Paisley. At first the coal was mined on a very local scale, mainly for use in the manufacture of salt from sea water in large open iron pans. It was not until the invention of the steam engine for pumping, drainage, lifting and transporting coal that deep deposits could be exploited.

Oil and gas are also present in the Carboniferous rocks in the eastern part of the Midland Valley, including the oil shales (see next section). Cannel coal (from the Scots pronunciation of "candle", so-called because of the long yellow flame; it was once used for lighting) was used by James Young to produce oil, before he turned to the much more abundant oil shales. Sandstones in the Carboniferous acted as reservoirs for oil and gas that migrated from the oil shales and then became trapped in anticlinal fold structures, beneath a cap of impermeable shale beds. Near Edinburgh, at Cousland (Midlothian), natural gas was discovered in 1937 in the Cousland anticline, and 20 years later

commercial production started, with gas being transported north by pipeline to Musselburgh on the coast, where it was mixed with coal gas to supply the local town. Production continued for 10 years, until 1965. Oil was first found in 1922 in the same anticlinal structure, and by 1965 a total of 30 000 barrels had been obtained. Other attempts at exploiting oil and gas in the Carboniferous rocks of the Midland Valley have been unsuccessful.

Coals younger than the Carboniferous are present also, and some of these, especially the Jurassic coals at Brora in Sutherland (165 million years old), were also exploited in the past, until the 1970s. The Brora coal was used for domestic heating in the Highlands and to fire the local brickworks. However, its pyrite (iron sulphide) and ash content made it unsuitable to exploit

The origin of coal

In order for coal to form, certain requirements have to be met. First, there needs to be abundant plant material (wood, roots, leaves, stems, spores), which is rapidly buried to prevent total decomposition at the surface. Secondly, the raw material has to be buried deeply enough for the subsurface temperature to reach a certain level and the pressure of overburden has to be enough to expel excess water and gases (volatiles). Finally, these natural processes take place only very slowly, so millions of years are required to allow coal to form. The Carboniferous is the most important geological period for coals. Before then, dense forests did not exist, whereas more recent deposits have not yet had enough time to allow organic matter to be converted to high-quality coal – although younger low-grade coal deposits are certainly found around the world, but none occurs in Scotland save for a tiny field at Brora in eastern Sutherland which has Jurassic coal (165 million years old).

The greatest coal-forming event in Earth history took place about 300 million years ago in the Carboniferous period. Carboniferous vegetation included giant clubmosses, tree ferns, horsetails and ancestors of modern conifers, growing rapidly in vast luxuriant forests. Almost all of these plant types are now extinct, although much of the basic structure and composition of the plant material remain essentially unchanged. The main organic compounds in plants are carbohydrates such as cellulose, and hydrocarbons (resins, oils and waxes) that serve to protect plants and also resist decay. Plants will normally decay quickly at the Earth's surface after death, yielding mostly water and carbon dioxide that return to the atmosphere. However, depending on certain circumstances, some plant materials can be preserved and transformed to coal. Economic coal deposits are of humic coal, which forms from plant material growing in place, in swampy marshy ground. Plant debris – twigs, roots, leaves, and so on – is saturated by stagnant water, and only limited decomposition occurs at shallow depths. On the surface, new plant growth continues at a rate that keeps pace with death and decay, so that thick peaty layers build up to form humus. This is compressed to form peat.

Breakdown of plant material produces methane, carbon dioxide and some hydrogen – gases that escape into the atmosphere. Over time, the percentage of carbon increases and the percentage of gases (volatiles) decreases. A second environment of plant decay, known as putrefaction, exists where fine plant materials, including pollen and spores, are brought by rivers and the wind into enclosed basins of deep stagnant water, totally devoid of oxygen. Here, algae and bacteria mix with the dead vegetation to form black organic mud known as sapropel. The end product, on consolidation, is cannel coal, usually of limited extent and less important than the humic coal formed from peat by slow oxidation.

Coal forms from peat under the influence of increased compaction and temperature caused by burial. There is a progressive series of coal types, from peat through brown coal or lignite, to bituminous or black coal, and finally anthracite (the highest-ranking coal), which is very high in carbon content and very low in volatiles (see table below). Higher-ranking coals burn with a hotter, clearer flame and produce very little ash, because they have low mineral (inorganic) content. Low-ranking bituminous coals were once used in the production of coal gas and coke. Because anthracite requires high pressure to form, it is usually found as the most deeply buried coal, often in the cores of synclinal fold structures. Higher-grade coals are also shinier (more reflective), harder and cleaner to handle.

Coal is still an extremely important source of energy, and global reserves are estimated to be sufficient for the next 200 years, based on current production levels. It will remain in great demand as oil reserves decline, but it is unlikely that deep mining for coal will resume on a large scale in Scotland.

Rank	Moisture (%)	Volatiles (%)	Carbon (%)	Heat value
Peat	75–90	65–80	60	Very low
Lignite	35–75	55–65	70	Low
Bituminous	10–25	25–50	85	Medium–high
Anthracite	0	0–5	95	Very high

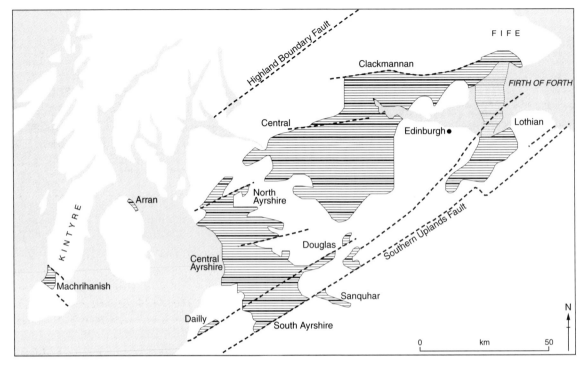

Figure 9.14 Carboniferous coalfields of Central Scotland.

further. The Brora coal was first mined at the end of the sixteenth century and was worked intermittently until 1975. Annual production was 6000 tonnes from the 1 m-thick seam; shafts were sunk up to 100 m below ground level. Current estimates of the remaining reserves are put at 12 million tonnes.

Younger still are the soft brown lignite coals of the Inner Hebrides, which are found as thin seams beneath the basalt lavas in Tertiary lake sediments (60 million years old). These have low heat value and are very smoky, but they were exploited locally on Mull and Skye and used by blacksmiths at the beginning of the eighteenth century.

Oil shale

Scotland was the first country in the world to produce oil on a commercial scale. Production began in 1851 with the pioneering work of James Young, who obtained oil by distilling coal in a factory at Bathgate in West Lothian. The material he used was local, from the Carboniferous Coal Measures (315 million years old) at Boghead on Torbane Hill. Technically, the material is known as sapropelic coal, meaning that the

organic matter that makes up the coal was derived from the decay of plants (floating algae, spores, leaves and weeds) in shallow-water lakes and lagoons. Local varieties of this coal are boghead coal and torbanite, named after local places (Boghead and Torbane Hill in West Lothian). Boghead coal is of very limited lateral extent and was quickly depleted. Oil shale was then used to produce oil, after large reserves were discovered in 1858. The oil shales are of Lower Carboniferous age and are older than the Coal Measures, at about 340 million years old. They were deposited in cyclical sequences where mudstone, limestone, shale, oil shale, sandstone, coal are repeated in rhythmic fashion through hundreds of metres. Such sediments were deposited in a shallow stagnant lagoon, close to the sea, which slowly subsided and was occasionally flooded by the sea (limestone formed) and inundated by river deltas (sandstones formed). Boghead coal gave an oil yield five times greater than oil shale.

The remaining reserves of oil shale were put at almost 120 million tonnes in 1978, all in the Lothian region. About half was considered to be workable, but the yield would be equivalent to only three or four

210

Scotland's oil shale industry

James Young (1822–83), a chemist from Glasgow, discovered that oil could be extracted from coal by distillation; he was one of the first in the world to produce oil and refine it, and the world's oil industry was founded on the processes invented by Young (nicknamed Paraffin Young). This oil-shale industry peaked during the First World War, when about 2 million barrels per year were being produced. During the lifetime of the industry, 75 million barrels, or 10 million tonnes, of oil was obtained from 164 million tonnes of shale, equivalent to a small North Sea field. As much as 1 billion tonnes of shale could still remain, but this is not likely to be exploited, because of the high cost of doing so. At first, coal was used to produce the oil, and a refinery was opened at Bathgate in 1851. Oil for lighting and lubrication was produced, as well as ammonia, naphtha and paraffin wax for candles. From 1862 until 1962, the main source of the oil was oil shale as a result of a rise in price of the coal, which had become scarce, and the fact that coal distillation had been separately patented. At one stage, 10 000 people worked in the industry, and by 1865 there were 120 refineries in and around Bathgate. Oil was produced by heating the crushed shale to 500°C in a cast-iron retort, a vertical funnel surrounded by a brick heating oven. Steam was forced in to obtain the maximum yield of crude oil. Spent shale was removed and dumped in heaps, known as shale bings. These were up to 70 m high and flat, conical or pyramidal in shape – and they still mark the West Lothian landscape. Weathering converted the iron oxide in the shale to a bright red colour. Some of the spent shale was used in brick making, and the shale bings were rapidly denuded at the end of the twentieth century as the material found uses in the construction industry. The Five Sisters bing has been retained as a monument to industrial archaeology. (See Kerr 1999.)

months' output from a small field in the North Sea. There could be as much as a billion tonnes of oil shale still remaining (i.e. the total resource), but the costs of mining, processing, waste disposal and environmental reclamation are probably prohibitive and the material is unlikely to be extracted. Further details of the oil-shale industry and the geological conditions of formation can be found in the box (above). Outside the Midland Valley, oil shales are also found on the Inner Hebridean islands of Skye and Raasay. The shales are of Jurassic age (165 million years old), much younger than the Carboniferous varieties of West Lothian. Yields of oil are lower than the poorest Carboniferous types, and the deposits, which were surveyed in the interwar years, were never considered economic. Like their older equivalents, the Jurassic oil shales were deposited in a warm shallow stagnant lagoon that occupied the area of the Inner Hebrides and was separated by the Highland massif from the Jurassic seas of the east (the Moray Firth and North Sea Basins), where shales were deposited that later formed productive horizons for oil and gas.

Oil and gas

The discovery of oil and gas in offshore basins in the Moray Firth and North Sea in 1970 proved to be the most important find of natural resources in Scotland in the twentieth century, contributing significant revenues and providing employment. Investment in exploration has allowed geologists to interpret the structure of the North Sea and the oil and gas traps in ways that have made fundamental contributions to our understanding of the origin and evolution of sedimentary basins in general. In Scotland, just the right combination of conditions existed in the North Sea region to allow these valuable deposits to form. These crucial factors are considered in detail in the box; what follows is a brief summary.

Oil and gas, like peat and coal, are derived from organic matter – plants, algae, bacteria and planktonic animals – that is buried and preserved before being completely oxidized. In Scotland, the appropriate conditions have existed in two main periods: the Carboniferous, when dense coal swamps flourished 300 million years ago, and the Jurassic (140 million years ago), characterized by tropical seas and shallow lagoons with abundant and prolific life-forms. Plant remains tend to produce gas when deeply buried, and most of the gas in the southern North Sea is derived from Carboniferous coal beds. In the northern North Sea off Scotland and Shetland, the Jurassic rocks are the source of oil. The initial source rocks, mainly shales and muds of the Kimmeridge Clay formation, are rich in algae and bacteria, which yield oil, provided that depth of burial is no more than 4 km. Once formed by slow cooking, the oil then migrates slowly (a few millimetres per year) through permeable rock (generally sandstone with open pore spaces) until finally it becomes trapped in a reservoir rock within a structure (fold, unconformity, salt dome or fault trap) that prevents further migration.

Jurassic coals exist in the Moray Firth region. Onshore, the Brora coal was worked (see p. 136); the Dornoch and other coals off shore were the source of

North Sea oil and gas

Oil, natural gas (mostly methane) and gas condensate are together referred to as "petroleum". Nearly all petroleum consists of hydrocarbons, or compounds made of carbon and hydrogen, which may be solid, liquid or gas. In nature, these materials have come from the decay of bacteria, algae and plants. Deep burial and heating under pressure causes organic carbon (in the form of kerogen, which is solid bitumen) to decompose into oil and gas.

For oil and gas to form commercial deposits, certain conditions are necessary. First, there has to be an organic-rich source rock, such as fine mudstone with abundant algae, bacteria and plant debris derived from the land. The source rock then has to be buried rapidly by other sediments in a subsiding basin, until at depths of 3–4 km the temperature reaches 100–150°C. At this point, kerogen in the source rock decomposes to produce oil and gas, which migrate away from the source through a carrier bed, such as porous sandstone or fractured limestone, into a reservoir rock, to form a hydrocarbon accumulation. The deposit must be covered by a caprock, such as impermeable and impervious salt or clay, to form a seal that will trap the oil and gas, and prevent it from escaping to the surface. And finally, of course, the deposit has to be located and drilled to extract the petroleum. Such a combination of geological circumstances does not occur often in the crust, and Scotland is fortunate to have had the benefit of these factors.

Petroleum forms by a process similar to pressure cooking, and the region in the crust where oil forms is referred to as the "kitchen" by petroleum geologists. If the burial is not deep enough, and therefore pressure and temperature are too low, then the source rock will be immature (i.e. the temperature has not reached 100°C). Between 100°C and 150°C, the source rock matures to produce crude oil and gas (the gas being dissolved in oil). Above 150°C, any oil present breaks down (or "cracks") to produce only gas; nothing more can be produced once the temperature exceeds 200°C, beyond which point the petroleum is said to be over-mature.

Sedimentary source rocks rich in kerogen from algae produce mostly oil, whereas plant-rich kerogen tends to give only gas. Where Carboniferous or Jurassic coal is the source rock, again gas is obtained rather than oil. In the case of the North Sea, the southern part off the coast of England has a basement of Carboniferous Coal Measures, and most of the fields are of natural gas. However, in the Scottish sector the source rock is mostly Kimmeridge Clay, a thick fine organic-rich black sediment (derived from planktonic algae), laid down in deep oxygen-starved marine waters during the late Jurassic period (150–140 million years ago), and so the bulk of the petroleum deposits in the northern North Sea are of valuable light crude oil. The clay was buried deeply in the Central Graben and Viking Graben to the appropriate depths where maturity is assured. Much older rocks either do not contain enough organic matter, if any at all, or are too deeply buried and are therefore now incapable of producing hydrocarbons. On the other hand, very young rocks are either not buried deeply enough or have not had enough time to mature, and so do not produce oil or gas. The oldest source rock in Scotland is the Devonian fish bed from the Caithness Flags (Middle Old Red Sandstone), which, together with a Jurassic source, has yielded the oil in the Beatrice Field of the inner Moray Firth.

gas in various fields. Once natural gas (methane) was brought ashore in 1967, it quickly displaced coal gas as a fuel, with important economic consequences for Scotland's coal industry. Oil was first brought ashore in 1975, some ten years after the hundred-year-old oil-shale industry finally closed. Official estimates at the end of 2002 indicated that North Sea oil reserves could last until 2030. Untapped reserves were put at 42 billion barrels, which is more than the 32 billion barrels extracted since 1967. Future fields are likely to be smaller and daily production levels less than the 1999 record. In 2002, the North Sea oil and gas industry supported 265 000 jobs and employed 7 per cent of the entire Scottish workforce.

Alternative energy sources

Fossil fuels are of limited extent and are being rapidly depleted, hence the urgent need in most countries to develop alternatives. Thanks to Scotland's location in the North Atlantic region, the climate is generally cool, wet and windy, and for this reason there are considerable opportunities for the development of hydroelectric, wind, wave and tidal renewable-energy schemes. Of these, only hydroelectric power has been harnessed to any considerable extent. Here, the legacy of the most recent ice age has been utilized to great effect. Long narrow ice-scoured valleys and lochs in upland western regions are easily dammed. Corries too have been used, such as the Loch Awe pumped storage scheme, south of Oban (see Fig. 8.6). Here, Coire Cruachan has an elegantly constructed dam at its mouth, creating a reservoir whose waters are harnessed to even out the peaks and troughs of demand. This must be one of the most visually pleasing examples in the Highlands of a construction using natural geological features. The second pumped storage scheme is at Foyers, in the Great Glen. Here, the Falls of Foyers, a 60 m-high waterfall in the Foyers granite, were harnessed in 1896, to provide power for an aluminium smelter; this was the first hydroelectric scheme in Scotland. It remained in operation until 1970, when it was replaced by the pumped-storage generating station. In terms of the landscape, the main effect is one of visual impact and changes to river dynamics down stream of dams, where natural erosion, deposition and floodplain evolution are halted, and changing water levels create bare patches along lochsides.

Tidal schemes could have great potential along the western seaboard, with its deep rocky clefts where the sea has excavated joints, faults and shear zones in metamorphic rocks. Only one scheme is operating at present, on Islay; this is quite unobtrusive. Additional schemes have been proposed for the coast around Orkney.

Of more serious visual impact are the wind farms, which have been growing in number since the late 1990s. They have to be located on exposed hilltop sites to be effective and are therefore an intrusion in the landscape (unlike the picturesque Dutch windmills). Plans for the future massive expansion of renewable energy sources include siting wind installations off shore. Already there have been objections to a planned wind farm in the Solway Firth. On shore, plans were put forward at the end of 2002 for 500 wind turbines at 40 locations around Scotland.

Nuclear energy will always have a role to play in Scotland's energy needs for the foreseeable future. Nuclear power stations, like their coal- and gas-fired equivalents, are large, conspicuous and tend to be sited on coastal promontories, so their impact is considerable. Scotland's uranium resources are classed as strategic reserves, but are unlikely ever to be exploited for use as fuel in nuclear reactors; all fuel is imported. Scotland's climate is such that solar energy is not considered to be viable in any large-scale schemes. However, there is scope for local schemes, where solar panels are used to heat domestic water.

Several of the Caledonian granites in the Grampian Highlands were explored in the mid-1980s as possible sources of geothermal energy. Some of the bodies have very high heatflow values, including in particular the Bennachie granite near Huntly and Old Meldrum, but they are too far away from centres of population, and the investigations (known as the Hot Dry Rock Project) did not lead to any exploitation. The method would have involved fracturing the granite at depth, then pumping cold water down the cracks and extracting hot water from adjacent boreholes.

Resources and the landscape

People have always used rocks for building, ores for smelting and fuels for burning. These are natural materials, in limited and finite quantities, and gradually they are being depleted and can never be renewed except within the workings of the rock cycle, over many millions of years. Extracting these resources has had an impact on the environment and will continue to do so, especially as demand increases and the need for greater efficiency leads to ever-larger installations. Quarries and gravel pits are now much more extensive in scale than previously, since blasting, crushing, sorting and storing all take place on site, and access roads are wider, so that the visual impact overall is much greater. The roadstone quarry on Dunsinane Hill in Perthshire (made famous in Shakespeare's *Macbeth*) is visible from a great distance. Removal of sand and gravel from glacial outwash deposits or sand from coastal dunes destroys landforms or causes instability.

Once the resource is depleted, many extraction sites – quarries, mines and opencast pits – are filled with waste and landscaped. However, this process can actually destroy natural features, such as ridges and valleys, and can create new hazards in the form of groundwater contamination and possible eventual subsidence into old mine shafts and buried quarries, as happened in Edinburgh in 2001, with serious consequences for residents. Geological exposures may be lost for ever by restoration and reclamation work.

Disposal of radioactive waste from nuclear plants is likely to have an impact on the landscape in the future, assuming that a decision is taken to bury the waste in deep underground repositories. There are many obstacles yet to be overcome, including finding a suitable site with stable geological and engineering conditions; but much of the problem is tied up with political and economic considerations. It is proposed to treat high-level waste (which retains 90% of its radioactivity for thousands of years, during which time heat continues to be produced) by fusing it into artificial rock or embedding it in inert ceramic grains. This material would then be enclosed in copper cylinders, surrounded by inert clay, which in turn would be buried in vertical holes drilled through the floor of an impermeable concrete bunker 50 m underground.

Wherever repositories may be located in Scotland, and they may be close to existing nuclear power stations, the environmental impact will be considerable during the construction phase at least.

As climate changes, more rainfall, floods, higher sea levels and more storms will put pressure on river and coastal landscapes. Erosion will increase, and with it the demand for more defences. There have already been many attempts to tame rivers by straightening their courses, then building on floodplains, which sooner or later are flooded naturally, with serious economic consequences. Barriers to stop coastal erosion and cliff collapse inevitably result in changes to the coastal outline, and the problem is simply transferred along the coast. If planners, engineers and development agencies were to work more closely with Scottish Natural Heritage, for example, many potentially undesirable effects could be avoided. Working in harness with nature to use resources in a rational way that preserves the natural landscape features and geological heritage is likely to be more sustainable in the longer term, and more pleasing visually for those who love the Scottish landscape.

Agriculture and forestry can also obscure or destroy landforms, and create possibly unwanted visual impacts that destroy the naturalness or feeling of remoteness. The rapid growth of highly visible telecommunications masts and access roads in the Highlands, and the spread of fish farms in western sea lochs and freshwater hill lochs, means that one has to travel farther and look harder for relatively unspoilt landscapes.

Tourism and recreation also impact on the environment. Heather burning of upland moors (to allow young grass to grow as food for grouse) has created vegetation scars that are long lasting and highly visible. Cross-country tracks are bulldozed across hills and accelerate erosion. Natural features that are promoted as mass tourist attractions, such as Smoo Cave near Durness in northwest Sutherland (see Fig. 3.25), bring new constructions in their wake, including footpaths, stairways, fences, souvenir shops, notices, safety installations, viewing platforms, artificial lighting, covered walkways to prevent water dripping onto tourists (and it was dripping water that made the cave, after all), and five-minute boat trips (cancelled in times of flood, i.e. the thundering cascade of water tumbling down from the disappearing stream above). At Smoo Cave the forces of commercialization seem to be more powerful than the awe-inspiring forces of nature. Ski resorts, as in Glen Shee, Glen Coe, the Lecht and the Cairngorms, are also features with high visibility and an impact factor far beyond the confines of the slopes. The activities of collectors also have a deleterious impact, especially when they unscrupulously use powerful explosives to remove material in bulk. Not only are mineral and fossil sites plundered in this way (they inevitably involve rare and irreplaceable specimens), but the sites are destroyed and future research becomes impossible. All this suggests that careful planning, conservation and management are needed, and the agency set up to provide this service is Scottish Natural Heritage.

Summary – resources through time

Scotland lies at a geological crossroads, as it were, at the edge of continental plates, where activity has been intense and varied over billions of years. As a consequence, the geology is very complex, considering the small size of the country. Rocks of almost all geological periods are to be found, and examples of a great many different types of mineral and fuel resources serve to illustrate the variety of geological environments that existed in the past. In terms of ore deposits, pulses of igneous activity related to crustal stretching brought influxes of metals and heat from the upper mantle. These ores were subsequently recycled in the crust by the actions of the rock cycle, mainly the tectonic part of it (i.e. mountain building, in which shear zones have played an important role in localizing ore-bearing fluids). Sandstones and limestones for building, and coal, oil and gas in sedimentary rocks depend for their origin on sedimentary basins, mostly created by crustal stretching or sagging of the surface layers.

Table 9.1 is an attempt to put Scotland's resources into a framework of geological environments over time, and the types of useful materials that formed in such environments.

Table 9.1 Resources in time and place.

Period	Metal-ore deposits	Other natural resources
Quaternary	Gold placers in streams (Helmsdale); chromite in offshore sediments (Rum)	Sand and gravel for aggregates; clay for bricks
Tertiary	Chromite in layered ultrabasic rocks of Rum	Dolerite and basalt as local roadstone; skarns in marbles
Cretaceous		Glass sand (Lochaline); brick clay (Caithness)
Jurassic	Sedimentary ironstones (Raasay)	Coal; oil (North Sea); brick clay; sandstone for building; limestone
Triassic		Salt domes in North Sea (oil and gas traps); building stones (New Red Sandstone)
Permian		Red sandstone – building stone
Carboniferous	Ironstones (Midland Valley); lead–zinc–copper (Southern Uplands); silver (Ochil Hills)	Coal, oil shale, oil; sandstone for building; agricultural lime; brick clays; cement; glass sand; dolerite for roadstone
Devonian	Gold and silver in volcanic springs (Rhynie); copper in granites; uranium in flagstones	Old Red Sandstone – building stone; granite – building stone
Silurian	Chromite and platinum (Unst ophiolite);	Talc (Unst); greywackes (Southern Uplands) – building stone and host for lead–zinc ores (Leadhills)
Ordovician	Chrome–nickel and platinum (Huntly intrusions)	Greywackes (Southern Uplands) – building stone and host for lead–zinc ores; skarns in Durness Limestone (NW Sutherland and Skye)
Cambrian		Potash in Fucoid Beds; quartzite – roadstone (NW Sutherland)
Dalradian	Copper–zinc in volcanic rocks; baryte–zinc in sediment host	Greenstones – building slabs; roofing slates; agricultural lime
Moinian		Mica; building stones locally
Torridonian		Building stones locally
Lewisian	Iron–gold (Loch Maree Group); magnetite (Tiree), ironstone (Iona)	Mica; ceramic pegmatites; polished marble (Iona, Tiree)

People and the landscape

Scotland's natural landscape has been created by the action of weathering and erosion on a wide variety of rocks, over many years of Earth history – mountain ranges, hills, valleys, plateaux, fjords, lochs, seacliffs and beaches. Natural landforms have been used by people ever since the first settlers arrived shortly after the end of the most recent ice age, 10 000 years ago. Geological resources too have been exploited since earliest times, for tools, weapons, shelter and energy. These resources were the basis of industry: the availability close by of coal, iron ore, limestone, dolerite and sandstone in Central Scotland fuelled the industrial and agricultural changes of the eighteenth and nineteenth centuries. The human use of the landscape for various purposes has left a mark and contributed to its evolution in the form of constructions for defence and fortification, ritual and burial, settlement, transport and communication, trade and industry, agriculture, water supplies, fuel and recreation.

Landscape features have obviously played the major part in influencing settlement patterns in Scotland. There is relatively little low-lying flat land, and the Caledonian mountains were natural barriers to movement, with only the excavated fault lines affording through-routes, the Great Glen and Strathmore (immediately south of the Highland Boundary Fault) being the two most striking examples. Earliest settlements are clustered around the sea, principally the archipelagoes of Orkney, Shetland, the Western Isles and the Inner Hebrides. Transport was mainly by sea for several millennia. It is worth noting that Scotland's coastline is almost 10 000 km long (longer than the eastern seaboard of the USA).

Rock types and climate help determine soil type and therefore the potential for agriculture and settlement. Thus, the areas with sedimentary rock outcrops in the Midland Valley, around the Moray Firth shores, Orkney and the coastal lowlands of western Sutherland have provided the majority of the fertile well drained soils and arable landscapes of Scotland.

In addition to rocks and climate, the effects of the most recent glaciation have had a major influence on settlement. For example, excavated valleys allowed for shelter and communications. Corrie lochans provided summer grazing in the form of shielings. Glacial deposition provided well drained mounds of sand and gravel in the form of eskers, kames, drumlins and moraines. In lowland areas, these were the only patches of dry land with cultivation potential – the remainder was boggy ground and tidal marshes when sea level was higher. Uplift after the melting of the ice created raised beaches on which thick fertile soils formed. As the sea retreated, potentially fertile carselands were exposed in the east-coast firths and became available for cultivation, after the removal of postglacial peat and subsequent drainage. On the other hand, in the northern isles the land was inundated and river valleys became arms of the sea; coastal plains were lost and barren gneiss pavements were left behind.

Over time, people have re-used the same landforms for different purposes, so that much of the evidence for the earliest settlements has all but gone (aided in part by postglacial sea-level rise in the islands). Prehistoric landscape features include the stone circles, villages, burial cairns and brochs in Orkney, Shetland, Lewis and parts of Caithness, Sutherland and Argyllshire. The main legacy was the clearing of the native forest and the rapid encroachment of blanket peat from 5000 to 3000 years ago. Natural forest regeneration and tree-planting schemes have never re-created the natural wilderness of the immediate postglacial period.

When the Romans arrived 2000 years ago, they viewed Scotland as a hostile northern frontier land, and set about establishing defences in what was a military zone of occupation. Their surveyors and engineers made expert use of natural landforms as they built roads, camps, forts, signal stations and defensive walls. Many remnants are still visible in today's landscape, and some are still in use, such as Dere Street in the Borders and part of the A9 in Perthshire, which follow Roman roads. Signal towers were built on the three Eildon hills near Melrose (named Trimontium by the Romans) and along the Gask Ridge west of Perth. This ridge ends in the Ardoch fort near Crieff, and where the earthworks have remained as an impressive landscape feature. The Antonine wall has its foundations on an esker ridge at Callendar Park (Falkirk); the land on either side was marsh, thus creating an additional line of defence. The Antonine Wall was built in AD 143 by the Roman army on the orders of Emperor Antonius Pius. It stretched for 60 km between the Clyde (near Bearsden) and the Forth (Falkirk) and formed the northern frontier of the Roman Empire for 20 years. Even more famous than the Antonine Wall is Hadrian's Wall, built in AD 122 on the orders of Emperor Hadrian. Most of the 120 km wall is still intact. It runs from the Solway on the west to the Tyne on the east, almost along the Scotland/England border. The foundations of the eastern part are on the Great Whin Sill, a Carboniferous igneous intrusion (300 million years old), similar in age to the Midland Valley Sill to the north.

For a thousand years after the Romans left, the natives and the succeeding waves of invaders and settlers made their own impression on the landscape. Many of the fortified sites – hill forts, duns and brochs – were re-occupied and partly extended or built over. Local stone was used not only for construction but also for ritual purposes, including the carved stones and crosses in the west, made of fine-grain resistant greenstone (metamorphosed Dalradian lava), whereas in the east, around the Moray Firth, the Picts used the tough Devonian flagstones for their extremely intricate carved symbol stones. These treasures of the cultural heritage have remained intact for over twelve centuries, in testimony to great craftsmanship.

Medieval times saw gradual agricultural improvements and changes in land use, much of it promoted by the religious orders, who made their mark in the impressive monasteries, abbeys and cathedrals, built of durable local sandstone (Devonian, Carboniferous and Permian). The monks were also responsible for opening the first coal mines, to fuel the saltpans at the coast. Secular buildings include castles, walled towns and towers, most of which remain, some in ruins, others thriving, but all contributing to the landscape (Fig. 9.15). Castles were built on ancient volcanic plugs and sills – Dumbarton Rock, Edinburgh Castle and Stirling Castle – for visibility, deterrence and defence (Fig. 9.16).

From about the middle of the seventeenth century, landscape change increased rapidly, with the development of planned towns, clearance of Highland villages, enlarged field systems, drainage, water supply

Figure 9.15 Duffus Castle, Morayshire, between Elgin and Lossiemouth: a fourteenth-century castle built of local Hopeman Sandstone (Permian) on an artificial hill or motte; it failed because of weak foundations on loose soil, and was never occupied.

and sewage installations, the opening up of mines, pits and quarries for coal, oil shale, limestone, metal ores, sandstone, roadstone, roofing slate, sand and gravel, and brick clays. Much of the evidence of earlier land use was lost in the process. Remnants do remain, such as overgrown and abandoned Highland clearance villages, run rig (long narrow ridges of soil) field systems and high shielings (for summer pasture). Crofting townships continue to disappear, except on some of the Hebridean islands. The few remaining crofting settlements in Sutherland make a dramatic impact on the landscape along the shores of Loch Inchard, near the village of Rhiconich. Cattle drove-roads (from the West Highlands to Crieff and on to Stirling) are still in evidence, used today by ramblers.

The pace of change in the Highlands accelerated most rapidly after the Jacobite uprisings of 1715 and 1745, when military installations altered the landscape significantly through wholesale clearance of forests and villages, the construction of military roads and bridges in previously impenetrable lands, and the building of forts and garrison towns from which to control the population. Use was made of geological features such as the Great Glen Fault, with Fort William, Fort Augustus and Fort George along its length, Fort George being situated on a postglacial shingle promontory that guards the entrance to the inner Moray Firth. At least the architecture is pleasing, and they did use local stone. Like the Roman roads of two millennia ago, many of the military roads and bridges are still used today, thanks to superb engineering and durable local stone. Replacing people with deer and sheep in the Highlands had a lasting negative impact on the landscape, because of overgrazing, soil erosion, forest destruction and neglect of the now empty lands.

Developments since the end of the Second World War include the expansion of forestry and peat extraction, ever larger quarries, more housing and new industrial features such as oil installations (e.g. Sullom Voe in Shetland) and refineries (e.g. Grangemouth). Much of the heavy industry that existed in the Midland Valley has now gone, leaving scarcely a trace: shale bings (Gaelic: beinn, a hill) remain in West Lothian, but there is little to see of the coal mining that once dominated so much of Central Scotland (to say nothing of shipbuilding: a solitary crane stands on the Clyde at Glasgow). Now, at the beginning of the present century, the latest developments are designed to provide some degree of protection to the natural landscape, in the creation of Scotland's first natural parks at Loch Lomond and in the Cairngorms. Visitor surveys consistently show that it is the variety of striking landscapes that draws people back to the country.

I hope that you have enjoyed this trip through three billion years of Earth history in Scotland, and that you will be inspired to go out and visit these classic landscapes. Research continues into the Lewisian gneiss, the Assynt thrusts, the Glencoe Igneous Complex, structures in the Dalradian, and many more subjects. A cornucopia of geological research, Scotland continues to fascinate not only scientists but also many others who are attracted by its wonderful landscape and intrigued by what shaped it and what lies beneath.

Figure 9.16 Edinburgh Castle and the Royal Mile, forming a crag-and-tail feature; the crag is the Castle Rock and the tail is of sedimentary material, with Holyrood House at the end; deep hollows on either side of the crag were excavated by moving ice.

Appendix

Taking it further

Textbooks, websites and excursion guides will get you so far, but if you want to make significant progress in the study of geology, it is best to enrol in an evening class at a local university, college or community centre, or enrol in a course with the Open University (website www.open.ac.uk). It would also be a good idea to join a local geological society (in Scotland these are in Aberdeen, Edinburgh, Glasgow and Inverness; see the Bibliography for their website addresses). Winter lectures and summer field trips are organized, and members of the Edinburgh and Glasgow societies receive the *Scottish Journal of Geology* twice a year. Local museums may have a geology curator who can be approached for assistance with the identification of minerals, rocks and fossils. Every other year (in odd years), there is a national Scottish Geology Week, held in August, with events throughout the country; check details on the special website www.scottishgeology.com; this also has brief outlines of the geology of the main regions of Scotland and links to geological societies.

Fieldwork

This book deals with the broad features of Scotland's geology and landscape, and does not offer help with describing and identifying rocks. That is best done under the guidance of a trained specialist tutor, as is fieldwork, where different skills are required, and particularly in geological mapping where expert guidance is essential. Training in field techniques is a slow and painstaking process. However, it is worth remembering that all the major advances in our understanding of the geology of Scotland were done on the basis of the rocks and structures seen in the field. All available excursion guides to localities in Scotland are listed in the Bibliography, although the less experienced reader may find some of them challenging. If you undertake any fieldwork, it is essential to follow the Country Code and to take extra precautions when visiting working quarries (permission must be sought in advance, and protective headgear has to be worn). Disused quarries and mine shafts are full of hidden dangers (flooding, roof collapse, unstable faces, etc.) and are best avoided completely. Do not remove mineral and fossil specimens from rockfaces at outcrops, only from loose scree. At SSSIs, samples may be collected only with the prior permission of Scottish Natural Heritage (check their website for details, www.snh.org.uk; this site also gives details of the 2002 Natural Heritage Futures initiative, covering all aspects of geology, landscape, wildlife and conservation, with huge resources available on line).

In general, it is best to follow the countryside lover's advice: take nothing but photographs, leave nothing but footprints. Enjoy looking at Scotland's varied landscapes, which are the result of a billion years of geological evolution; they have inspired scientists for 250 years and more.

Geological maps

Geological maps at various scales are published by the British Geological Survey (www.bgs.ac.uk), and may be purchased from the bookshop at Murchison House, Edinburgh (tel. 0131–667–1000). The single sheet covering the whole of Scotland is the "Ten Mile" (i.e. ten miles to one inch scale, or 1:625000), North Sheet, Solid Edition, 2002. This is the fourth edition, but only the topography has been updated; towns, roads and rivers are shown, and this is the most useful map when touring. Detailed maps are available at a scale of 1:50000; the total is about 130 sheets for the whole country, but many are divided into west and east half-sheets, so that you would need to purchase about 165 maps for all of Scotland at that scale. Most of the newer maps have extensive notes in the margin, and cross sections to illustrate structure. Several have accompanying memoirs or brief descriptive accounts. Special sheets at the 1:50000 scale are available for Arran and Assynt. A broader overview of larger regions is contained in the set of maps that show offshore as well as land-based geology, at a scale of 1:250000. There are 17 sheets in this series. Three classic areas of British geology – Ballantrae, Moffatdale and Glencoe – have special sheets at 1:25000 scale, showing much detail; these have accompanying descriptive booklets. The Outer Hebrides are covered by four sheets at the scale of 1:100000. No geological excursion guides for the Outer Hebrides are available currently.

Glossary of technical terms

accretion   Growth of continents by collision; or movement together of wedges of rock; or theory of the origin of planets by capture of matter in space.

accretionary prism   A wedge-shape pile of sediments scraped off a subducting plate at a destructive plate boundary (subduction zone) and interleaved by overthrusting, resulting in strata being repeated.

acidic   Igneous rock with over 65% silica, e.g. granite.

agate   Colour-banded form of silica found lining cavities in lava.

agglomerate   Mixture of ash and broken lava in volcanic neck.

aggregate   Crushed rock, used as road stone and in concrete manufacture.

albedo effect   Reflection of Sun's heat by ice and snow back into space.

alluvial deposits   River sediments.

alluvial fan   River deposits laid down at the mouth of a valley that opens out onto a flat plain at the edge of a steep mountain slope; fan shape spreads outwards and sediments thin away from the valley mouth.

alluvium   Sand and gravel in river beds.

ammonite   Extinct fossil mollusc with coiled shell; important zone fossil, used for dating Jurassic rocks.

amphibole   Silicate of calcium, iron and magnesium (a ferromagnesian mineral); dark green to black; elongate crystals, double chain structure.

amphibolite   Hornblende–feldspar– biotite gneiss, often with garnet; formed from high-grade metamorphism of basic lavas.

amygdale   Almond-shape cavity in lavas, filled with minerals, e.g. agate, zeolite, calcite, quartz (rock crystal).

anaerobic   Environments in which oxygen is excluded, e.g. in deep oceans or in stagnant lakes.

andalusite   Aluminium silicate found in some schists.

andesite   Igneous rock, lava of intermediate composition; named after the Andes mountains.

anhydrite   Calcium sulphate, an evaporite mineral; similar to gypsum, but without water in its crystal structure.

anhydrous   Lacking in water.

anorthosite   Coarse-grain layered igneous rock containing mostly anorthite, a calcium-rich plagioclase feldspar.

anthracite   High-ranking coal, i.e. high in carbon and heat value, low in ash and volatiles; found in deeply buried Carboniferous coal fields.

anticline   Fold shape with limbs dipping away from axis; oldest rocks in centre or core of fold.

Archaean   Precambrian rocks older than 2500 million years.

arête   Glacial landform, a sharp ridge where two corries meet back to back.

arkose   Type of coarse sandstone with over 20% feldspar fragments; red or brown in colour; e.g. Torridonian Sandstone.

Armorican orogeny   Mountain-building episode in central Europe, Carboniferous to Permian, 360–290 million years ago; Gondwana collided with Laurussia and created the supercontinent of Pangaea; also known as Hercynian or Variscan orogeny.

arthropods   Invertebrate animals, with jointed limbs and a hard outer skeleton; group includes crustaceans and extinct trilobites.

asbestos   Fibrous silicate mineral found in metamorphosed ultrabasic igneous rocks.

ash   Fine volcanic material from explosive eruptions.

asthenosphere   Weak layer in upper mantle, beneath lithosphere, plastic, able to flow slowly.

augen gneiss   Gneiss with large eye-shape feldspar crystals surrounded by mica; a type of metamorphosed granite.

augite   Calcium–iron–magnesium silicate, in pyroxene group, found in basic igneous rocks (gabbro, basalt); black, stumpy crystals, single-chain structure.

aureole   Zone of country rock around an igneous intrusion, affected by heat and fluids.

Avalonia   Precambrian continent containing southern Britain, Ireland and western Europe; collided with Laurentia and Baltica to create the Caledonian mountain belt.

Baltica   Precambrian continent containing

Scandinavia; collided with Laurentia and Avalonia to create the Caledonian mountain belt.

banding   Layering of different minerals in metamorphic rocks, usually gneisses.

banded ironstone (banded iron formation) Precambrian sedimentary rock made of iron ore and chert.

Barrovian metamorphism   Type of regional metamorphism found in the Grampian Highlands, showing progressive grades from slate (originally shale) through schist to gneiss with increasing pressure and temperature.

Barrow's metamorphic zones Progressive sequence of minerals in regional metamorphic rocks, from chlorite to biotite, then garnet, kyanite and sillimanite.

baryte (barytes, barite; heavy spar)   Barium sulphate, dense, white, creamy or pinkish mineral used in oil drilling (heavy mud).

basalt   Dark fine-grain basic volcanic lava; contains olivine, pyroxene, calcium-rich plagioclase feldspar, iron ore.

basement   Older rocks on which sediments have been deposited; usually refers to a Precambrian craton, e.g. Lewisian Gneiss basement of the Northwest Highlands.

basic igneous rock   Rock with 44–52% silica; rich in ferromagnesian minerals (olivine, pyroxene) and calcium-rich plagioclase feldspar, no free quartz crystals.

basin   Broad area of slowly subsiding continental crust into which sediment has been transported by rivers; some basins are rift valleys or graben structures when bounded by two parallel normal faults.

batholith   Very large (over 100 km^2 exposed at the surface) plutonic igneous intrusion, usually granite.

bed   A stratum or layer of sediment or sedimentary rock.

bedding plane   Flat surface between two beds of sediment or sedimentary rock.

bedrock   Solid, firm rock on which other, younger, unconsolidated materials may be deposited, e.g. soil, peat, sand and gravel; boulders and scree are not bedrock, but instead are described as superficial materials.

bentonite   Light coloured, soft clay rock

derived from weathering of volcanic ash; can absorb large volumes of water.

biotite Black or dark brown flaky mica (silicate mineral) with sheet structure.

bitumen Solid type of petroleum, same as asphalt, pitch and tar (alternative names).

bituminous coal Normal coal, intermediate in rank between lignite (brown coal) and anthracite; high in volatiles.

bivalve Marine animal (mollusc) with shells in two parts; clams, cockles, mussels, oysters, scallops.

blanket bog Broad, flat expanse of peat.

bog iron ore Soft form of limonite, hydrated iron oxide; porous; formed in freshwater bogs and streams in peaty areas; no longer used as an ore.

bole Fossil clay-rich soil formed on top of weathered lava flows; usually red due to presence of iron ores.

boudinage Black-pudding-shape structure in deformed rocks, due to thinning and stretching of layers.

boulder clay Glacial deposit formed by melting ice sheet; contains unsorted material.

brachiopod Lamp shell, a marine invertebrate animal with two symmetrical shells (valves) of unequal size.

braided stream Stream or river with many fast flowing interweaving channels that criss-cross over a wide area and carry a high load of sediment.

breccia Rock consisting of broken fragments; formed in volcanic explosions, or along fault lines, or as scree formed by collapse of rock slopes.

brittle Property of a rock that breaks by fracturing.

bryozoans Colonial marine invertebrate animals with calcareous skeleton; grew as mats on the sea floor.

Buchan type metamorphism High-temperature regional metamorphism, found in Dalradian schists of Northeast Scotland, adjacent to large basic–ultrabasic igneous intrusion.

bulk minerals Crushed rock (aggregate) and sand and gravel – natural resources used in the construction industry; the term used in this book is industrial materials.

calcite Calcium carbonate; mineral found in limestone and marble; white or colourless, soft, dissolves in weak acids.

calcrete Type of soil cemented by calcium carbonate (calcite); found in desert regions.

caldera Large circular volcanic structure with steep walls, usually marking a ring fault, and a flat, depressed downfaulted floor containing volcanic vents.

Caledonian orogeny Mountain building episode 500–400 million years old that created the mountain chain in Ireland, Scotland, Wales and western Norway.

Cambrian period Geological time period, from 545 to 495 million years ago; above the Precambrian and below the Ordovician; named from Cambria, Roman name for Wales.

cannel coal Fine-grain coal made of spores and fine plant fragments; burns with a long, bright yellow candle-like flame ("cannel" is a local pronunciation of "candle"); related to boghead coal.

carbon cycle Exchange of carbon in the crust, biosphere, atmosphere and hydrosphere in a continuous process.

carbonate minerals Common minerals in sedimentary rocks, e.g. limestone; examples are calcite (calcium carbonate), dolomite (calcium–magnesium carbonate) and siderite (iron carbonate); dissolve easily in weak acids; often form cement in sedimentary rocks, especially sandstones.

Carboniferous period Geological time period, from 354 to 290 million years ago; above Devonian and below Permian; named from the coal content in the Coal Measures, the upper part of the Carboniferous; in USA the terms Mississippian (lower) and Pennsylvanian (upper) are used instead.

carse Floodplain of a river beside its estuary; contains alluvial and marine deposits, sometimes also glacial sands and post-glacial peat; carse soils are often heavy, clay-rich and poorly drained, and were originally covered by low, thick scrubby vegetation; also known as carseland.

cauldron subsidence Collapse of a cylindrical volcanic crater and sub-surface igneous intrusion into deeper levels of the crust, with the formation of a ring fracture; this is followed by upwelling of magma to form a ring intrusion near the surface.

cement Mineral that binds grains together in sedimentary rocks; grows in pore spaces and along grain boundaries; commonest cements are silica, calcite and iron ores.

cementstone Rock that produces commercial cement when crushed; contains lime, silica and alumina in correct proportions.

cephalopods Marine animals (molluscs) with a large head, well developed eyes and tentacles, in a coiled or straight shell with internal chambers; examples are octopus, squid and cuttlefish; ammonites are extinct cephalopods with intricate chamber walls.

chalcopyrite Copper–iron sulphide, an important copper ore mineral.

chalk Pure white limestone made of shells of shallow-water marine carbonate plankton (foraminifera) and calcareous algae.

Chalk Name given to the upper Cretaceous period, in which chalk formed.

chemical weathering Weathering (decay or decomposition) of rocks to produce minerals stable at the Earth's surface.

chert Hard, compact form of silica found in chalk; formed of silica-rich organisms; similar to flint and jasper.

chilled margin Zone of fine-grain rock at edge of an igneous body, formed due to rapid cooling.

chlorite Green flaky silicate mineral with sheet structure, related to mica; found in low-grade metamorphic rocks, slate and phyllite.

chromite Ore mineral, source of chromium; chromium–iron oxide; black or dark brown

and very dense, with cubic symmetry.

cirque French for corrie.

cladding Cut and polished stone slabs used to cover external face of a building.

clastic Used to describe sedimentary rocks made of broken fragments.

clay minerals Silicates with a sheet structure, derived from weathering of feldspars and other minerals.

cleavage Ability of a mineral or rock to break along a plane of weakness.

Coal Measures Upper part of the Carboniferous, in which coal seams predominate.

cobbles Rounded rock fragments, 64–256 mm diameter.

cold-based ice Continental ice sheet, frozen to bedrock.

collision zone Boundary between two tectonic plates; continent–continent collision produces a mountain chain; ocean–ocean collision produces a volcanic island arc and trench (subduction zone); ocean–continent collision produces a trench and mountain chain with earthquakes and volcanoes.

columnar jointing Closely spaced cooling joints in igneous rocks, e.g. lava flows.

concretion Nodule of hard material with no internal structure, found in sedimentary rocks, e.g. chert in chalk, ironstone in sandstone, or limestone in sandstone.

cone sheet Igneous intrusion, circular at the top and tapering downwards to a point; cone sheets are often found stacked one inside the other.

conformable Continuity of beds of sediment, laid down in sequence without time gaps or breaks in sedimentation.

conglomerate Sedimentary rock made of large and small rounded fragments.

conodonts Extinct group of small marine animals, possibly related to worms; only the feeding apparatus is preserved; lived from Cambrian to Triassic times.

constructive plate boundary Zone of oceanic crust where new material is being added by injection of lava at mid-ocean ridge.

contact aureole Zone of thermal metamorphism around an igneous intrusion.

contact metamorphism Heat affected zone of country rock surrounding an igneous intrusion; produces a contact aureole.

continental crust Top part of the continental lithosphere above the Moho; average 35 km thick, up to 80 km in mountain chains such as the Himalaya.

continental drift Notion that continental blocks could move apart; superseded by plate tectonics (by the sea-floor spreading mechanism).

continental shelf Part of the continental crust overlain by up to 200 m sea water; also known as continental platform.

continental shield Extensive area of ancient Precambrian rocks that forms the basement in continental regions; also known as a craton.

convection currents Motion within a fluid whereby hot material rises, spreads outwards then sinks down when cool; thought to operate in the upper mantle.

corals   Marine invertebrate animals that build calcium carbonate skeletons; mostly live in reefs as colonies; some are solitary and live on the sea floor; corals require clean, well aerated, warm sea water and live only at shallow depths on the continental shelf.

cordierite   Magnesium–iron–aluminium silicate mineral in high-grade metamorphic rocks.

core   Innermost part of the Earth from 2900 km beneath the mantle to the centre of the Earth at 6370 km; mainly iron–nickel alloy; outer core is liquid and the source of the magnetic field, inner is solid.

correlation   Comparing sedimentary rock sequences, usually on the basis of their fossil content, and may be combined with rock type and environment.

corrie   Glacial landform on a mountain side, caused by ice dislodging large, steep-walled segment of bedrock; also cirque (French) or cwm (Welsh).

country rocks   Rocks around an igneous intrusion or mineral vein.

cover rocks   Sedimentary rocks deposited on older basement rocks.

crag and tail   Glacial landform, crag being formed of hard igneous rock forming steep face and tail being made of softer sedimentary rock or glacial sediment behind crag; ice movement direction was from crag to tail, giving streamlined feature.

craton   Large stable area of ancient continental crust unaffected by earthquakes or volcanic activity since the Precambrian; also known as shield.

Cretaceous period   Geological time division, 142 to 65 million years; above Jurassic and below Tertiary; named after chalk (creta in Latin); in Scotland, occurs mainly in the North Sea.

crinoids   Sea lilies, related to echinoderms (starfish, sea urchins); with stem and head made of calcite plates; common as fossils in Carboniferous limestones.

cross bedding   Structure in sedimentary rocks formed by water currents depositing sand at different angles on sloping surfaces; also known as cross stratification.

crust   Outermost layer of the Earth; continental crust is 35 km thick on average, and up to 4 billion years old; oceanic crust is less than 10 km thick and made of basalt, up to 180 million years old; lies above the Moho, with the mantle beneath.

crustaceans   Arthropod animals, mainly marine invertebrates, with several pairs of jointed legs and a hard, protective outer shell.

crystal   Solid, natural inorganic material, with a regular internal structure and flat faces arranged in different symmetry classes.

crystalline rocks   Rocks with a texture of interlocking crystals; igneous and metamorphic rocks are crystalline; some sedimentary rocks such as certain limestones and evaporite deposits are made of crystals, but the term is not usually used

cwm   Welsh for corrie.

cyclothem   Regular, repeated arrangement of sedimentary layers; especially common in Coal Measures: sandstone, shale, fossil soil (fireclay), coal, and limestone.

Dalradian   Sequence of late Precambrian sedimentary and volcanic rocks in Scotland and Ireland, 750–600 million years old, folded and metamorphosed during the Caledonian orogeny.

density currents   Fast flowing submarine currents consisting of mud and small rock fragments mixed into a turbid slurry with sea water; very dense and highly erosive; greywackes are sedimentary rocks produced by density currents (or turbidity currents).

destructive plate boundary   Zone of the Earth's crust where one plate is consumed beneath another by subduction along an ocean trench; such boundaries are marked by earthquakes and volcanoes.

detritus   Weathered rock fragments transported and deposited elsewhere, eventually to form a sedimentary rock, e.g. conglomerate or sandstone.

Devonian period   Division of the geological column, 417 to 354 million years ago; above Silurian and below Carboniferous; named after Devon; in Scotland, the Old Red Sandstone continental beds were deposited then.

dextral   Rightwards-directed movement, e.g. of a fault.

diagenesis   Conversion of loose sediment to a sedimentary rock by burial, compaction and cementation.

diatomite   Soft, powdery, porous rock made from diatom shells; used as a filter and purifying agent.

diatoms   Microscopic single-cell planktonic algae with silica cells or shells.

dimension stone   Building stone, dressed and shaped into blocks; can usually be cut in any direction.

dip   Maximum inclination of a plane (e.g. a bed or a fault), at right angles to the strike.

displacement   Distance of physical movement.

dissected peneplain   Originally flat land surface cut by valleys.

disseminated   Description of ore minerals that are widely dispersed throughout a body of rock; the ore is therefore usually of low grade.

dolerite   Medium-grain basic igneous rock, found in dykes and sills; sometimes referred to as whinstone; used as road stone.

dolomite   The mineral calcium–magnesium carbonate; also a form of limestone made of that mineral.

downthrow   Vertical displacement of one side of a fault.

drift   Unsorted loose material plastered over the surface when ice sheets melt.

drumlin   Smooth, rounded hill of glacial sediment, formed by moving ice.

ductile   Deformation of rock by thinning,

stretching and flow (opposite of brittle).

dune bedding   Form of cross bedding found in sand dunes; common in the New Red Sandstone (Permian to Triassic) of Scotland.

dyke   Igneous intrusion in the form of a vertical sheet; may have formed a feeder to surface lava flow.

dyke swarm   Closely spaced set of vertical dykes radiating out from an igneous centre.

echinoderms   Marine invertebrate animals having a calcite skeleton with fivefold symmetry; includes sea urchins, star fish and crinoids (sea lilies).

elastic deformation   Non-permanent changes in the shape of a rock, which will return to its original shape when the stress is removed (opposite of plastic deformation).

era   Division of geological time that contains several periods, e.g. the Palaeozoic era ("era of ancient life") is made of the Cambrian, Ordovician, Silurian, Devonian, Carboniferous and Permian periods.

erratics   Boulders carried by moving ice and dropped far from its source.

esker   Long, sinuous ridge of gravel, a landform created by deposition of sediment in a tunnel inside an ice sheet.

essexite   A type of gabbro; found at Lennoxtown (Campsies) and Crawfordjohn (Sanquhar); highly distinctive glacial erratic.

estuary   Mouth of a river where it flows into the sea, and where fresh water and salt water meet and mix.

Eurasia   Giant continent consisting of Europe, India and Asia.

eurypterid   Extinct arthropod fossil, water scorpion; Ordovician to Permian; found in Carboniferous shales of the Midland Valley.

eustatic sea level changes   Worldwide rise or fall in sea level; related to growth and decay of continental ice sheets or growth of mid-ocean ridges.

evaporite   Sedimentary rock formed by chemical precipitation of salts in shallow seas and lagoons during extremely hot climatic conditions; e.g. gypsum (calcium sulphate) and halite (sodium chloride, rock salt) are evaporite minerals.

exhumed topography   Buried ancient landscape exposed by erosion of overlying rocks; e.g. Torridonian on top of Lewisian in Northwest Highlands.

extinct volcano   Volcano that has not erupted in historical times.

extinction   Complete disappearance of a species of animals or plants in nature.

extrusive   Igneous lava poured out at the surface.

facies   Total set of characteristics in a rock; in sedimentary rocks this will include grain composition and size, sedimentary structures, etc. and can be used to interpret past environment of deposition, e.g. river bed, delta, desert floor, shallow sea, lake or lagoon; in metamorphic rock the facies is a combination of pressure and temperature conditions that produced a particular set of minerals under those conditions.

facing stones   Thin slabs of decorative stone (may be polished) attached to rubble walls.

fan   A slope of rock detritus that widens down the slope and spreads out like a fan.

fault   Break in rocks where there has been physical movement (displacement) between two blocks.

fault breccia   Broken, angular fragments of wall rock, surrounded by crushed rock, in a fault zone.

fault displacement   Relative movement of one block of rock against another.

feldspars   Silicate minerals, with three-dimensional lattice structure; commonest minerals in the crust; plagioclase feldspars contain calcium and sodium; orthoclase feldspar contains potassium.

felsite   Red or black very fine-grain igneous rock, originally a volcanic glass.

ferromagnesian minerals   Silicate minerals containing iron and magnesium; e.g. olivine, pyroxene (augite), amphibole (hornblende).

fireclay   Fossil soil beneath coal seams, made of clay minerals and very heat resistant (refractory).

fissile   Describes a fine-grain rock that splits easily along bedding planes, e.g. shale (sedimentary), or along cleavages, e.g. slate (metamorphic).

flagstones   Finely laminated siltstone, used as paving slabs.

flint   Form of non-crystalline silica, found in chalk; similar to chert.

flood basalt (plateau basalt)   Basalt lava flows covering an extensive area, and usually forming stepped or trap landscape feature.

floodplain   Low part of a river valley, which is liable to flooding; formed of alluvium deposited by meandering river.

flow banding   Structure formed in igneous rock by lava being streaked out during flow.

fluorite   Calcium fluoride, a purple mineral, forming cubic crystals; used as a flux in smelting; common gangue mineral in lead–zinc ore deposits; also called fluorspar.

fluvial sediments   Sediments transported by rivers.

fluvioglacial   Sediment carried by meltwater streams from glaciers; also glaciofluvial.

fold   Structure formed where beds have been bent or buckled.

fold axis   Line separating the two limbs of a fold.

fold limb   Part of a fold lying between fold hinges.

foliation   Planar fabric in metamorphic rocks, usually used to describe schist or gneiss.

fool's gold   Pyrite or iron pyrites, a cubic brassy yellow mineral and a common iron ore; iron sulphide.

foraminifera   Microscopic single-cell planktonic marine animals of wide distribution; extremely important for correlating sedimentary rocks from Devonian to the present.

formation   Unit of rocks with consistent internal structure and composition, recognizable over a wide area and which can be mapped.

fossil fuel   Peat, coal, oil and gas, of organic origin, preserved in sedimentary rocks.

fracture   A break in rocks, e.g. joint or fault; or the way in which a mineral breaks.

gabbro   A coarse-grain basic igneous rock containing olivine, pyroxene and plagioclase (calcium-rich feldspar); usually black speckled with cream or white.

galena   Lead sulphide, a soft, dense shiny grey mineral, and the main ore of lead metal.

gangue mineral   Waste mineral (e.g. quartz, fluorite, calcite) found in association with useful metal ores in a mineral deposit.

garnet   A cubic silicate mineral (of iron, magnesium, aluminium), usually deep red in colour, found in metamorphic rocks that formed at medium to high pressure, e.g. schist or gneiss; tangerine-coloured garnets at Elie, Fife, are called "Elie rubies" – these were transported from the upper mantle in volcanic pipes and deposited in ash layers.

gastropods   Molluscs with coiled shell, e.g. snails; from Cambrian to present day; now at their most abundant.

geological column (stratigraphic column)   Division of geological time into different periods; the column is shown with the oldest formations or periods at the bottom.

glacial erratics   Boulders of rock carried by ice and deposited far from their source; useful for deducing ice movement direction.

glacial till   Boulder clay, unsorted, mixed sedimentary material deposited directly from a melting ice sheet.

gneiss   Very coarse-grain, banded, high-grade metamorphic rock with foliation often marked by biotite mica.

Gondwanaland   Supercontinent formed 2000 million years ago by the accretion of the southern continents, Australia, Antarctica, South America, India and Africa; began to rift apart 180 million years ago.

graben   Rift valley formed by downfaulting of crustal block between two parallel normal faults.

graded bedding   Structure in sedimentary rocks (usually deposited in water), with coarser material at the base of a bed, gradually becoming finer towards the top.

Grampian orogeny   Deformation and metamorphism of Moine and Dalradian rocks in the Highlands, 510–480 million years ago; gabbros and granites in Aberdeenshire formed in this event; caused by collision of edge of Laurentia with an island arc.

granite   Coarse-grain acidic igneous rock, intruded as plutons into folded country rocks; contains quartz, plagioclase feldspar (sodium-rich) and orthoclase feldspar (potassium-rich), in nearly equal amounts, often also biotite mica; pink, white, grey or silver overall.

granular   Rock texture in which mineral grains are of even size.

granulite   Coarse-grain, high-grade regional metamorphic rock formed under very high pressure and temperature conditions; granular texture, lacks banding or foliation; common in the Lewisian Complex, especially around Scourie and Gairloch; contains pyroxene, plagioclase feldspar, garnet, quartz.

graphite   Soft, platy, black carbon; may be found in graphite schist, a metamorphosed sedimentary rock.

graptolite   Extinct fossil animals that lived as floating colonies; tiny polyps occupied small cups connected along branching arms; Cambrian to Devonian; extremely important for dating, especially in the Ordovician and Silurian; resemble pencil marks on slate or fine mudstone.

Grenville orogeny   A series of mountain building events (folding, metamorphism, migmatite formation) in the Precambrian that led to the formation of the Rodinia supercontinent, comprising Laurentia, Baltica, Siberia and Gondwana, 1000–850 million years ago, and affecting the Moinian of the Northern Highlands.

greywacke   Medium-grain sedimentary rock, a type of sandstone, containing fragments of igneous and metamorphic rocks, angular quartz and clay minerals; formed by slumping of sediment on the continental shelf, then carried to the deep ocean by turbidity currents or mud slurries; grey, black, purple or dark green, tough, weakly metamorphosed; common in the Southern Uplands; locally known as whinstone; quarried used as road stone.

grit   Coarse, angular sedimentary rock, usually referring to metamorphic quartzite in the Dalradian.

groundmass   Tiny fine-grain crystals in an igneous rock that also contains larger crystals; matrix.

gypsum   Calcium sulphate, an evaporite mineral; soft, transparent crystals with flat, tabular habit and pronounced cleavage; alabaster is a white form of gypsum; anhydrite is another form, without water in its structure; used in plaster.

half-life   Time taken for a mass of a radioactive element to decrease by half.

halite   Sodium chloride, rock salt; an evaporite mineral, found in salt domes, which are important oil traps.

hanging valley   Glacial landform feature caused by small tributary stream falling into a main valley that has been deepened by ice.

hæmatite (hematite)   Iron oxide, a common red iron ore.

Hercynian orogeny   Mountain building episode in central Europe, Carboniferous to Permian, 360–290 million years ago; Gondwana collided with Laurussia and created the supercontinent of Pangaea; also known as Variscan or Armorican orogeny.

Highland Border Complex   Series of sedimentary and igneous rocks, including ocean floor pillow lavas, of Cambrian to Ordovician age (530–450 million years old) found patchily in a narrow belt along the Highland Boundary Fault.

hinge   Part of a fold where two limbs meet and the fold curves over to form a line.

Holocene   The most recent geological period, from 10 000 years ago when the last ice melted; subdivision of the Quaternary.

hornblende   Calcium–iron–magnesium–aluminium silicate, the commonest of the amphibole family; black to dark green elongate crystals, found in hornblende schist, amphibolite and banded gneiss.

hornfels   Fine-grain, hard splintery rock found in thermal aureoles around large igneous intrusions; may contain spots of new mineral growth.

host rock   The country rock into which magma is injected, or which is affected by mineralization.

hummocky moraine   Glacial landform produced by irregular mounds of unsorted mixed material dumped at the front of melting glaciers.

humus   Organic matter contained in soil.

hybrid rocks   Igneous rocks formed from the mixing of acidic and basic magmas to give rocks of intermediate composition, often with unusual minerals; may also form by country rocks around an intrusion being melted and incorporated into the (basic) magma.

hydrated   Mineral altered by the addition of water.

hydrocarbons   Natural materials containing hydrogen and carbon, e.g. methane gas.

hydrological cycle (water cycle)   Natural system that connects all the water in the atmosphere, seas, oceans, clouds, rivers, lakes, soils, plants and glaciers on Earth, constantly circulating by evaporation, precipitation and transpiration.

hydrothermal alteration   Changes to minerals and rocks caused by the action of hot water-rich fluids adjacent to igneous intrusions or mineral veins.

hydrous   Said of minerals containing water molecules.

Iapetus Ocean   Wide (up to 5000 km) ocean between Laurentia (North America, Greenland and Scotland), Gondwanaland (Europe and the southern continents) and Baltica (Scandinavia and Russia) that existed from 600 to 420 million years ago; continental collision led to the closure and disappearance of Iapetus by subduction and the growth of the Caledonian mountain chain; the line of collision between Scotland and England along the Solway Firth is known as the Iapetus Suture.

ice age   Period of time marked by rapid fluctuations between extremely cold and mild climatic conditions; there have been 20 such events in the last 2.5 million years.

ice cap   Small glacier in mountainous region.

ice sheet   Continental glacier covering the entire landscape; shaped like a flattened dome.

igneous rock   Rock formed by cooling and crystallization of molten magma in underground intrusions or on the surface as lavas.

ignimbrite   Igneous rock formed by ash and broken crystals in a violent volcanic eruption, so hot that clouds of material are welded together.

index fossil   A distinctive fossil which is abundant in a particular sedimentary horizon.

index mineral   Characteristic mineral in a metamorphic rock which marks a metamorphic zone.

inlier   Older rock surrounded by younger.

inselberg   Landscape feature formed by steep, isolated mountain rising above low, flat terrain.

interglacial   A short and relatively warm climatic phase between glacials.

intermediate igneous rock   Rock with 52–66 percent silica and less than 10% quartz crystals; usually contains 75% feldspars and 25% ferromagnesian minerals (hornblende, biotite, pyroxene); diorite is coarse grained (plutonic) and andesite fine grained (lava).

interstadial   Short, relatively warm phase in a glaciation.

intrusion   Igneous body forced into country rocks at depth; minor intrusions are dykes (vertical), sills (parallel to beds, usually horizontal), cone sheets; major intrusions are granite and gabbro plutons.

intrusive igneous rocks   Rocks formed by physical intrusion or emplacement at depth.

island arc   Arc-shape chain of volcanic islands formed when one oceanic plate is subducted beneath another at a destructive plate boundary.

isostatic rebound (isostatic uplift)   Recovery of initial level of the land surface to a state of equilibrium by removal of temporary local load of ice; uplift may also result from erosion of mountains and gradual uplift of mountain root.

isotopes   Forms of the same element with different masses; chemically identical but with different physical properties; some are radioactive and decay, most are stable.

isotopic dating (radiometric dating)   Method of dating rocks by measuring amounts of radioactive isotopes and their decay products; the rate of decay (half-life) is known and the time since decay began (= the age of a rock or mineral) can be calculated, assuming that the rock or mineral was a closed system.

jasper   Microcrystalline form of silica (quartz) stained red by hæmatite (iron oxide).

joint   Natural crack or fracture in a rock, along which there has been no movement; cooling joints are common in igneous rocks.

jökulhlaup   Icelandic name for catastrophic flood caused by sudden melting of ice.

Jurassic period   Division of the geological column, 206–142 million years; above Triassic and below Cretaceous; named after Jura Mountains in France and Switzerland; North Sea oil deposits occur in Jurassic

rocks; also found in Skye and Mull.

kame   Glacial landform; irregular mound of bedded sand and gravel formed at edge of melting ice sheet.

kaolinite   Sedimentary rock made of the clay mineral kaolin; derived from the weathering of feldspar.

karst   Landform in limestone area; includes caves, dry river valleys, swallow holes, underground streams and limestone pavements; caused by carbonate being dissolved by rainwater.

kerogen   Organic matter (bitumen) found in sedimentary rocks such as oil shale; carbon-rich, derived mostly from plant debris.

kettle hole   Glacial landform feature created by block of stagnant ice melting to form a lake-filled depression in glacial drift (till).

knock and lochan   Topography formed in Lewisian Gneiss by glacial erosion of the landscape; low, bare, rounded rocky hills (cnoc in Gaelic) with small, shallow lochs and peat bogs between.

kyanite   Aluminium silicate mineral found in high-grade (high pressure) metamorphic schists; forms flat, blue blades.

laccolith   Igneous intrusion, shaped like a mushroom, with flat floor and domed top; e.g. Traprain Law in East Lothian and the Eildon Hills at Melrose in the Borders.

Lake Orcadie   Devonian (380 million years old) shallow-water lake or inland sea that occupied the area of the present-day Moray Firth, Caithness, Orkney and south Shetland, where flagstones formed and in which primitive fish lived.

lamellibranchs   Bivalves, marine or freshwater molluscs, invertebrate animals with soft body inside two shells joined by a hinge, e.g. mussels, oysters, clams.

lamination   Very fine layering of sedimentary rocks, with layers up to 1 mm thick.

landform   A natural physical feature on the surface, e.g. valley, mountain, plateau.

laterite   Soil rich in iron and aluminium left behind as a residue in regions of deep tropical weathering.

lattice   Regular arrangement of the atoms in a crystal.

Laurasia   Supercontinent in the northern hemisphere, formed in the Triassic by accretion of North America, Greenland, Europe and Asia.

Laurentia   Precambrian continent, consisting of North America, Greenland and Northwest Scotland, collided with Baltica and Avalonia to create the Caledonian mountain belt.

lava   Stream of molten rock erupted onto the surface by a volcano (from a crater or a fissure), to form a lava flow.

layered intrusion   Large basic to ultrabasic igneous intrusion in which crystals have settled into layers, under gravity, during cooling and crystallization; each layer usually has a distinctive composition, possibly including ore mineral horizons.

leaching Removal of soluble materials in soil by groundwater movement.

limb One half of a fold; both limbs meet at the fold hinge.

limestone Sedimentary rock made mostly of calcite (calcium carbonate), often of organic origin; common in the lower Carboniferous.

limonite Rusty-brown hydrated iron ore (iron oxide).

lineation Linear structure in a rock; usually refers to parallel arrangement of needle-shape minerals (e.g. hornblende) lined up in metamorphic rocks.

lithification Process that changes loose sediment into rock by compaction, squeezing out of water from pore spaces, and growth of cement.

lithosphere Strong, rigid, solid outer part of the Earth, made of crust and upper mantle; above the weak asthenosphere.

liver stone A quarrying term for a compact massive sandstone with thick beds and no fine-scale internal laminations. Very large blocks of such rock can be removed from a quarry in one piece. For example, the six columns forming the pillars at the entrance to Old College, University of Edinburgh (built 1789–1828), are almost 7 m high and weigh 9 tonnes each. Each one was transported from Craigleith quarry on a special carriage pulled by 16 horses. At the peak of its production in the early 1800s, the quarry was over 100 m deep.

Loch Lomond re-advance (Loch Lomond Stadial) Cold period at the end of the most recent ice age, 11 000–10 000 years ago; re-establishment of a mountain ice cap from Torridon to Loch Lomond, after the melting of the main ice sheet.

machair Coastal landform in western Scotland, consisting of shell sand covered by rich, fertile soil; sand dunes and white sandy beaches at sea level.

magma Molten rock, usually containing dissolved gases; cools to form crystalline igneous rocks.

magma chamber Body of magma in the crust which feeds lava to surface volcanoes; crystals forming in the chamber may become stratified to form a layered intrusion.

magnetite Magnetic iron oxide; cubic mineral and important iron ore; very common as a minor mineral in basic igneous rocks.

magnetic striping Alternating stripes of normal and reversely magnetized ocean floor arranged symmetrically on both sides of mid-ocean ridges.

magnetic reversals (geomagnetic reversals) Changes in the Earth's magnetic field between normal polarity (magnetic pole in the north) and reversed polarity.

mantle Layer of the Earth below the crust and above the core; made predominantly of peridotite; upper mantle contains the weak, partially molten asthenosphere; boundary between crust and mantle is marked by the Moho.

mantle convection Slow movement of plastic material in the upper mantle; hot material rises, spreads laterally and cools, then cold material sinks back into the mantle; the mechanism that is thought to control plate movements.

mantle hot spot Unusually hot part within the upper mantle, resulting in partial melting of mantle peridotite and creation of magma chambers and volcanic provinces in the crust.

mantle plume Vertical up-rise of very hot mantle material into the lower crust, producing a mantle hot spot.

marble Metamorphosed limestone containing calcite (calcium carbonate), dolomite (calcium–magnesium carbonate) and sometimes other minerals; has sugary or granular texture.

marker horizon A layer of rock that has certain distinctive features (e.g. composition, fossils) that allow it to be used over a wide area as a point of reference.

marl Very fine lime mudstone.

matrix Groundmass of finer crystals or particles in a rock (not the same as cement in a sedimentary rock).

meander Bow-shape loop of a river in its floodplain.

Mesozoic Era of geological time between Palaeozoic and Cenozoic; contains the Triassic, Jurassic and Cretaceous periods, total span 248–65 million years ago.

metallogenic province A region of the crust in which there are many similar ore mineral deposits.

metamorphic aureole Heat-affected zone of country rock surrounding a large igneous intrusion; rocks may be converted to hornfels by contact metamorphism (thermal metamorphism).

metamorphic grade Relative intensity of pressure and temperature conditions during metamorphism; slate is a low-grade rock, gneiss is high-grade.

metamorphic rocks Rocks which have recrystallized in the solid state by the action of heat and pressure on pre-existing rocks at depth within the crust.

metamorphism Changes in the mineral content and texture of a rock brought about by increased pressure, temperature (above about 300°C) and fluid activity in the crust.

metasediments Informal term for metamorphosed sedimentary rocks.

metasomatism Chemical changes brought about in rocks by the action of hot fluids (water and carbon dioxide in the main), usually penetrating from large igneous intrusions into reactive country rocks (host rocks); useful mineral deposits may result from chemical reactions and precipitation.

methane Hydrocarbon gas, a natural compound; found with crude oil, and the main constituent of natural gas; also produced in marshes and by agricultural activities; a greenhouse gas; present in coal mines as firedamp.

mica Sheet-like complex potassium–aluminium silicate mineral with good cleavage; muscovite is colourless, biotite is black or dark brown (because of iron content); muscovite is common in schist, biotite is common in granite, schist, gneiss and amphibolite.

microfossils Fossils of plants and animals, usually plankton, spores and seeds, that can be seen only with the aid of a microscope; very important in correlating oil-bearing strata in the North Sea.

mid-ocean ridge Continuous underwater mountain range snaking around the Earth's oceans; source of new oceanic crust: basalt lava is erupted onto the ocean floor at the ridge and spreads out sideways; Iceland sits astride the Mid-Atlantic Ridge.

migmatite High-grade metamorphic rock of mixed composition (usually granite and amphibolite) and patchy appearance, resulting from partial melting and injection of granitic veins into more basic rock; common in parts of the Lewisian Gneiss Complex.

Milanković cycles Regular variations in the amount of heat received at the Earth's surface from the Sun as a result of motions of the Earth's orbit and the tilt of its axis; three separate cycles operate on time scales of 96 000, 40 000 and 21 000 years, and their combination has been used to explain climatic changes, especially the Pleistocene glaciations.

mineral Naturally occurring crystalline solid with a regular internal structure; minerals (especially silicates) are the building blocks of rocks.

mineralization Impregnation of country rocks (host rocks) with fluids rich in ore minerals, to create an ore deposit; mineralization may be widely disseminated, or localized along narrow shear zones.

Moho Boundary separating the crust from the mantle of the Earth, marked by a sharp change in seismic (earthquake) wave velocity due to changes in composition (mantle peridotite); abbreviation for Mohorovicic discontinuity.

Moinian Precambrian metamorphosed sediments (mica schists and quartzites), 1000–800 million years old, north of the Great Glen Fault and east of the Moine Thrust Zone in the Northern Highlands of Scotland; also called Moine Schist.

mollusc Invertebrate animal with soft body, protected usually by a shell; e.g. snail, clam, squid, octopus; first appeared in the Cambrian period.

monadnock Prominent, isolated mountain standing up above a low, flat plain.

moraine Mound of unstratified and unsorted glacial drift deposited by a glacier; terminal moraine forms at the front or snout of a glacier, lateral moraine along valley walls, and medial moraine where two glacier flows meet and their lateral moraines merge.

mud cracks Structure formed in fine sediment when ponds or river beds dry up and mud or clay dries and shrinks; useful way-up structure.

mudstone Very fine-grain massive sedimentary rock which does not split easily;

consists of clay and fine silt; often black due to presence of organic matter (carbon).

muscovite   White mica, a sheet silicate; splits easily into very thin flakes; common in schist.

mylonite   Fine-grain high-pressure metamorphic rock with banded structure, from extreme crushing and recrystallization of pre-existing rocks in narrow shear zones and thrusts; from Greek word for "mill".

nappe   Large sheet of rock transported above a thrust fault; from French for "tablecloth"; nappe structures are common in the Northwest Highlands, in the Moine Thrust Zone.

nautiloid   Free-swimming marine mollusc with a coiled, chambered shell; common as fossils in Ordovician and Silurian rocks; *Nautilus* is a modern form.

New Red Sandstone   Permian and Triassic rocks, deposited in desert conditions.

normal fault   Fault with downward movement of one side relative to the other.

normal magnetism   Orientation of the Earth's magnetic field with the north magnetic pole near the north geographic pole, as today.

North Atlantic Drift   Northeastern part of the Gulf Stream that brings warm equatorial waters north to Scotland, Iceland and northern Norway; responsible for higher winter temperatures than other places on the same latitude.

North Atlantic Volcanic Province   Broad area of igneous activity during the Tertiary, 60–50 million years ago, embracing East Greenland and Northwest Scotland that produced large volumes of basalt lava from central complexes, e.g. Mull, Skye, Rum; resulted from a mantle plume.

nose   Part of a fold where the hinge curves, same as hinge zone.

nunatak   Isolated peak of bedrock protruding above an ice cap; Greenlandic word.

obduction   Process by which isolated slices of oceanic crust are detached and thrust upwards into folded rocks of the continental crust during mountain building; the deformed oceanic rocks in such situations are referred to as ophiolites.

obsidian   A black glassy volcanic rock, formed from the rapid chilling of lava at the surface; acidic to intermediate composition; widely used in prehistoric times to make tools; pitchstone is a recrystallized form of obsidian.

ocean floor spreading   Mechanism in plate tectonics theory whereby new oceanic crust is formed by the injection of dykes and pillow basalt lavas along ocean ridges; also known as sea-floor spreading.

ocean ridge   Narrow zone (1–4 km wide and 2–3 km high) in an ocean where new crust forms; the Mid-Atlantic Ridge is an example.

ocean trench   Long narrow V-shape depression of the oceanic crust above a subduction zone where an oceanic crustal

plate is deflected down into the mantle; may be up to 10 km deep; most occur around the Pacific Ocean and are associated with deep and shallow earthquakes and volcanic zones.

oceanic crust   Crust that underlies the ocean basins; made of basalt, 7 km thick on average; created at ocean ridges by dykes injecting pillow lavas into rift zones.

oil shale   Fine-grain black shale rich in kerogen, a hydrocarbon; when heated, liquid oil is produced; common in the Carboniferous of the Midland Valley.

Old Red Sandstone   Continental sedimentary rocks formed in valleys and flood plains during the Devonian period by rivers flowing off the Caledonian mountains; the red colour is due to iron staining in the cement.

olivine   Iron–magnesium silicate mineral found in basic and ultrabasic igneous rocks; dark green or black and very dense, weathers easily at the surface; alters to serpentine during metamorphism.

oolith   Small (up to 2 mm), round, egg-shape particle in a sedimentary rock, usually a sand grain or tiny piece of shell surrounded by fine layers of calcite or iron compound.

oolitic   Description of a sedimentary rock made of ooliths, e.g. oolitic limestone, oolitic ironstone.

ophiolite   Slice of old oceanic crust (basalt lava) thrust into folded continental rocks during mountain building; often metamorphosed to serpentinite.

Ordovician period   Division of the geological column, 495–443 million years ago, above Cambrian and below Silurian; named after the ancient Welsh tribe, the Ordovices.

ore body   Large continuous mass of metal ore.

ore deposit   Accumulation of a useful mineral that can be worked economically.

ore mineral   Metal compound or native element found in host rocks; the useful part of an ore, in contrast to the waste or gangue material.

orogeny   Mountain building episode, a sequence of events including folding, metamorphism, thrusting, faulting, granite intrusion and crustal thickening, occupying several tens of millions of years.

orthoclase   Potassium–aluminium silicate, a type of feldspar; common mineral in granite and pegmatite; pink, white or cream in colour.

ostracod   Tiny marine crustacean animal with a carbonate bivalve.

outcrop   Bedrock which appears at surface exposures.

outlier   Isolated body of younger rock surrounded by older rocks and situated at a distance from the main outcrop.

outwash plain   Sand and gravel deposited by glacial meltwater in braided streams flowing across a broad, flat plain in front of the snout of a glacier.

overturned fold   Fold which has been tilted beyond the vertical.

Pangaea   Name of a supercontinent that is thought to have existed from 300 to 200 million years ago and included all the world's continents; rifted into the northern continents or Laurasia and the southern continents or Gondwana, separated by the Tethys Sea; surrounded by the ocean Panthalassa.

Panthalassa   Name of the ocean that surrounded the supercontinent of Pangaea.

partial melting   Process in the upper mantle and lower crust that generates molten magma; pressure at depth in the Earth is too great to allow complete melting, but if pressure is slightly released, and small amounts of water are present (1–5%), then mantle peridotite can melt partially and copious amounts of basalt may be produced; in the case of crustal rocks, e.g. thick folded sediments in a mountain chain, temperatures may be reached that will allow granite to form by partial melting of the lower crust; melts are lighter than the surrounding solid rocks and therefore will rise closer to the surface; partial melting is a key stage in the rock cycle.

patterned ground   Symmetrical shapes such as polygons found at the surface in arctic regions as a result of frost action (repeated freeze–thaw cycles).

pebbles   Rounded rock fragments, 4–64 mm in diameter; sedimentary rock containing pebbles is known as conglomerate.

pegmatite   Extremely coarse-grain igneous rock, usually granite in composition; contains large well formed crystals of quartz, pink feldspar and mica over 25 mm in diameter; formed in the final stages of cooling and crystallization of magma containing water-rich fluids.

peneplain   Extensive, low, flat or gently undulating landscape produced by erosion; when elevated and cut into by rivers, the feature is referred to as a dissected peneplain, where all the hill tops are at the same height.

peridotite   Very coarse-grain dense, black ultrabasic igneous rock containing olivine and pyroxene (iron–magnesium silicates); the material that makes up the mantle; also found in large layered igneous intrusions such as in Rum and Morven–Cabrach, Huntly; crystals may have settled out of the magma to form separate layers; may be altered to serpentinite during metamorphism in the presence of water-rich fluids.

periglacial   Cold climatic conditions, usually around ice sheets and large glaciers, and the features that are produced in such environments.

period   Main division of geological time, spanning tens of millions of years, e.g. Cambrian period (also spelled with capital P; colloquially, "the Cambrian").

permafrost   Permanently frozen soil in arctic regions.

Permian period   Division of geological timetable, 290–248 million years; above Carboniferous and below Triassic; named after Perm in eastern Russia; Permian and

Triassic desert sandstones are named the New Red Sandstone in Britain.

phenocryst   Large crystal in an igneous rock, surrounded by finer groundmass.

phyllite   Green coloured low-grade metamorphic rock (between slate and schist in grade), typically with closely spaced tiny folds, crinkles and corrugations, giving a wavy appearance; phyllite splits easily but irregularly; green colour derives from the presence of chlorite, a silicate with a sheet structure resembling mica.

pillow lava   Form of basalt erupted from submarine volcanic vents and chilled to form an outer fine-grain skin.

pitchstone   Volcanic glass, usually black; formed by very rapid cooling of lava so that crystals had no time to grow; forms extremely sharp fragments when broken; similar to obsidian.

placer deposits   Accumulation of useful minerals by sorting and concentrating in moving water (rivers and beaches); minerals are usually dense, e.g. gold or chromite.

plagioclase   Group of the feldspar silicate minerals, containing calcium and sodium; the commonest minerals in igneous and metamorphic rocks.

plankton   Mass of tiny floating animals and plants at the surface of a sea, ocean or lake.

plastic deformation   Permanent changes to the structure of a rock, e.g. folding and stretching, which cannot be undone; opposite of brittle deformation (e.g. faulting and fracturing).

plate tectonics   Theory of global tectonics that states that the lithosphere is divided into a number of large rigid plates, which are capable of moving across the surface of the globe and which interact at their boundaries.

plateau   Landform feature consisting of extensive, elevated flat region with relatively steep sides.

plateau lavas   Thick series of lava flows, usually basalts, forming an extensive flat-topped plateau across thousands of square kilometres.

Pleistocene   Subdivision of the Quaternary period, 1.82 million to 10 000 years ago; followed by the Holocene, the most recent period; noted for its numerous climatic fluctuations and the ice ages.

plug   Landform feature caused by upstanding vertical mass of igneous rock in the vent of an ancient volcano.

pluton   Large, deep-seated igneous intrusion.

plutonic igneous rock   Coarse-grain igneous rock formed inside a pluton.

Polar Front   Boundary between cold polar air or water and warm tropical air or water; presently located between Newfoundland and Iceland; during glacial episodes it is pushed south, to the latitude of Spain and Portugal.

pollen analysis   Study of pollen grains from plants, found in peat and lake deposits; used for dating sediments and for interpreting past climate on the basis of tree distribution.

pore spaces   Open spaces between grains

in a sediment or sedimentary rock; pore spaces may be interconnected, allowing fluids to move through.

porphyry   Type of igneous rock containing larger crystals (phenocrysts) amid finer-grain matrix or groundmass.

porphyry ore deposit   Very large, low-grade ore deposit, usually in igneous rocks such as granites or andesite lavas, in which the ore minerals are scattered or dissemi-nated in the host rock; e.g. porphyry copper deposits, which may also contain minor amounts of molybdenum, gold and silver.

Precambrian   Division of geological time, older than Cambrian, 545 million years; generally devoid of fossils with hard parts; usually subdivided into Archaean (4560–2500 million years) and Proterozoic (2500–545 million years).

precipitate   Mineral formed by deposition, usually from sea water as a result of a chemical reaction.

pyrite   Iron sulphide, a brassy yellow iron ore that forms cubes; also known as fool's gold; common in slate, sometimes also in coal.

pyroclastic   Volcanic rock formed of broken crystals, rock and ash during explosive eruptions.

pyroxene   Family of calcium–iron–magnesium silicates with single chain structure; dense, dark in colour; very common in basic and ultrabasic igneous rocks; augite is one of the commonest pyroxenes.

quartz   Silicon dioxide, one of the commonest rock-forming minerals, especially in granite; sandstone consists mostly of quartz grains; harder than glass; colourless or white; perfect hexagonal crystals are known as rock crystal.

quartzite   Metamorphosed sandstone in which quartz grains and silica cement have recrystallized to form a rock with a granular texture in which the grains are evenly shaped; very hard, white and tough rock; the Cambrian Quartzite of Northwest Scotland is a form of sedimentary quartzite with a recrystallized silica cement.

quartzo-feldspathic   Used of metamorphic rocks made of quartz and feldspar, e.g. granite gneiss.

radiocarbon dating   A method used to date organic remains (wood, peat, bones, teeth, cloth) that are younger than about 40 000 years; based on the decay of carbon–14.

radiolaria   Microscopic single-cell planktonic animals with spherical silica shells; Cambrian to present day; their remains contribute to deep-sea ooze and to the formation of chert.

radiometric age dating   Laboratory method used to determine the age of a mineral or rock by measuring and comparing amounts of radioactive isotopes and their decay products; using the known decay rate (half-life), it is possible to calculate the time at which an event occurred in the past; the

minute quantities of isotopes are measured in a sensitive mass spectrometer instrument.

raised beach   Landform created around the coast by uplift or rebound of the land surface and fall in sea level, after the melting of a continental ice sheet; raised beaches are now well above the level of the modern beach; fossil cliffs and caves may be present.

regional metamorphism   Whole-scale alteration of rocks over a wide area, involving increases in pressure, temperature and fluid activity and resulting in recrystallization during deformation within a continental mountain chain.

reserves   Amount of a useful mineral resource that can be removed by mining.

reservoir rock   A rock with pores or fractures in which oil and gas have accumulated.

resources   Total amount of metal ores, coal or oil, etc., including known reserves as well as uneconomic deposits and those not yet discovered.

reverse fault   Steep fault in which one block of rock is pushed up over another.

reversed magnetism   Orientation of the Earth's past magnetic field opposite to the present orientation, i.e. magnetic North Pole was adjacent to geographical South Pole.

rhyolite   A fine-grain acid lava, usually grey or white in colour and rich in quartz and feldspar; the surface equivalent of granite.

ring dyke   Minor igneous intrusion in the form of a vertical cylindrical wall.

ripple marks   Marks left on the surface of sediment by currents washing back and forth; a useful way-up indicator.

roche moutonnée   A landform created by glacial erosion when bedrock outcrops are streamlined by moving ice that climbs up a smooth whaleback and plucks material away from the sheltered steep side.

rock-forming minerals   Minerals that are sufficiently abundant to make up rocks; the commonest groups are the silicates in igneous and metamorphic rocks, and carbonates, halides, etc. in sedimentary rocks.

Rodinia   Supercontinent that formed 1000 million years ago by the collision of Laurentia, Baltica, Siberia and Gondwana in the Grenville orogeny, when parts of the Northern Highlands formed; Rodinia began to rift apart about 750 million years ago, when the Iapetus Ocean formed as a result; Rodinia is derived from the Russian for "motherland".

salt   Natural rock salt, sodium chloride, the cubic mineral halite; forms as an evaporite mineral; common in the Permian and important in oil trap structures in the North Sea (salt domes).

sandstone   Medium-grain sedimentary rock made of quartz fragments cemented by silica, iron ore or calcium carbonate; usually displays bedding.

sandur   Icelandic term for an extensive outwash plain in front of a melting glacier;

braided streams transport large volumes of sand and gravel.

sapropel   Organic-rich mud formed in stagnant lakes.

sapropelic coal   Fine-grain, massive coal formed from spores and fine plant debris; examples are boghead coal and cannel coal.

schist   Medium- to coarse-grain metamorphic rock rich in mica, foliated and roughly banded; result of high grade regional metamorphism; the main or significant mineral is often attached, e.g. mica schist, garnet schist, hornblende schist.

schistosity   Foliation structure in schist due to parallel alignment of platy mica.

scree   Accumulation of loose, angular rock fragments at the foot of a steep mountain slope, usually caused by frost action.

seat earth   Fossil clay-rich soil formed beneath coal seams and usually containing fossil tree roots.

sediment   Grains of loose material deposited by water, wind or ice or precipitated from solution.

sedimentary basin   Large flat depression in the continental crust into which rivers flow and deposit their sediment load; usually long-lived and with thick sequences of sedimentary rocks resulting from subsidence of the basin floor.

sedimentary breccia   Coarse angular sedimentary rock formed usually by the rapid removal and deposition of rock fragments close to the source area.

sedimentary rock   Rock formed on the Earth's surface by processes of weathering, erosion, transportation, deposition, burial, compaction and cementation (clastic or fragmental rocks); or by biological activity such as reef-building animals to create limestones; or as chemical precipitates, such as rock salt (halite, by evaporation of sea water).

serpentine   Group of hydrated iron–magnesium silicate minerals with a sheet structure; usually green and may be fibrous; result from metamorphism or hydrothermal alteration of olivine and pyroxene, sometimes also amphibole.

serpentinite   Metamorphic rock made mainly of serpentine group minerals, often also with chlorite, carbonate minerals, talc and magnetite; formed from olivine and pyroxene in basic and ultrabasic igneous rocks; dark green, streaked with black and red, greasy feel.

shale   Thinly bedded fine-grain sedimentary rock rich in clay minerals.

shear   Deformation caused by one part of a body of rock sliding sideways against another.

shear zone   Narrow belt of highly deformed rock in which recrystallized minerals are often strongly aligned; important for the movement of fluids in the crust and may be the sites of ore deposits.

shelf sea   Shallow sea above the continental shelf, usually less than 250 m deep; the North Sea is a shelf sea.

shield   Large stable ancient area of

Precambrian rocks at the centres of the major continents; lacking in earthquake and volcanic activity and usually surrounded by linear mountain chains.

siderite   Iron carbonate, reddish brown in colour; an important ore of iron.

silica   Silicon dioxide; chemically resistant; forms the mineral quartz and is a common cement in sandstone; jasper, opal, agate, chert and flint are forms of silica; combines with metals to form silicate minerals.

silicate mineral   Natural compound with the basic structure of 4-sided silicon–oxygen building blocks in various combinations with metals, to form groups with single or double chains, sheets or 3-dimensional frameworks, e.g. olivine, pyroxene, amphibole, garnet, feldspar, mica; extremely common as rock-forming minerals.

sill   Sheet of magma intruded into sedimentary rocks parallel to bedding planes.

sillimanite   Aluminium silicate mineral found in high-temperature metamorphic rocks.

siltstone   Fine-grain sedimentary rock, low in clay minerals and without closely spaced partings.

Silurian period   Division of geological time scale, 443–417 million years, above Ordovician and below Devonian; named after the ancient Welsh tribe, the Silures.

sinistral   Left-directed movement of a fault.

sink hole   Feature in limestone areas where a stream disappears underground, due to solution of the bedrock by rainwater; also known as swallow hole.

skarn   Mineral deposit found in limestone adjacent to large igneous intrusions; minerals are calcium-rich silicates, and ores of iron, copper, lead, zinc and many other metals can be found as skarn deposits.

slate   Fine-grain low-grade metamorphic rock rich in platy minerals and with a pronounced cleavage that allows it to be split into thin slabs.

slaty cleavage   Structure in slate caused by parallel alignment of platy minerals such as mica and chlorite.

slickensides   Parallel grooves and linear scratch marks on a fault plane caused by slippage and grinding of one block against another.

sole marks   Sedimentary structures found as casts on the underside of a bed and caused by objects rolling down slope, or slumping of heavy coarse sediment into the layer beneath; useful way-up indicators.

source rock   Sedimentary rock, usually organic-rich shale or mudstone, in which oil and gas were originally formed.

stadial   Brief cold interlude within a glaciation during which advance of glaciers took place, e.g. the Loch Lomond stadial (re-advance) at 11 000–10 000 years ago.

stockwork   An extensive, interconnected network of fractures around a plutonic igneous intrusion; the fractures are then filled with vein minerals; an important type of ore deposit situation in which ore minerals are

disseminated within the stockwork.

strata   plural of stratum, Latin for a bed or layer of sediment.

strata-bound deposit   A mineral deposit confined to a particular stratigraphic unit.

stratiform   A mineral deposit of large extent lying parallel to bedding in sedimentary rocks, or to mineral layering in igneous rocks.

stratigraphic column (geological column)   Diagram showing the subdivisions of geological time, in a sequence from oldest at the bottom to youngest at the top.

stratigraphy   Study of the relative ages of sedimentary rocks, based usually on their fossil contents.

stratum   Bed or layer of rock or sediment; plural is strata.

striations (striae)   Scratches or parallel marks caused by glaciers (glacial striae) or during faulting (slickensides).

strike   direction or trend of a structure, at right angles to the dip direction of a bed.

strike-slip fault   Fault in which the movement is sideways, parallel to the strike of the fault; e.g. the Great Glen Fault; sometimes called transcurrent fault, tear fault or wrench fault; blocks move relatively to the right (dextral fault) or left (sinistral fault).

stromatolite   Pillow-shape or mushroom-shape mounds of carbonate algal growth forming large reef-like structures in shallow water; the oldest known fossils belong to this group.

subduction   Tectonic process whereby an oceanic lithospheric plate descends below another plate into the upper mantle.

subduction zone   Long narrow belt at a destructive plate margin where subduction takes place; marked by an ocean trench; accompanied by shallow and deep earthquakes.

subsidence   Sinking or settling of the Earth's surface; may be caused by cooling and relaxation of crust after volcanic activity (thermal subsidence); or collapse of a large volcanic structure around ring faults (cauldron subsidence); or collapse due to subsurface workings (mining subsidence).

supercontinent   Amalgamation of a large group of continents into one unit, e.g. Pangaea.

suture   A line marking the join of two separate plates at a continent–continent collision zone.

swallow hole (sink hole)   Landform feature in limestone country, where a surface stream disappears underground due to dissolution of limestone by rainwater.

syncline   A V-shape fold in which the two limbs slope in towards each other, so that younger rocks are in the centre or core of the fold; opposite of an anticline.

tear fault   A strike-slip fault, with a vertical fault plane, where one block slides past the other to produce dextral (right-directed) or sinistral (left-directed) fault; sometimes known as a wrench fault or transcurrent fault; e.g. the Great Glen Fault.

tectonic   Related to the formation of large-scale structures in the Earth.

tectonic plates   Large, rigid segments of the Earth's lithosphere (crust and upper mantle).

tephra   Fine volcanic ash, glass, dust and broken crystals carried by the wind and deposited as layers on the surface some distance from an explosive eruption.

terrace   Flat strip of land with steep edges, often above a river bed in a relatively wide valley, and usually caused by uplift of the land.

terrane   Large segment of the crust, usually bounded by major faults, in which the geological evolution is distinct from adjacent terranes; also spelled terrain.

Tertiary period   Division of geological time, 65–2 million years; above Cretaceous and below Quaternary.

Tethys Sea   Former sea that existed between Laurasia and Gondwana; obliterated by the creation of the Alpine–Himalayan mountain chain; the Mediterranean Sea is the last remnant of Tethys.

texture   The way in which the grains, particles or crystals in a rock are held together, and relate to each other; e.g. crystalline texture, fragmental texture.

thermal metamorphism   Process of recrystallization of country rocks around a large igneous intrusion in a thermal aureole (metamorphic aureole), due to heat exchange.

throw   The amount of movement or displacement along a fault.

thrust fault   Reverse fault with movement plane shallower than 45°; horizontal compression forces one block across another.

till   Loose glacial drift, unsorted, unstratified; boulder clay, with mixed sizes and compositions of the material; formed by deposition under an ice sheet.

tillite   Sedimentary rock made of fossilized glacial till.

tombolo   Narrow spit of land, usually sand and gravel ridge, joining a small island to the mainland.

topaz   A colourless mineral, hydrous aluminium silicate containing fluorine, found in granite; very hard and glassy looking; may be used as a gem.

tourmaline   Complex silicate mineral containing boron; with hexagonal symmetry, forming elongate pencil-like crystals; black form is schorl; found in some pegmatites and granites.

trace fossil   Structure in a sedimentary rock created by an animal whose remains are not preserved, e.g. worm tubes, burrows, feeding trails, footprints.

transcurrent fault   A strike-slip fault (wrench fault) in which blocks on opposite side of a vertical fault plane move sideways past one another.

transform fault   Strike-slip fault in oceanic crust, along which two segments of an ocean ridge are displaced relative to one another.

trap   Structure in which oil and gas are found, formed by a reservoir rock and an impermeable cap rock.

trap topography   A step-like landscape feature formed by horizontal basalt lava flows, usually with a fossil soil horizon (bole) between each flow which is preferentially weathered out; typically seen in the Campsie, Kilpatrick and Gargunnock Hills and on Skye and Mull.

trench   Long, deep, linear segment of oceanic crust above a subduction zone; 2–10 km deep and thousands of km long.

trilobite   Extinct fossil arthropod, with head, segmented body and tail; Cambrian to Permian.

tsunami   Giant destructive wave originating in the ocean and reaching the shore with high speed and enormous force to create a catastrophic flood; caused by submarine earthquakes or by sediment slumping down the continental shelf into the deep ocean; Japanese word meaning "harbour wave".

tuff   Consolidated volcanic ash, often bedded.

turbidite   Sediment or sedimentary rock formed by material slumping down continental shelf; beds are usually graded; many of the greywackes in the Southern Uplands are turbidites.

turbidity current   Dense slurry of sediment and water that moves extremely rapidly downslope towards the ocean floor and capable of gouging out a narrow channel on the continental slope; may be triggered off by earthquakes.

type locality   Where a sequence of rocks of a particular age is defined, or a place where the first find of a new fossil has been made.

ultrabasic igneous rock   Igneous rocks with less than 44% silica; made almost entirely of olivine and pyroxene, plus some iron ore, i.e. the ferromagnesian silicates, with no feldspars and no quartz; dense, black rocks, usually very coarse grained, found in large intrusions, sometimes with crystals in layers; also known as ultramafic rocks.

unconformity   A break in the geological record, sometimes expressed as younger rocks lying on top of older rocks that have been uplifted and folded; represents a time gap.

upthrow   Side of a fault that has moved upwards relative to the opposite side.

vein   Narrow intrusion of igneous rock or mineral ore, or a fracture filled with minerals.

valve   One shell of a bivalve animal.

Variscan orogeny   Mountain building episode in central Europe, Carboniferous to Permian, 360–290 million years ago; Gondwana collided with Laurussia and created the supercontinent of Pangaea; also known as Hercynian or Armorican orogeny.

vesicle   Gas bubble in lava, later filled with secondary minerals.

vesicular   Bubbly texture of volcanic rocks caused by escape of gases.

volatiles   Materials which are easily boiled and driven off by heat and which escape from a rock.

volcanic rocks   Igneous rocks formed from the cooling of lava and ash erupted onto the Earth's surface.

volcanic vent   Opening in the Earth where lava has been extruded onto the surface.

wadi   River bed or steep-sided canyon in desert regions, where large amounts of sediment are transported in intermittent flow during flash floods.

wall rock   Country rock adjacent to an igneous intrusion or mineral vein; often affected by some sort of alteration.

warm-based ice   Ice at the base of a thick glacier with a thin film of water separating it from the bedrock; can move rapidly across the surface, causing extensive erosion.

water cycle (hydrological cycle)   Natural system that connects all the water in the atmosphere, seas, oceans, clouds, rivers, lakes, soils, plants and glaciers on Earth, constantly circulating by evaporation, precipitation and transpiration.

way-up structure   Feature in a sedimentary rock or metasediment that indicates the top and bottom of the beds, e.g. graded bedding, cross bedding, ripple marks.

welded tuff   A fine-grain volcanic rock resulting from the compaction of ash and crystal shards due to the intense heat in a burning, explosive cloud.

whinstone   Informal name for black, fine-grain rock that breaks into angular fragments; used as a road stone; greywacke, dolerite and basalt have all been called whinstone; named from the fact that whin or gorse grows on the thin, poor soils overlying such rocks.

wrench fault   Strike-slip fault, tear fault, transcurrent fault; fault plane is vertical and blocks slide past each other laterally, to the right (dextral) or left (sinistral).

xenolith   Block of country rock caught up in an igneous rock; Greek for "foreign rock".

Zechstein Sea   Shallow, warm, salty inland sea that existed in Germany, Denmark and the North Sea during the late Permian period (255–250 million years ago); evaporite deposits formed (thick beds of halite, gypsum, anhydrite).

zeolite   Large group of minerals, hydrated silicates of sodium, calcium and aluminium; usually white or colourless, rather soft; form in cavities in lavas; can lose and gain water easily; used in water softening, filtering and purification plants.

zinc blende   Zinc sulphide, a cubic mineral and the major ore of zinc; also called sphalerite.

zircon   Common mineral in granite and gneiss; colourless silicate of zirconium, forming tiny elongate crystals, often inside biotite mica; has very high melting temperature and forms early in the crystallization of igneous rocks; often contains small amounts of radioactive uranium and hence is useful for age-dating purposes.

Gaelic terms

abhainn river
ach, **achadh**, **auch** field
allt stream
àrd high
avon anglicized form of abhainn

baile township, scattered settlement, town
bàn white
beag, **bheag** small
bealach mountain pass
beinne mountain
ben anglicized form of beinne
binnein pinnacle, hill summit
bog soft
buidhe yellow

cairn anglicized form of càrn
càm bent, crooked
camas bay, inlet
caol narrow (usually inlet of the sea)
càrn heap of stones
ceann head
ceum step
cille chapel, monk's cell
cìr comb, crest of a hill
clach stone
cnoc rounded hill (pronounced "krok")
cnocan small hillock
coire corrie (glacial landform; originally
 means pot or cauldron)
còinich mossy
crag anglicized form of creag
creag stone, rock, cliff
cruach, **cruachan** pile of stones
cùl round hill (shaped like a person's back)

dearg red
drum, **druim** ridge
dubh black
dùn prominent hill, fort

eas waterfall

fada long
fiacal tooth

garbh rough
glas greyish green
gleann valley
glen anglicized form of gleann
gorm green, bluish green

inbhir mouth of a river
inver anglicized form of inbhir

kil- anglicized form of cille
kin anglicized form of ceann
knock anglicized form of cnoc
kyle anglicized form of caol

lairig high mountain pass
loch lake
lochan small lake

maol bare, round hill
meall shapeless hill
mòine peat, peat bog
mór, **mhór** large
mullach headland

òb bay
òrd high place or round hill

poll pool
port harbour, bay

rannoch bracken
rinn point of land, promontory, headland
 (hence Rhinns)
ros, **ross** promontory
ruadh red
rubha promontory, peninsula

sàil heel, end of a hill ridge
sgurr peak, rocky summit
sithean little round hill, fairy knoll
 (pronounced "she-an")
spidean sharp peak, pinnacle
sròn nose (pronounced "stron")
sruth stream (pronounced "stroo")
stac steep conical hill
stòr, **storr** steep high cliff, pinnacle
srath, **strath** wide river valley
stron anglicized form of sròn

tòrr round hill
tràigh shore, beach

uamh cave
uig bay (from Norse vík)
uisge water

Bibliography

General geology

Barnes, J. W. 1995. *Basic geological mapping*. Chichester: John Wiley.

Pocket-size guide to field mapping.

Dryburgh, P. M. 1996. *Assynt: the geologists' Mecca*. Edinburgh Geological Society, Edinburgh.

Booklet detailing the early history of research in Assynt, and the famous Highland geological controversy of 1860.

Duff, P. McL. D. 1993. *Holmes' Principles of Physical Geology*. London: Chapman & Hall.

One of the greatest and most influential geology textbooks of all time, Arthur Holmes' masterpiece was updated first by his wife Doris Reynolds, then by Donald Duff.

Dunning, F. 1995. *Britain's offshore oil and gas*. London: Geological Museum and United Kingdom Offshore Operators' Association.

This booklet is available in a more up-to-date edition on-line; see p. 235.

Emeleus, C. H. & M. C. Gyopari 1992. *British Tertiary Volcanic Province*. London: Chapman & Hall.

One in a series of definitive volumes on geological conservation sites in Britain, published by the Joint Nature Conservation Committee.

Fry, N. 1991. *The field description of metamorphic rocks*. Chichester: John Wiley.

Pocket-size guide to identifying and describing metamorphic rocks and minerals in the field.

Gordon, J. 1997. *Reflections on the Ice Age in Scotland: an update on Quaternary studies*. Glasgow: Scottish Natural Heritage & Scottish Association of Geography Teachers.

Also includes a summary of the geology of Scotland, and a treatment of pre-Quaternary landform evolution.

Hancock, P. L. & B. J. Skinner 2000. *The Oxford companion to the Earth*. Oxford: Oxford University Press.

An encyclopaedic approach to topics in modern geology: 800 entries.

Harding, R. R., R. J. Merriman, P. H. A. Nancarrow 1984. *St Kilda: an illustrated account of the geology*. British Geological Survey Report Vol. 16, No. 7. London: HMSO.

1:25 000 scale geological map with excursion guide.

Kerr, D. 1999. *Shale oil: Scotland: the world's pioneering oil industry*. Edinburgh: Kerr.

Illustrated history of the mining and processing of oil shales.

McClay, K. 1991. *The mapping of geological structures*. Chichester: John Wiley.

Pocket-size guide to identifying, describing and mapping small-scale folds, faults and shear zones in the field.

McIntyre, D. B. & A. McKirdy 1997. *James Hutton, the founder of modern geology*. Edinburgh: HMSO.

Colour booklet about Hutton, published to coincide with the bicentenary conference; includes many photographs of places he visited and sketches by his illustrator, Sir John Clerk of Penicuick.

McKirdy, A. & R. Crofts 1999. *Scotland: the creation of its natural landscape: a landscape fashioned by geology*. Perth: Scottish Natural Heritage.

Booklet giving an overview of a series of related booklets on different regions: Edinburgh, Borders, Cairngorms, Skye, Arran, Northwest Highlands, Orkney and Shetland, Stirling to Loch Lomond, Fife and Tayside, published jointly with the Geological Survey.

McMillan, A. 1997. *Quarries of Scotland*. Edinburgh: Historic Scotland (Technical Advice Note 12).

Excellent pictorial record of stone quarries, especially Caithness Flagstones, sandstones and granite; including the geology of building stones, rock properties, quarrying methods, etc.

Oldroyd, D. R. 1990. The Highlands Controversy. Chicago: University of Chicago Press.

Explains how geological knowledge advanced through fieldwork in the nineteenth century.

Ross, S. 1992. *The Culbin Sands – fact and fiction*. Aberdeen: Centre for Scottish Studies, Aberdeen University.

Expert study of the geomorphology of the sand dunes and the changing shape of the coastline between the rivers Nairn and Findhorn, with detailed maps and excellent aerial photographs.

Scottish Natural Heritage. 2000. *Minerals and the natural heritage in Scotland's Midland Valley*. Perth: Scottish Natural Heritage.

Deals with environmental impact of quarrying in the Midland Valley.

Thorpe, R. E. & G. Brown 1991. *The field description of igneous rocks*. Chichester: John Wiley.

Pocket-size guide to identifying and describing igneous rocks and minerals in the field.

Toghill, P. 2000. *The geology of Britain: an introduction*. Shrewsbury: Swan Hill Press.

Large format, colour illustrations.

Trewin, N. (ed.) 2003. *Geology of Scotland*, 4th edn. London: Geological Society.

Comprehensive coverage by 30 authors.

Tucker, M. E. 1996. *Sedimentary rocks in the field*. Chichester: John Wiley.

Pocket-size guide to identifying and describing sedimentary rocks and fossils in the field.

Walsh, J. A. 2000. *Scottish slate quarries*. Edinburgh: Historic Scotland (Technical Advice Note 21).

Detailed study of slates, with many photographs, maps and diagrams, together with information on the history of slate quarries.

Whittow, J. 1992. *Geology and scenery in Britain*. London: Chapman & Hall.

Separate chapters for each region in Britain.

Wickham-Jones, C. R. 2001. *The landscapes of Scotland: a hidden history*. Stroud: Tempus Publishing.

An account of the human influence on landscape development, from earliest times.

Wilson, R. C. L., S. A. Drury, J. L. Chapman 2000. *The great Ice Age: climate change and life*. London: Routledge.

Comprehensive treatment of theories of ice ages.

Woodcock, N. & R. Strachan 2000. *Geological history of Britain and Ireland*. Oxford: Blackwell Science.

An advanced textbook by expert geologists.

British Geological Survey publications

British Geological Survey 2002. *Geological Survey ten-mile map: north sheet (solid)*. Scale 1:625 000.

A folded one-sheet geological map of Scotland; a companion drift sheet is available, showing unconsolidated sediments only.

Cameron, I. B., P. Stone, J. Smellie 1986. *Geology of the country around Girvan*. Explanation of 1:50 000 sheet 7. London: HMSO/British Geological Survey.

Short description of an area of complex geology, well illustrated.

Chapman, G. R. 1999. *Coal resources map of Britain, 1:1 500 000 scale*. Keyworth, British Geological Survey and the Coal Authority.

Map of coalfields, with notes on resources and reserves onshore and offshore, including geological columns and notes on methane gas.

Colman, T.B. 1996. *Metallogenic map of Britain and Ireland, 1:1 500 000 scale*. Keyworth: British Geological Survey.

Geological map overprinted with metal ore deposit occurrences.

Emeleus, C. H. 1997. *Geology of Rum and the adjacent islands (Eigg, Muck and Canna)*. London: The Stationery Office.

British Geological Survey Memoir for 1:50 000 scale Geological map, Sheet 60 (Scotland).

Highley, D.E. 1996. Industrial mineral resources map of Britain, 1:1,000,000 scale. Keyworth: British Geological Survey.

Geological map overprinted with useful mineral deposit occurrences.

Kokelaar, P. & I. Moore 2004 (in preparation). *The Glencoe area*. Classic Areas of British Geology. London: HMSO/British Geological Survey.

A new interpretation of the geology, to accompany a 1:25 000 scale geological map.

Lott, G. K. 2001. *Building stone resources map of Britain, scale 1:1 000 000*. Keyworth: British Geological Survey.

Geological map showing stone quarries, photographs of the rocks and buildings around the country, and notes on stone properties.

Regional handbooks are published for Orkney and Shetland, the Northern Highlands, the Grampian Highlands, the Midland Valley, Tertiary volcanic districts, Southern Scotland.

Provide detailed accounts of geology and mineral resources. The Orkney and Shetland volume includes suggested field excursions.

Stone, P. & J. L. Smellie 1988. *The Ballantrae area*. Classic Areas of British Geology. London: HMSO/British Geological Survey.

Detailed description (1:25 000 map) of the geology, with suggested excursion itineraries.

Webb, B. C., A. W. A. Rushton, D. E. White 1993. *Moffatdale and the Upper Ettrick Valley*. Classic Areas of British Geology. London: HMSO/British Geological Survey.

Detailed description of an important area in Scotland; excellent treatment of Lapworth's graptolite zones; maps, sections, glossary. Description of a 1:25,000 map.

Field excursion guidebooks

Allison, I., May, F. & R. A. Strachan 1988. *An excursion guide to the Moine geology of the Scottish Highlands*. Edinburgh: Scottish Academic Press.

Includes A9 sections, Fort William, Glenfinnan, Morar, Ullapool, Moine Thrust zone, Eriboll, Bettyhill.

Auton, C. A., C. R. Firth, J. W. Merritt 1990. *Quaternary of Beauly to Nairn field guide*. London: Quaternary Research Association.

Barber, A. J., A. Beach, R. G. Park, J. Tarney, A. D. Stewart 1978. *The Lewisian and Torridonian rocks of Northwest Scotland*. London: Geologists' Association, Guide No. 21.

Covers Scourie, Laxford, Stoer, Assynt, Lochinver, Gairloch, Torridon, Dornie.

Bell, B. R. & J. W. Harris 1986. *An excursion guide to the geology of the Isle of Skye*. Glasgow: Geological Society of Glasgow.

Includes an extensive introduction to the Tertiary and pre-Tertiary geology of Skye.

Birnie, J., J. Gordon, K. Bennett, A. Hall 1993. *Quaternary the Shetland Isles field guide*. London: Quaternary Research Association.

Browne, M. & C. Gillen 2003. *Stirling geology: an excursion guide*. Edinburgh: Edinburgh Geological Society.

Covers Stirling, Falkirk, Alva, Fintry, Kilsyth, Keltie Water, Aberfoyle and Perth.

Burton, C. J. & J. J. Doody 1995. *Excursion guide to the geology of southern Kintyre*. Glasgow: Geological Society of Glasgow.

Covers the Kintyre Peninsula south of Campbeltown–Machrihanish.

Dawson, A. G. & S. Dawson 1977. *Quaternary of Islay and Jura field guide*. London: Quaternary Research Association.

Emeleus, C. H. & R. M. Forster 1979. *Field guide to the Tertiary igneous rocks of Rum, Inner Hebrides*. Newbury, Nature Conservancy Council.

Faithfull, J. & B. Upton 2003 (in preparation). *Excursion guide to the geology of Mull*. Edinburgh: Geological Societies of Edinburgh and Glasgow.

Glasser, N. F. & M. R. Bennett 1996. *Quaternary of the Cairngorms field guide*. London: Quaternary Research Association.

Gribble, C. D. 1976. *Ardnamurchan: a guide to geological excursions*. Edinburgh: Edinburgh Geological Society.
  Includes a full colour 1:40 000 scale map of the Tertiary Complex.

Hall, A. M. 1996. *Quaternary of Orkney field guide*. London: Quaternary Research Association.

Hambrey, M. J., I. J. Fairchild, B. W. Glover, A. D. Stewart, J. E. Treagus, J. A. Winchester 1991. *The Late Precambrian geology of the Scottish Highlands and Islands*. Geologists' Association Guide No. 44. London: Geologists' Association.
  Torridonian of the northwest, and Dalradian of the southwest and Islay and the Garvellachs.

Hill, J. & D. Buist 1994. *A geological field guide to the Island of Bute, Scotland*. Geologists' Association Guide No. 51. London: Geologists' Association.
  Straightforward trails around the whole island, north and south of the Highland Boundary Fault.

Johnson, M. R. W. & I. Parsons 1979. *Geological excursion guide to the Assynt district of Sutherland*. Edinburgh Geological Society, Edinburgh.
  A new edition is in preparation, with a new Moine Thrust website and geological map.

Land, D. H. & R. F. Cheeney 1996. *Discovering Edinburgh's volcano*. Edinburgh: Edinburgh Geological Society.
  A fold-out map and geological excursion guide to Arthur's Seat and Holyrood Park; including the famous Hutton's sections.

Lawson, J. A. 1981. *Building stones of Glasgow*. Glasgow: Geological Society of Glasgow.

Lawson, J. D. & D. S. Weedon 1992. *Geological excursions around Glasgow and Girvan*. Glasgow: Geological Society of Glasgow.
  Covers Aberfoyle, Loch Lomond, Cumbrae, Largs, Ayr, Lesmahagow, Dob's Linn

Lawson, T. J. 1995. *Quaternary of Assynt and Coigach field guide*. London: Quaternary Research Association.

MacDonald, J. G. & A. Herriot 1983. *Macgregor's excursion guide to the geology of Arran*. Glasgow: Geological Society of Glasgow.
  Includes extensive geological introduction and notes on archaeology.

MacGregor, A. R. 1996. *Fife and Angus geology: an excursion guide*. Durham: Pentland Press and Edinburgh Geological Society.
  Covers Stonehaven, Glen Esk, Arbroath, Dundee, Perth, Comrie, St Andrews.

McAdam, A. D. & E. N. K. Clarkson 1986. *Lothian geology: an excursion guide*. Edinburgh: Scottish Academic Press.
  Includes Edinburgh, North Berwick, Dunbar, Siccar Point, Lammermuirs, North Esk, Pentlands, Bathgate.

McAdam, A. D., E. N. K. Clarkson, P. Stone 1992. *Scottish Borders geology: an excursion guide*. Edinburgh: Scottish Academic Press.
  Includes St Abb's, Berwick-on-Tweed, Border abbeys, Eildon Hills, Biggar, Leadhills and Wanlockhead.

McKerrow, W. S. & F. B. Atkins 1985. *Isle of Arran*. London: Geologists' Association.
  More compact than the guide by MacDonald & Herriot; contains questions for each locality.

McMillan, A. A., R. J. Gillanders, J. A. Fairhurst 1999. *Building stones of Edinburgh*. Edinburgh: Edinburgh Geological Society.
  Extensive coverage of properties of building stones and notes on famous quarries.

Merritt, J. W., E. R. Connell, D. R. Bridgland 2000. *The Quaternary of the Banffshire coast and Buchan field guide*. London: Quaternary Research Association.
  This part of Scotland has an unrivalled sequence of Quaternary features, and is internationally important.

Nicholas, C. J. 2000. *Exploring geology on the Isle of Arran*. Cambridge: Cambridge University Press.
  A field notebook format, encouraging the visitor to observe and make deductions.

Pattison, D. R. M. & B. Harte 2001. *The Ballachulish Igneous Complex and aureole: a field guide*. Edinburgh: Edinburgh Geological Society.
  With full colour map and excellent field photographs of Dalradian metamorphic rocks.

Peacock, J. D. & R. Cornish 1989. *Glen Roy area field guide*. Cambridge, Quaternary Research Association.
  Includes Glen Spean and an excursion to the famous Parallel Roads.

Roberts, J. L. 1998. *The Highland geology trail*. Edinburgh: Luath Press.
  Car excursions from Inverness to Thurso, Cape Wrath, Ullapool, Skye, Fort William and Oban; summary of the geology of the north.

Roberts, J. L. & J. E. Treagus 1977. Dalradian rocks of the Southwest Highlands. *Scottish Journal of Geology* **13**(2), 87–184.
  This is the only available source of excursion guides to classic areas of Dalradian geology in Cowal, Kintyre, Tayvallich, Jura, Loch Awe and Loch Leven districts, and may be available in local libraries.

Stone, P. 1996. *Geology in south-west Scotland: an excursion guide*. Keyworth: British Geological Survey.
  Covers Dumfries and Galloway, with extensive notes on ore mineralization and graptolite zones.

The accompanying 1:30 000 scale map has been superseded by the BGS 1:50 000 map (sheet 60).

Tipping, R. M. 1999. *Quaternary of Dumfries and Galloway field guide.* London: Quaternary Research Association.

Trewin, N., B. Kneller, C. Gillen 1987. *Excursion guide to the geology of the Aberdeen area.* Edinburgh: Scottish Academic Press/Geological Society of Aberdeen.

Trewin, N. H. & A. Hurst 1993. *Excursion guide to the geology of east Sutherland and Caithness.* Edinburgh: Geological Society of Aberdeen/Scottish Academic Press.

Walker, M. J. C., J. M. Grey, J. J. Lowe 1992. *Quaternary of the Southwest Scottish Highlands field guide.* London: Quaternary Research Association.

Covers Elgin, Moray Firth coast, Fraserburgh, Collieston to Stonehaven coast, Glen Esk, Deeside, Aberdeenshire granites.

Includes Devonian fish beds, Golspie to Helmsdale coast, Kildonan gold deposit.

Websites

Aberdeen Geological Society:
http://www.abdn.ac.uk/geology/ags/ags.htm

British Geological Survey:
http://www.bgs.ac.uk/ — Has links to other surveys; and an "ask the expert" service.

Dept of Geology & Petroleum Geology, University of Aberdeen:
http://www.abdn.ac.uk/geology/ — Has links to the Rhynie chert research pages.

Division of Earth Sciences, University of Glasgow:
http://www.earthsci.gla.ac.uk/

Dynamic Earth, Edinburgh:
http://www.dynamicearth.co.uk/ — Website for the visitor centre in Holyrood Park.

Edinburgh Geological Society:
http://www.edinburghgeolsoc.org/ — Has virtual tour of Edinburgh's geology, and lists of publications.

Geological Society of London:
http://www.geolsoc.org.uk — Many links to geological societies worldwide.

Glasgow Geological Society:
http://www.geologyglasgow.org.uk/ — Features field-trip photographs.

Grant Institute, University of Edinburgh:
http://www.glg.ed.ac.uk — Has many good field-excursion archives.

Leeds University (Moine Thrust):
http://earth.leeds.ac.uk/moine/ — An extensive resource of maps, photographs, diagrams and text written by experts.

Quaternary Research Association:
http://www.qra.org.uk/ — Lists all their excursion guides.

Rhynie chert:
http://www.abdn.ac.uk/rhynie/ — A teaching and learning resource.
http://members.tripod.com/Lyall/Rhynie/rhynie.htm
http://www.uni-muenster.de/GeoPalaeontologie/Palaeo/Palbot/erhynie.html — With many excellent colour photographs and reconstructions of the environment at Rhynie.

University of St Andrews, School of Geography & Geosciences:
http://www.st-and.ac.uk/~www_sgg/schoolpage.html

Professor Chris Scotese (University of Texas):
http://www.scotese.com — Includes computer animations of past and possible future arrangements of the continents, with plate tectonics and 3-D world globes, as well as geological maps of the Earth for each of the past periods.

Scottish Geology:
http://www./scottishgeology.com/ — Devoted to the geology of Scotland, for non-specialists; there is a section on each of the geological regions, with field photographs and many useful links.

Scottish Natural Heritage:
http://www.snh.org.uk — Has links to SNH "Futures", detailing geology, landscape and conservation for 25 areas.

United Kingdom Offshore Operators' Association:
http:// www.oilandgas.org.uk/education/ — Information on North Sea oil and gas, and resources for students and teachers.

United States Geological Survey:
http://www.usgs.gov — One of the best sites for general geology and plate tectonics.

Mr Wood's Fossils: http://www.mwfossils.pwp.blueyonder.co.uk/ — Stan Wood discovered *Lizzie* the fossil reptile.

Index of names

Index of topics